RNA BINDING PROTEINS:

New Concepts in Gene Regulation

ENDOCRINE UPDATES

Shlomo Melmed, M.D., Series Editor

---

1. E.R. Levin and J.L. Nadler (eds.): Endocrinology of Cardiovascular Function. 1998. ISBN: 0-7923-8217-X
2. J.A. Fagin (ed.): Thyroid Cancer. 1998. ISBN: 0-7923-8326-5
3. J.S. Adams and B.P. Lukert (eds.): Osteoporosis: Genetics, Prevention and Treatment. 1998. ISBN: 0-7923-8366-4.
4. B.-Å. Bengtsson (ed.): Growth Hormone. 1999. ISBN: 0-7923-8478-4
5. C. Wang (ed.): Male Reproductive Function. 1999. ISBN 0-7923-8520-9
6. B. Rapoport and S.M. McLachlan (eds.): Graves' Disease: Pathogenesis and Treatment. 2000. ISBN: 0-7923-7790-7.
7. W. W. de Herder (ed.): Functional and Morphological Imaging of the Endocrine System. 2000. ISBN 0-7923-7923-9
8. H.G. Burger (ed.): Sex Hormone Replacement Therapy. 2001. ISBN 0-7923-7965-9
9. A. Giustina (ed.): Growth Hormone and the Heart. 2001. ISBN 0-7923-7212-3
10. W.L. Lowe, Jr. (ed.): Genetics of Diabetes Mellitus. 2001. ISBN 0-7923-7252-2
11. J.F. Habener and M.A. Hussain (eds.): Molecular Basis of Pancreas Development and Function. 2001. ISBN 0-7923-7271-9
12. N. Horseman (ed.): Prolactin. 2001 ISBN 0-7923-7290-5
13. M. Castro (ed.): Transgenic Models in Endocrinology. 2001 ISBN 0-7923-7344-8
14. R. Bahn (ed.): Thyroid Eye Disease. 2001 ISBN 0-7923-7380-4
15. M.D. Bronstein (ed.): Pituitary Tumors in Pregnancy ISBN 0-7923-7442-8
16. K. Sandberg and S.E. Mulroney (eds.): RNA Binding Proteins: New Concepts in Gene Regulation. 2001. ISBN 0-7923-7612-9

RNA BINDING PROTEINS:

New Concepts in Gene Regulation

Edited by

Kathryn Sandberg, PhD
Departments of Medicine and Physiology & Biophysics
Georgetown University School of Medicine
Washington, DC

and

Susan E. Mulroney, PhD
Department of Physiology & Biophysics
Georgetown University School of Medicine
Washington, DC

Springer Science+Business Media, LLC

Library of Congress Cataloging-in-Publication Data

RNA binding proteins : new concepts in gene regulation / edited by Kathryn Sandberg and Susan E. Mulroney.
p. cm. -- (Endocrine updates ; 16)
Includes bibliographical references and index.
DOI 10.1007/978-1-4757-6446-8

1. Genetic regulation. 2. RNA-protein interactions I. Sandberg, Kathryn. II. Mulroney, Susan E. III. Series.

QH450 .R63 2002
572.8'65--dc21

2001054415

Originally published by Kluwer Academic Publishers in 2002
MyCopy version of the original edition 2002

Printed on acid-free paper.
www.springer.com/mycopy

Contents

Contributors ... vii

Preface ... xiii

Acknowledgements ... xv

Section I: Translational Control

1. The role of RNA-binding proteins in IRES-dependent translation
Sung Key Jang and Eckard Wimmer ... 1

2. Translational regulation of masked maternal mRNAs in early development
Nancy Standart ... 35

3. Poly-C binding proteins: Cellular regulators of mRNA fate and function
Andrea V. Gamarnik and Raul Andino ... 53

4. Autoregulation of protein synthesis by translation
Guim Kwon, Guang Xu, Wilhelm S. Cruz, Connie A. Marshall and Michael L. McDaniel ... 71

5. Translational initiation of ornithine decarboxylase mRNA
Lo Persson and Koichi Takao ... 87

Section II: mRNA Metabolism

6. Regulation of mRNA stability by AUF1
Gerald M. Wilson and Gary Brewer ... 101

7. RNA binding by members of the 70-kDa family of molecular chaperones
Christine Zimmer, Eszter Nagy, John R. Subjeck, and Tamás Henics ... 119

8. Post-transcriptional control of type-1 plasminogen activator inhibitor mRNA
Joanne H. Heaton and Thomas D. Gelehrter ... 135

9. Post-transcriptional control of the GAP-43 mRNA by the ELAV-like protein HuD
Nora Perrone-Bizzozero and Rebecca Keller 157

Section III: Hormonal and Homeostatic Regulation

10. RNA-dependent protein kinases
Raymond Petryshyn, Sergie Nekhai, and Evelio D. Perez-Albuerne 175

11. Regulation of messenger RNA-binding proteins by protein kinases A and C
Richard A. Jungmann 193

12. Post-transcriptional regulation of iron metabolism
Tracey A. Rouault 213

13. Cytochrome P450 RNA-protein interactions
Matti A. Lang and Françoise Raffalli-Mathieu 225

14. Site-specific cleavage of insulin-like growth factor II mRNAs
Erwin L. van Dijk and P. Elly Holthuizen 239

15. Hormonal regulation of the EGF/receptor system
Stephen W. Spaulding and Lowell G. Sheflin 255

16. The role of RNA binding proteins in tumorigenesis
Sreerama Shetty 271

17. Regulation of G-protein coupled receptor cytosolic mRNA binding proteins
Kathyrn Sandberg, Zheng Wu, Hong Ji, Eric Hernandez and Susan E. Mulroney 285

Index 307

Contributors

Raul Andino, Ph.D.
Associate Professor
Department of Microbiology and Immunology
University of California at San Francisco, Box 0414
San Francisco, CA 94143-0414

Wilhelm S. Cruz, Ph.D.
Instructor
Biology Department
Campus Box 8118
Washington University
Rebstock 130
St Louis, MO 63110

Gary Brewer, Ph.D.
Associate Professor
Department of Molecular Genetics and Microbiology
UMDNJ-Robert Wood Johnson Medical School
675 Hoes Lane
Piscataway, NJ 08854

Andrea V. Gamarnik, Ph.D.
Group Leader
Instituto de Investigaciones Bioquimicas
Fundacion Campomar
Av Patricias Argentinas 435
Capital Federal, Buenos Aires, Argentina

Thomas D. Gelehrter, M.D.
Professor and Chair
Department of Human Genetics
University of Michigan Medical School
4909 Buhl, Box 0618
1241 E. Catherine Street
Ann Arbor, MI 48109-0618

Tamás Henics, M.D., Ph.D.
Senior Staff Scientist
Department of Molecular Biology
INTERCELL, Inc.
Rennweg 95B
A-1030 Vienna, Austria

Assistant Professor
Dept. of Medical Immunol and Microbiol
Faculty of Medicine
University of Pecs, Szigeti u. 12
7643 Pécs, Hungary

Joanne H. Heaton, M.S.
Department of Human Genetics
University of Michigan Medical School
4909 Buhl, Box 0618
1241 E. Catherine Street
Ann Arbor, MI 48109-0618

Eric Hernandez
Graduate Student
Department of Microbiology and Immunology
Georgetown University
4000 Reservoir Road, NW
Washington, DC 20007-2145

P. Elly Holthuizen, Ph.D.
Senior Scientist
Department of Physiological Chemistry
University Medical Center Utrecht
Universiteitsweg 100
3584 CG Utrecht
The Netherlands

Sung Key Jang, Ph.D.
Associate Professor
Department of Life Sciences
Pohang University of Science and Technology
Hyoja-Dong San 31, Pohang
Kyungbuk 790-784, Korea.

Hong Ji, M.D.
Assistant Professor
Department of Physiology
Georgetown University
4000 Reservoir Road, NW
Washington, DC 20007-2145

Richard A. Jungmann, Ph.D.
Professor
Department of Cell and Molecular Biology
Northwestern University Medical School
Olson Pavilion 8321
303 East Chicago Avenue
Chicago, IL 60611-3008

Rebecca Keller, Ph.D.
Research Assistant Professor
Department of Chemistry
Clark Hall 104
University of New Mexico School of Medicine
Albuquerque, NM 87131

Guim Kwon, Ph.D.
Research Instructor
Department of Pathology and Immunology
Washington University School of Medicine
660 S. Euclid Avenue
St Louis, MO 63110

Matti A. Lang. Ph.D.
Professor and Chair
Department of Biochemistry
Faculty of Pharmacy
Biomedical Centrum Box 578
University of Uppsala
Uppsala, Sweden

Connie A. Marshall, B.S.
Research Associate
Department of Pathology and Immunology
Campus Box 8118
Washington University School of Medicine
660 S. Euclid Avenue
St Louis, MO 63110

Michael L. McDaniel, Ph.D.
Professor
Department of Pathology and Immunology
Campus Box 8118
Washington University School of Medicine
660 S. Euclid Avenue
St Louis, MO 63110

Susan E. Mulroney, Ph.D.
Associate Professor
Department of Physiology & Biophysics
Georgetown University School of Medicine
3900 Reservoir Rd, NW Rm 253 Basic Science Building
Washington, DC 20007

Eszter Nagy, M.D., Ph.D.
Senior Staff Scientist
Department of Molecular Biology
INTERCELL, Inc.
Rennweg 95B
A-1030 Vienna, Austria

Assistant Professor
Dept. of Clinical Biochemistry
Faculty of Medicine
University of Pecs, Szigeti u. 12
7643 Pécs, Hungary

Sergie Nekhai, Ph.D.
Assistant Professor
Sickle Cell Center
Howard University
2121 Georgia Avenue, NW
Washington, DC 20059

Evelio D. Perez-Albuerne, M.D., Ph.D.
Attending Physician
Department of Hematology/Oncology
Children's National Medical Center
111 Michigan Avenue, NW
Washington, DC 20010

Nora Perrone-Bizzozero, Ph.D.
Professor
Department of Neuroscience
University of New Mexico School of Medicine
915 Camino de Salud NE
Albuquerque, NM 87131-5223

Lo Persson, Ph.D.
Professor
Department of Physiology and Neuroscience
University of Lund
BMC F:13
S-22184 Lund, Sweden

Raymond A. Petryshyn, Ph.D.
Scientific Review Administrator
Grants Review Branch
Division of Extramural Activities
National Cancer Institute
Room 8133
6116 Executive Blvd. MSC 8328
Bethesda, MD 20892

Françoise Raffalli-Mathieu, Ph.D.
Department of Biochemistry
Faculty of Pharmacy
Biomedical Centrum Box 578
University of Uppsala
Uppsala, Sweden

Tracey A. Rouault, M.D.
Chief
Section on Human Iron Metabolism
Cell Biology and Metabolism Branch
National Institute of Child Health and Human Development
Building 18, Room 101
Bethesda, MD 20892

Kathryn Sandberg, Ph.D.
Associate Professor
Division of Nephrology
Department of Physiology & Biophysics
Georgetown University School of Medicine
3800 Reservoir Rd, NW   Rm 394 Building D
Washington, DC 20007

Lowell G. Sheflin, Ph.D.
Research Assistant Professor of Medicine
SUNY at Buffalo
VA Western New York Healthcare System
Veterans Administration Medical Center
3495 Bailey Avcenue
Buffulo NY 15215

Sreerama Shetty, Ph.D.
Associate Professor
Division of Medicine
Department of Specialty Care Services
University of Texas Health Center at Tyler
11937 US Hwy 271
Tyler, TX 75708-3154

Stephen W. Spaulding M.D., C.M.
Professor of Medicine and Physiology & Biophysics, SUNY at Buffalo
Associate Chief of Staff for Research & Development
VA Western New York Healthcare System
Veterans Administration Medical Center
3495 Bailey Avcenue
Buffalo NY 15215

Nancy Standart, Ph.D.
Senior Lecturer
Department of Biochemistry
University of Cambridge
80 Tennis Court Road
Old Addenbrook Site
Cambridge CB2 1GA, UK

John Subjeck, Ph.D.
Professor
Department of Molecular and Cellular Biophysics
Roswell Park Cancer Institute
Elm and Carlton Streets
Buffalo, NY 14263

Koichi Takao, Ph.D.
Research Associate
Laboratory of Cellular Physiology
Department of Clincial Dietetics and Human Nutrition
Faculty of Pharmaceutical Sciences
Josai University
1-1 Sakado
Saitama, 350-0295 Japan

Erwin L. van Dijk, Ph.D.
Post-Doctoral Fellow
Centre de Genetique Moleculaire
CNRS-UPR2167
Avenue de la Terrasse, Batiment 24
F-91198 Gif-sue-Yvette Cedex
France

Gerald M. Wilson, Ph.D.
Research Associate
Department of Molecular Genetics and Microbiology
UMDNJ-Robert Wood Johnson Medical School
675 Hoes Lane
Piscataway, NJ 08854

Eckard Wimmer, Ph.D.
Professor
Department of Molecular Genetics and Microbiology
School of Medicine
State University of New York at Stony Brook
Stony Brook, NY 11794-5222

Zheng Wu, Ph.D.
Research Associate
Department of Physiology
Georgetown University
4000 Reservoir Road, NW
Washington, DC 20007-2145

Guang Xu, Ph.D.
Post-Doctoral Fellow
Department of Internal Medicine
Washington University School of Medicine
1021 NT, Box 8069
St Louis, MO 63110

Christine Zimmer
Graduate Student
Department of Molecular Biology
INTERCELL, Inc.
Rennweg 95B
A-1030 Vienna, Austria

Preface

RNA binding proteins are an exciting area of research in gene regulation. A multitude of RNA-protein interactions are used to regulate gene expression including pre-mRNA splicing, polyadenylation, editing, transport, cytoplasmic targeting, translation and mRNA turnover. In addition to these post-transcriptional processes, RNA-protein interactions play a key role in transcription as illustrated by the life cycle of retroviruses. Unlike DNA, the structure of RNA is highly variable and conformationally flexible, thus creating a number of unique binding sites and the potential for complex regulation by RNA binding proteins.

Our enthusiasm for compiling this book came from several sources. Predominant factors included the timeliness and importance of the topic, as well as from the willingness of the contributors to add one more task to their already overloaded schedules. We took this as a good sign, and feel very fortunate to have so many leaders in this important area contribute chapters to this book. There is no way to adequately thank them for their time, experience, and patience. For us, it was a wonderful collaboration, and we were delighted to interact with each of them to produce this work.

Although there is a wide range of topics included in this book, general themes have been repeated, highlighting the overall integrative nature of this field. We have separated the chapters into three different sections; *Translational Control, mRNA Metabolism, and Hormonal and Homeostatic Regulation*, and although each chapter could integrate to other areas, this helps focus the general themes. RNA binding proteins is an exciting new area of research in gene regulation. A multitude of RNA-protein interactions are used to regulate gene expression. In eukaryotes, protein synthesis does not occur until the newly transcribed RNA is extensively modified and exported into the cytoplasm. In the cytoplasm, a variety of RNA-protein interactions are employed by the cell to regulate the levels of protein expression. Unlike DNA, the structure of RNA is highly variable and conformationally flexible, thus creating a number of unique binding sites and potential regulatory mechanisms for RNA binding proteins.

In the first section on translational control, we focus on RNA binding proteins that interact with internal ribosome entry sites (chapter 1), maternal mRNAs in early development (chapter 2), pyrimidine stretches, which play an important role in translation and in mRNA stability (chapter 3), and with translation initiation factors (chapters 4 and 5). In the second section on mRNA metabolism, RNA binding proteins that specifically regulate mRNA stability are examined including regulation of the angiotensin AT_1 receptor by AUF1 (chapter 6), the molecular chaperone family (chapter 7), control of the type-1 plasminogen activator inhibitor mRNA and regulation of GAP-43 mRNA by HuD (chapter 9). The final section on hormonal and homeostatic regulation includes chapters on signaling by RNA-dependent protein kinases (chapter 10) and protein kinases A and C (chapter 11). Also included are chapters on homeostatic regulation of gene expression by iron-dependent (chapter 12), cytochrome P450-dependent (chapter 13) and insulin dependent (chapter 14) RNA binding proteins. The role of RNA binding proteins in regulating epidermal growth factor receptors is examined in chapter 14 and the importance of RNA binding proteins in tumorigenesis is addressed in chapter 15. Lastly, the emerging field of RNA binding proteins in regulating G-protein coupled receptors is presented in chapter 17.

These chapters were written with the seasoned investigator as well as the student in mind. To facilitate the students' experience, we have included summaries of the key concepts reviewed within each chapter, as well as guiding questions that could be used to stimulate class discussions. Considering the importance of this field to biochemistry, cellular and molecular biology, this book is appropriate for science curricula at advanced undergraduate and graduate levels as well as for scientists who discover that their gene of interest is regulated post-transcriptionally. We hope this compendium educates, enthralls, and stimulates the readers to look to the future possibilities in this rapidly evolving field.

Kathyrn Sandberg *Susan E. Mulroney*

Washington, DC

Acknowledgements

We wish to thank all the authors that contributed chapters to this book. They have done an excellent job in summarizing the major advances on each topic. We would also like to thank our colleagues at Kluwer Academic including Jill Almeida and Melissa Ramondetta for their patience and assistance in preparing this book. In particular, we thank Barbara Murphy at Kluwer, who supported this project from the beginning. Lastly, we would like to acknowledge the support of the NIH, the National Kidney Foundation and the American Heart Association who have funded our research in this important area of science.

1

THE ROLE OF RNA-BINDING PROTEINS IN IRES-DEPENDENT TRANSLATION

Sung Key Jang and Eckard Wimmer*

*Pohang University of Science and Technology, Pohang, Hyoja-Dong, Korea and. *SUNY Stony Brook School of Medicine, Stony Brook, NY*

Picornaviridae are a large family of human and animal RNA viruses whose best known members are poliovirus (PV) belonging to genus Enterovirus, human rhinovirus (HRV) of Rhinovirus, encephalomyocarditis virus (EMCV) of Cardiovirus, hepatitis A virus (HAV) of Hepatovirus, and foot-and-mouth disease virus (FMDV) of Aphthovirus. It has been estimated that the incidence of human infections by picornaviruses exceeds 6 billion per year. Fortunately, the vast majority of these infections are self limiting, with no serious sequelae. However, the diseases range from the very serious (poliomyelitis, meningitis, heart disease, hepatitis) to the benign (common cold) and together, they cause enormous hardship in the human population. The genome of picornaviruses is single stranded and of plus strand polarity. That is, it functions as mRNA after the virus has entered the host cell. Molecular biologic studies of these viruses have revealed numerous important mechanisms, including the internal ribosomal entry site (IRES) (1,2). The description of picornaviral IRES elements broke the dogma that all eukaryotic translation begins with a mechanism of cap-dependent ribosome "scanning" from the 5' end of mRNA. This chapter will discuss the role of RNA binding proteins in IRES-dependent translation of picornaviruses. The existence of viral IRESs has led to the discovery of IRES elements in numerous cellular mRNAs. These concepts regarding IRESs have revolutionized the general perception of the control of gene expression by translation in eukaryotic cells.

BACKGROUND

The discovery of the internal ribosomal entry site (IRES) resulted from studies investigating the mechanisms of gene expression in viral genomes. The mRNAs from the picornaviruses, encephalomyocarditis virus (EMCV) and poliovirus (PV), were the first eukaryotic mRNAs shown to contain IRES elements (1,2, and see below). An IRES was next discovered in the Hepatitis C virus (HCV) (3), which is the sole member of the genus Hepacivirus of Flaviviridae. Flaviviruses, just like picornaviruses, are plus strand RNA viruses. It has been estimated that 1% of the global population is infected with HCV of whom 70% will carry the virus for life. 30% of the persistently infected individuals will eventually develop serious (fatal) liver disease. Member viruses of the genus Pestivirus (bovine diarrhea virus and classical swine fever virus) of Flaviviridae that cause devastating disease in live stock, also carry IRES elements (4). Not surprisingly, IRES containing animal viruses have gained notoriety and their study is of widespread interest.

Most recently, certain insect viruses (Plautia stali intestine virus and cricket paralysis virus) have been found to contain IRES elements. These viruses are particularly fascinating as their genomes are dicistronic, and the expression of their ORFs is being controlled by two IRES elements (5,6). The genomes of these viruses resembles that of an artificially made dicistronic poliovirus (7) but the strategy of internal initiation of translation is very different (8,9).

The Discovery of Picornaviral IRES Elements

The discovery of the 5'-terminal "cap" and 3'-terminal poly(A) in eukaryotic mRNAs launched a new era in studies of eukaryotic gene expression. Suddenly, much effort was expended to decipher the function of these novel structures in cellular RNA. The fundamentally important role of the cap structure as an essential element in the translation of eukaryotic mRNA was quickly established (10,11). It has subsequently been shown that the cap structure binds to the cap-binding protein (eIF4E) that, in turn, forms a complex with other eukaryotic initiation factors (eIFs). The complex then recruits the 40S ribosomal subunit, itself in complex with several cellular components, to the 5' end of mRNA (12). From there, a search ensues 5' to 3' ("scanning") to identify a suitable AUG codon. A comparative analysis of cellular mRNAs combined with mutation studies of the 5' nontranslated region (5'NTR) of preproinsulin mRNA have led Marilyn Kozak to suggest that the 40S ribosomal subunit can "scan" from the cap through the 5'NTR until it encounters the initiation codon (13). Selection of an AUG codon is favored if it occurs in a special context of surrounding nucleotides, that is, an A or G residue at the –3 position and an A or G at the +1 position relative to the AUG ($^{A}/_{G}$XXAUG$^{G}/_{A}$). Here, the 40S ribosomal subunit complex will pause and attract the 60S ribosomal subunit, forming the complete initiation complex for translation (10,12).

The role of the 3'-terminal poly(A) of eukaryotic mRNAs, on the other hand, is less well understood and it is definitely not absolutely required for translation since several eukaryotic mRNAs (e.g. histone mRNAs, reovirus mRNA) lack this structure. However, if present, it can promote circularization of mRNA via the poly(A)-associated poly(A) binding protein (PABP), an event that stimulates the formation of the initiation complex formation (14).

The studies leading to these fundamental mechanisms in eukaryotic translation have been carried out mainly in cell-free translation systems of which the micrococcal nuclease-treated rabbit reticulocyte lysates (RRL) was (and still is) the preferred medium. Indeed, the RRL that was developed by Pelham and Jackson (15) played a pivotal role in all studies of eukaryotic protein synthesis. For certain viral mRNAs (e.g. poliovirus genomic mRNA), however, the RRL is deficient as it does not offer all cellular factors required to consistently initiate translation (see below). In this special case, cell-free HeLa cell extracts are superior over the RRL (7).

The "scanning hypothesis" provides an explanation for events preceding the first peptide bond formation in translation of the vast majority of eukaryotic mRNAs. There are many exceptions, most notably the genomes of a selected group of animal, plant, and insect "plus strand RNA viruses". The genome of this group of viruses within plus stranded RNA viruses lack the pivotal cap-structure yet it is efficiently used as mRNA. Satellite tobacco mosaic virus (STNV) was the first eukaryotic plus strand RNA virus whose genome was found to lack a cap structure (its 5' end is ppN; (16)).

The elucidation of the complete sequences of picornavirus RNAs revealed even more striking features. The 5' end of the genomes of these viruses is covalently linked to the small protein (VPg). VPg is removed from all newly synthesized viral RNA destined to engage in protein synthesis, yielding a 5'-terminal pU of the viral mRNA (17-19). Moreover, the initiating AUG is located several hundred nucleotides (nt) downstream of the 5' end and the intervening sequence may harbor multiple non-used AUG triplets, sometimes in good 'Kozak context'. Finally, computer-aided studies on RNA structures of the picornaviral 5'NTRs predicted very stable stem-and-loop structures which would greatly interfere if not completely abolish, scanning of the ribosomal subunit (20-23). Nevertheless, the 5'NTR of EMCV was known to very efficiently direct translation of reporter genes in RRL. These features of picornaviral mRNAs led us to investigate whether the picornaviral 5'NTR recruits ribosomes directly into the polynucleotide chain near the initiation codon without scanning from the 5' end. Artificial dicistronic mRNAs were generated to probe the possibility of internal ribosomal binding to the 5'NTR of EMCV (Fig. 1A). The dicistronic mRNAs contained the 5'NTR of PV which is inefficient in RRL (see below), followed by a reporter gene A, a part of the EMCV 5'NTR known to direct excellent translation in RRL, followed by reporter gene B (Fig. 1A). In translations directed by this dicistronic mRNA *in vitro*, the product of gene B appeared earlier than the product of gene A, and translation efficiency of gene B was not affected by the preceding gene A (1). These and other experiments clearly indicated that translation of gene B was 5' end independent. It was correctly concluded that the 5'NTR of EMCV was necessary and sufficient for translation of the reporter gene B. The most plausible explanation for this phenomenon is that the 5'NTR of EMCV contains an element recruiting ribosomes to the initiation codon independently from the 5' end. This entity in the 5'NTR of EMCV directing cap-independent translation was named 'internal ribosomal entry site (IRES)'. Pelletier and Sonenberg (2) showed independently that the 5'NTR of PV contains a genetic element functioning similarly to the EMCV IRES. Soon followed the discovery of IRES elements from other picornaviruses, such as foot-and-mouth disease virus (FMDV), human rhinovirus (HRV), and hepatitis A virus (HAV) (23). Ultimate proof that IRES elements promote internal ribosomal entry *in vivo* was provided by the construction and genetic analysis of a dicistronic PV (7) and *in vitro* by translation of a circular mRNA containing the EMCV IRES element (24).

The Discovery of Cellular IRES Elements

It has been known that PV infection inhibits protein synthesis directed by cellular mRNAs. Cellular mRNA translation in PV-infected cells may cease 3 hours post infection, whereas translation of PV mRNAs may continue for several more hours. Efforts to understand the mechanism of the discriminatory inhibition of cellular mRNA translation in PV-infected cells revealed a correlation between the blockage of translation of cellular mRNAs and the cleavage of a cellular protein of apparent molecular weight of 220 kDa that was historically called p220 (25,26). p220, now known as eukaryotic initiation factor 4G (eIF4G), appears to function as a scaffolding protein that links mRNA with the 40S ribosomal subunit. This is accomplished by its interaction with both RNA-binding proteins, such as eIF4E, eIF4A, eIF4B, and poly(A)-binding protein (PABP), and the 40S ribosomal subunit via the eIF3 complex. eIF4G is a member of eIF4F complex along with eIF4E, the cap binding protein, and eIF4A, a helicase (12).

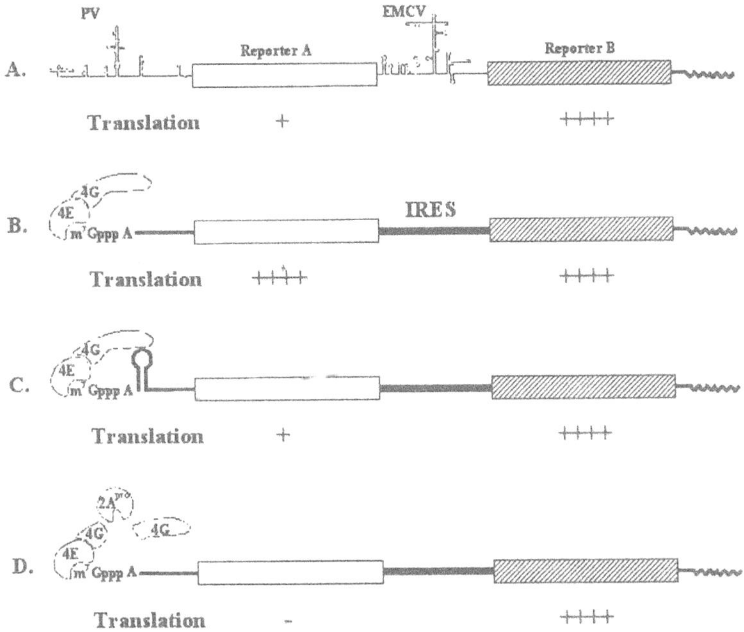

Figure 1. Constructs used for identifying IRES elements. *The assays utilize dicistronic mRNAs containing two reporter genes, A and B. Panel A depicts an artificial dicistronic mRNA that led to the discovery of the IRES element in the 5'NTR of EMCV. In this case, the dicistronic RNA was translated in the RRL, a cell-free system that is poor for the function of the poliovirus IRES but excellent for the function of the EMCV IRES. In most studies, the first gene is expressed by cap-dependent translation via the cap-structure indicated by m^7 Gppp, while the second gene is translated under the control of an RNA segment of interest in the intergenic region, depicted by thick lines (panels B to D). A hairpin structure of RNA is introduced at the 5' end of the dicistronic mRNA (panel C) to block cap-dependent translation. Panel D represents the inactivation of eIF4G by viral proteases or by chemicals. 4E, 4G and $2A^{pro}$ stand for eIF4E, eIF4G and protease $2A^{pro}$ of PV or HRV, respectively.*

Members of the genera *Enterovirus* (e.g. PV) and *Rhinovirus* (e.g. HRV) of the *Picornaviridae* family encode a proteinase $2A^{pro}$ that cleaves eIF4G (27,28). Similarly, members of the genus *Aphthovirus* (e.g. FMDV) of *Picornaviridae* encode a proteinase L that is different from $2A^{pro}$ yet efficiently cleaves eIF4G (29).

eIF4G is composed of the N-terminal domain containing eIF4E- and PABP-binding sites, the central domain containing eIF3- and eIF4A-binding sites, and the C-terminal domain containing the second eIF4A- and Mnk1 (a protein kinase)-binding sites. Proteolytic cleavage of eIF4G by the viral proteases results in separation of the N-terminal domain from the central and C-terminal domains. The C-terminal two-thirds of eIF4G generated by virus infection supports translation of cap-independent translation through IRES elements. Cap-dependent translation via scanning, on the other hand, is impaired due to the lack of the eIF4E-binding site in cleaved eIF4G that retains the pivotal eIF3-binding site. In other words, PV infection inhibits translation of cap-dependent mRNAs mediated by scanning but not IRES-dependent translation. It should be noted that EMCV does not encode a proteinase capable of cleaving eIF4G. Instead, EMCV infection leads to dephosphorylation of 4E-BP1, an eIF4E binding protein. Underphosphorylated 4E-BP1 binds to eIF4E and prevents the interaction between eIF4E and the cap-structure. As a consequence, cap-dependent translation is inhibited (30).

Selective inhibition of cap-dependent translation by PV has been used to identify cellular mRNAs containing IRESs. Sarnow and his colleagues found that a cellular mRNA encoding immunoglobulin binding protein BiP was immune from the translational inhibition by PV infection (31). Translation assays with artificial dicistronic mRNAs containing the BiP 5'NTR in the intergenic space revealed that translation of the second cistron occurs without a requirement for ribosomes to traverse the first cistron (32). This result strongly suggested that the BiP 5'NTR functions as an IRES.

Different experimental approaches have been used to provide evidence for the presence of an IRES element in an mRNA of interest (Fig. 1). In these experiments, RNA transcripts generated *in vitro* are translated in an *in vitro* system or transfected into suitable host cells, or DNAs containing the sequence of a dicistronic mRNA are transfected into suitable cells transcribing the plasmid DNAs so that the expression of reporter genes can be measured *in vivo*. The following criteria define an IRES.

First, the candidate RNA segment should support translation of the reporter gene B placed downstream of the putative IRES in an artificial dicistronic messenger (Fig. 1, panel B). Second, translation of the reporter B should remain unchanged even if translation of the first cistron (reporter A), which may use the scanning mechanism, is inhibited by a stable stem-and-loop structure between the cap and the initiating AUG (Fig. 1, panel C). Third, translation of the reporter B should be resistant to inhibition of cap-dependent translation either by PV super infection of the transfected cells (Fig. 1, panel D), or by chemicals specifically inhibiting cap-dependent translation such as rapamycin (33). Many cellular IRESs have been discovered satisfying some or all of these criteria (see below).

The Structure of IRES Elements

It is apparent that the viral IRES elements consume huge segments of the viral genomes (7% of the PV polynucleotide) that are folded into higher order structures (Fig. 2). At first glance, it appears as if these viral IRES elements contain much more genetic information than that would be required for attracting a ribosomal subunit (34). It is apparent that viral IRESs share little sequence or structural homology (Fig. 2). Therefore IRESs are defined by function, not by structure. This has been demonstrated by exchanging the cognate IRES of PV with that of other viruses, such as EMCV or HCV, thereby generating viable chimeric viruses (35-37).

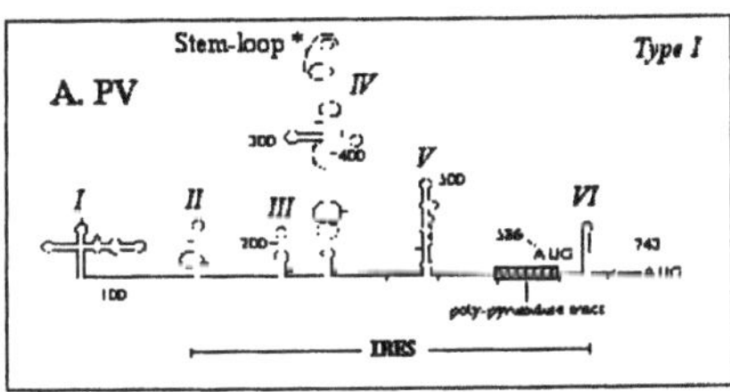

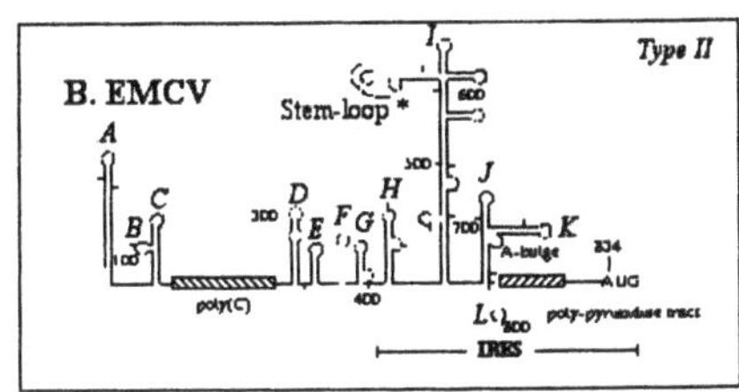

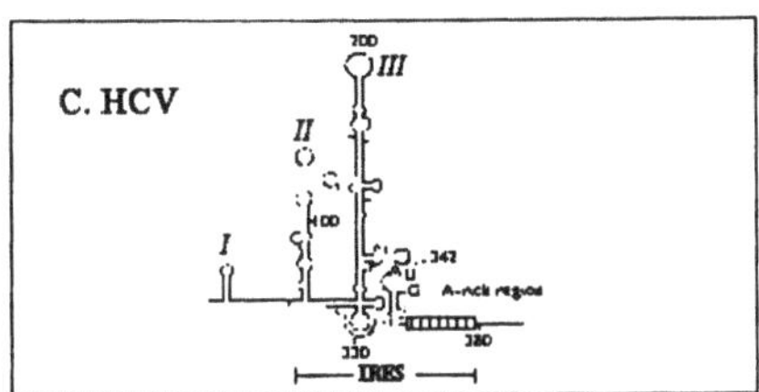

Figure 2. Schematic presentation of secondary structures of viral IRESs. *Panel A depicts a type I picornavirus IRES, represented by PV-1; panel B, a type II picornavirus IRES represented by EMCV, and panel C a flavivirus IRES represented by HCV. Numbers indicate nucleotides starting from the 5' end of genomic RNAs, and small bars mark 20 nucleotides. Structural domains of the RNAs have been arbitrarily identified by upper case alphabets (panel B) or by roman characters (panels A and C) (see text for references). A poly-pyrimidine tract, conserved throughout picornavirus IRESs is shown by a hatched box, and a stem-loop marked with an asterix (*), that is conserved in types I and II IRESs except HAV IRES, is indicated by a half circles. The poly(C)-tract in EMCV, the A-bulge at the J-K junction of EMCV, and the A-rich region in the 5'-terminal region of the core-coding sequence in HCV are also indicated. The areas required for IRES activities of the viral mRNAs are marked by a bar. The pictures are not drawn to scale, and some of the structures may have to be refined.*

The borders of IRES elements of picornaviruses have been defined by analyzing deletion mutants of the 5'NTRs in standard dicistronic mRNAs (22,38). Computer-aided prediction, phylogenetic analysis and biochemical studies of RNA secondary structure of picornaviral IRESs have led to structures shown in Fig. 2 (20,21,39,40). Schematic secondary structures of PV and EMCV are shown on panels A and B in Fig. 2, respectively. It has been subsequently suggested to divide the picornaviruses into two types (23): type I IRESs are carried by enteroviruses (e.g. PV) and rhinoviruses (e.g. HRV), while type II IRESs are carried by aphthoviruses (e.g. FMDV), cardioviruses (e.g. EMCV) and hepatoviruses (e.g. HAV), although the latter form a subgroup of type II IRESs.

The importance of higher order structures or single-stranded motifs in IRES function has been demonstrated largely by genetic approaches (making mutations, compensatory mutations, linker scanning mutations, etc.) (see 40). Remarkably, it has been reported that defective IRESs can be complemented in transfected cells *in trans* (41-43). It has been postulated that complementation may involve RNA-RNA interactions between different domains of the IRES. Complementation, however, has not been achieved in cell-free translation mixtures.

A striking feature shared by all picornaviral IRESs is the presence of a poly-pyrimidine tract (Yn tract; n = 5 to 7 nucleotides) at about 20 nucleotides (a spacer designated by Xm) upstream of an AUG triplet (panels A and B in Fig. 2). This Yn-Xm-AUG motif (m = approximately 20 nucleotides of unspecified sequence)(44) is important for IRES function with respect to the length of both the Yn tract and Xm spacer (22,23,45,46). Moreover, a GNRA tetra loop (stem-loop * on panels A and B in Fig. 2) is conserved in most of the picornaviral IRESs. The importance of this sequence was investigated by using random mutagenesis of the tetra loop and selection of the translationally active mRNAs (47). Optimal IRES activity was obtained from RNAs with the hairpin loop sequence fitting a RNRA consensus (R = purine, N = any nucleotide, A = adenine) (47). All functional IRES elements had a tetra loop with a 3' terminal A residue. However, the 5' terminal G residue does not seem to be essential for IRES function even though it is conserved in all known cardiovirus IRESs.

The IRES of hepatitis C virus (HCV; (3; for the latest structural arrangement see (48) and references therein) is also shown in Fig. 2C. Its structure differs from type I or type II IRESs of picornaviruses, and the recruitment of the 40S ribosomal subunit to the initiation codon occurs via a different mechanism.

Where is a Ribosome Loaded onto the IRES?

Typically picornaviral IRESs are about 450 nucleotides long. Where would the ribosome bind within this large RNA structure? As mentioned above, there are conserved Yn tracts about 20 nucleotides upstream of an absolutely conserved AUG, forming the Yn-Xm-AUG motif (23,44). The sequences between the Yn tract and the AUG triplet (Xm) vary among picornaviral IRESs, but the distance between the Yn tract and the AUG triplet is conserved. When the distance between the Yn tract and the AUG is artificially increased, translational efficiency of the mRNA decreased gradually (46). On the other hand, when the distance was reduced, translational efficiency of the mRNA dropped significantly in the case of PV mRNA (46). Alternatively, an AUG downstream of the original AUG was used as initiation codon when the length of Xm of EMCV was reduced significantly (22). These observations suggested that an intact Yn-Xm-AUG motif aids in efficient ribosome loading onto the IRES element.

In the commonly studied EMCV R strain, which contains a type II IRES, translation initiates mostly at the 11th AUG at nucleotide 834. A small proportion of translation commences at the 12th AUG located 12 nucleotides downstream of the 11th AUG. Remarkably, little if any translation was detected from the 10th AUG that is located just 8 nucleotides upstream of the 11th AUG, despite the fact that the 10th AUG is in good 'Kozak context'. Moreover, this 10th AUG is competent for translation via scanning when 5' deleted EMCV IRES segments were assayed (49,50). This strongly suggests that in the EMCV IRES the ribosome is loaded very near or directly onto the 11th AUG.

Whereas the ribosome loading site in type II picornavirus IRESs appears to be located close to, or at, the initiating AUG of the Yn-Xm-AUG motif, the situation in type I IRESs is much more complex. In poliovirus, the Yn-Xm-AUG596 motif is upstream of, and separated from, the initiating AUG747 codon by a "spacer" of 154 nucleotides (in rhinoviruses, the spacer is only 19 nucleotides long). In contrast to type II IRESs, the AUG of an IRES I motif is cryptic and it is rarely, if ever, used as initiation codon. In poliovirus, the AUG of Yn-Xm-AUG596 is in poor Kozak context (CxxAUG596) and out of frame with the ORF of the polyprotein. Translation (*in vitro*) from AUG596 of the Yn-Xm-AUG596 was observed only when the context of its AUG was upgraded (yielding the expected small protein of 6 kDa; (51,52)). Curiously, this genetic modification did not significantly decrease translation from AUG747. Nevertheless, the importance of the Yn-Xm-AUG596 motif in poliovirus has been supported by numerous genetic analyses: changes of the motif led to replication phenotypes of mutant viruses; efficient replication was restored only through pseudo reversions in which a Yn-Xm-AUG motif was reestablished (23,46,53). These data strongly indicate that loading of the 40S ribosomal subunit into type I IRESs occurs at their Yn-Xm-AUG motif, just as in type II IRESs.

How does the 40S ribosomal subunit reach the initiating AUG747 codon? A plausible transfer would be scanning. This mechanism was supported by Kuge and his colleagues who inserted a segment of 72 nucleotides containing an AUG codon into the spacer of 154 nucleotides (54). The insertion resulted in replication phenotypes. Pseudo reversions either deleted the sequence or changed the inserted AUG codon (54). This was contrasted by experiments in which a stable hairpin or an AUG codon was introduced into the spacer near domain VI. The insertions only modestly impaired translation of the polyprotein (55). It appears as if the effect of insertions into the spacer on ribosomal transfer crucially depends on the location of the insert. A mechanism of ribosome shunting (56) for the transfer of the ribosome from Yn-Xm-AUG596 to AUG747 cannot be excluded at present.

Remarkably, deletion of nt 564 to 726 of the poliovirus 5'NTR which included the AUG596 codon and the spacer, was not lethal for viral replication but merely yielded a replication phenotype (23,46,57). This surprising result can be explained as follows: in spite of the massive deletion, a new Yn-Xm-AUG motif was formed by the initiating AUG747 such that the original Yn-Xm-AUG596 motif was replaced by a new Yn-Xm-AUG747 motif. The IRES in these deletion mutants thus resembles a type II IRES. It remains a mystery, however, why entero- and rhinoviruses have retained (or acquired) the spacer between the important Yn-Xm-AUG motif and the AUG initiating polyprotein synthesis.

Requirement of Canonical Translation Factors in IRES Dependent Translation

Numerous cellular factors are required for the recruitment of 40S ribosomal subunit and, subsequently, the 60S ribosomal subunit to the initiation site of translation. These factors have been mostly identified by using cap-dependent mRNAs and cell-free translation systems. Is the entire set of cellular factors that is required for cap-dependent mRNA (often called canonical initiation factors) also essential for IRES-dependent translation?

Available evidence suggests that, indeed, most of canonical translational factors are required for translation of IRES-dependent translation (for diminished requirements of the HCV IRES, see below). The exception is eIF4E that binds specifically to the cap-structure. As mentioned, some picornaviruses produce proteinases that cleave eIF4G and liberate the N-terminal domain of eIF4G where eIF4E binds. Although the cleaved eIF4G can no longer direct cap-dependent translation, the central domain of its cleavage fragment can bind eIF3 and eIF4A that is necessary for picornavirus IRES-dependent translation (58). It has been reported that cleavage of eIF4G by picornavirus proteinases inhibits the function of a type II IRES of hepatitis A virus (59-61). The reason for this phenomenon (apparent requirement of intact eIF4G for IRES-dependent translation), exceptional amongst picornaviruses, remains to be elucidated.

The role of canonical translational factors and their binding to IRES elements have been systematically analyzed with RNA foot- and toe-printing methods by using purified translational factors and ribosomal subunits. In this respect, the IRESs of hepatitis C virus (HCV; Fig. 2C), which belongs to family *Flaviviridae*, EMCV, and FMDV have been studied most extensively. Remarkably, it was found in a reconstituted binding assay that the 40S ribosomal subunit can bind to the HCV IRES in the immediate vicinity of the initiation codon (resulting in a 48S complex) without assistance of any of the canonical initiation factors (62-64). The additional presence of eIF2/GTP/Met-tRNAi ternary complex was then required for precise positioning of the 40S ribosomal subunit to the initiation site (62). The bottom half of stem-loop III, including the pseudoknot structure and the region surrounding the initiation codon, are necessary and sufficient for the 40S ribosomal subunit binding (64). A direct contact of S6, a 40S ribosomal subunit protein, to this segment of the HCV IRES was detected by UV cross-linking (62). However, the minimal region required for 48S complex formation (the lower part of domain III) is not sufficient for IRES function, since domain II as well as most of the apical part of domain III are also required for IRES activity. Therefore, the formation of the 48S complex is probably only one of several steps required for commencement of IRES-dependent translation.

Initiation factor eIF3 is usually required for association of the 40S ribosomal subunit to mRNA. In the case of the HCV IRES, eIF3 is dispensable for this step but, curiously, it is absolutely required for 60S ribosomal subunit joining to the 48S complex along with a 50-70% ammonium sulfate subfraction (62). Toe printing and UV cross-linking assays have revealed that eIF3 bind to the apical half of domain III (65,66). It is conceivable that eIF3, while not directly associated with the 40S subunit but positioned closely to it through an interaction with the apex of domain III, somehow facilitates interaction between 40S and 60S ribosomal subunits. Other initiation factors such as eIF4A, 4B, 4E, and 4G are required neither for 48S complex formation nor for 80S complex formation (62). In accordance with this conclusion, a dominant negative mutant of eIF4A had no effect on translation of HCV RNA (62). The process of 80S ribosomal complex formation onto the classical swine fever virus (CSFV) IRES, which is related to HCV IRES, is overall very similar to that of the HCV IRES (62).

Other examples of IRES elements directly interacting with the 40S ribosomal subunit are the IRESs in the insect viruses Plautia stali intestine virus (5) and cricket paralysis virus (CrPV) (6). The genome of these insect viruses is dicistronic, that is, it consists of two ORFs that are separated by an internal IRES element. Surprisingly, the IRES element in the genome of CrPV can form 80S ribosome complexes with a CCU triplet in the ribosomal P site and a GCU triplet in the A site (9). As a consequence, translation of the second ORF encoding structural proteins starts with an alanine residue instead of usual methionine residue and no Met-tRNA is required (see below).

Unlike the HCV IRES, the EMCV IRES does not bind a salt-washed ribosome. Binding of the denuded 40S subunit requires eIF2, -3, and -4F, while eIF4B stimulates 48S complex formation about 2 fold (67). The eIF4G component of eIF4F is functionally replaceable with the central domain of eIF4G (58). A specific RNA-binding protein called pyrimidine tract-binding protein (PTB) further stimulates 48S complex formation (see below).

Toe- and foot-printing assays revealed that eIF4G binds to the basal stem and the A-bulge in the J-K domain of the EMCV IRES close to the initiation site (58,67,68). eIF4G associated with eIF4A binds to the EMCV IRES by 2 orders of magnitude stronger than eIF4G alone (69). Interaction of eIF4G with the FMDV IRES correlates well with the activity of the IRES (70). The region in FMDV equivalent to the J-K-L domain of EMCV binds to eIF4B independently of PTB (70-72).

The Identification of Cellular RNA-Binding Proteins in IRES Activity

It has been known over 20 years that PV mRNA is translated inefficiently and inaccurately in RRL, but that accurate translation can be achieved by the addition of a HeLa cell extract (73-75). Similar results have been reported for translation of HRV RNA (59,76). Indeed, the activities of IRES elements have been reported to be different in different cell lines (77). This indicates that at least some IRESs require different cellular factor(s) other than canonical initiation factors. UV cross-linking experiments using different IRES RNA probes contributed to identifying several cellular proteins interacting with specific IRESs.

Human antigen La: UV cross-linking of PV RNA revealed a cellular protein of 52 kDa specifically interacting with nucleotides 559-624 of PV IRES (78). This protein was subsequently identified as human antigen La that is recognized by antibodies from patients with autoimmune disorders such as systemic lupus erythematosus and Sjogren's syndrome (79). Addition of La protein to RRL, which contains a limited amount of La, stimulates translation of PV RNA and reduces aberrant translation (79). Interestingly, the La protein, which resides in the nucleus, is redistributed to the cytoplasm in the PV-infected cells. This is the consequence of the removal of the nuclear retention signal at the C-terminal end of the protein by the proteinase activity of PV $3C^{pro}$ (80).

It is now known that the La protein binds to several IRES elements and enhances translation. La-mediated stimulation of translation *in vitro*, however, requires relatively large amounts of protein. In any event, La binds to the initiation codon of HCV IRES and enhances translation when it is supplemented to RRL (81). Addition of RNA which was generated by La/SELEX, into RRL reduced translation of HCV mRNA, while addition of purified La protein along with the competitor RNA restored translation of the mRNA. Moreover, overexpression of the competitor RNA in the cell inhibited translation of HCV RNA (82). This strongly suggests that La protein may play a positive role in HCV mRNA translation.

La protein also binds to the cellular IRES of X-linked inhibitor of apoptosis (XIAP) (83) that is up-regulated by low dose ionizing irradiation (84). A dominant negative deletion mutant of La (amino acids 226-348) reduced translation of XIAP mRNA *in vitro* and *in vivo*. This suggests that La protein is also involved in cellular mRNA translation. Intriguingly, La mRNAs that are composed of at least two isoforms (La1 and La1') through alternative splicing, themselves contain IRES elements in the 5'NTRs (85). This may be a survival strategy of cells under stress conditions that suppresses cap-dependent translation. Even under stress conditions, La protein may be translated continuously through the IRES elements. La protein, in turn, may assist translation of some mRNAs such as XIAP that are required for coming out of the stress response when environmental conditions become normal.

PTB, unr and GAPDH: A cellular protein of 57 kDa (originally named p57) strongly binds to the 5' border of EMCV IRES (stem-loop H) (22). Binding of the protein to the IRES element correlated well with the IRES activity of IRES mutants (22). Polypeptide p57 is polypyrimidine tract binding protein (PTB) (86), also known as hnRNP I, and its cellular function is to modulate alternative splicing of certain pre-mRNAs. Depletion and repletion experiments of PTB in cell-free extracts showed that it is required for translation of a certain strain of EMCV (87,88). Similar experiments revealed that PTB is necessary for translation of FMDV, another picornavirus containing a type II IRES (89).

PTB is also required for translation directed by type I IRESs such as in HRV and PV. The type I IRESs, however, demand more PTB protein for translation than type II IRESs because the latter have higher affinity to PTB. Supplementing PTB to RRL enhances translation of HRV mRNA (90) and PV mRNA (91). An additional cellular protein called unr (upstream of N-ras), which contains five cold-shock domains and is essential for embryo development (92-94), is required for efficient translation of HRV mRNA in RRL (91). The enhancing effect of HRV mRNA translation by UNR and PTB is additive or sometimes synergistic (91).

PTB binds to the HCV IRES and is required for the IRES function (95), although Kaminski and her colleagues presented evidence contradicting this conclusion (87). However, a requirement of PTB for HCV IRES function was further supported by using the SELEX method similarly to the experiment of La protein described above (96). The necessity of PTB in translation under the direction of PV, HAV, and HCV IRESs was systematically investigated by using artificial dicistronic mRNAs containing the PTB gene as the first cistron, different IRESs at the intercistronic region, and the CAT reporter gene as the second cistron. Upon transfection to BS-C-1 cells, containing limited amount of PTB, expression of PTB stimulated activities of HCV, PV, and HAV IRESs by 5-, 12-, and 37-fold, respectively (97). All published data together strongly suggest that PTB is generally required for, or may at least enhance, IRES activities of all three major types of viral IRESs shown in Fig. 2.

The accessibility to PTB seems to play a role in the regulation of translation of some IRES-dependent mRNAs. For instance, glyceraldehyde 3-phosphate dehydrogenase (GAPDH) competes with PTB for the binding to stem-loop IIIa in the IRES element of HAV that resides at the equivalent position of stem-loop D of EMCV ((98); Fig. 2B). GAPDH suppresses translation of HAV mRNA possibly by changing the secondary structure of the IRES element (98,99). Binding of PTB, on the other hand, enhances translation of HAV mRNA in transient expression systems (99). Interestingly, HAV adapted to cell-culture contained a mutation in the IRES element reducing affinity to GAPDH consistent with the inhibitory role of GAPDH in translation (99).

PTB itself shows an inhibitory effect on translation of certain cellular mRNAs containing an IRES element. PTB binds to the cellular IRES of BiP, and overexpression of PTB inhibits translation directed by the BiP IRES (100). The mechanism of the translational inhibition remains to be elucidated. PTB also binds to the cellular IRES of vascular endothelial growth factor (VEGF) (101), but the effect of this interaction remains obscure.

PCBP: Poly(rC) binding protein 2 (PCBP 2), also known as hnRNP E2, was identified as a cellular protein specifically interacting with domain IV of PV (Fig. 2A) (102). Depletion of PCBP2 from HeLa cell-free extracts using a stem-loop RNA affinity column resulted in inefficient translation of PV mRNA in the system. Translation was restored by addition of recombinant PCBP2, but not by PCBP1, a closely related member of the protein family (103). Moreover, it was reported that type II IRESs do not require PCBP2 for its activity by using the same experimental strategy although the EMCV IRES could bind the protein (104). Among three K homology (KH) domains in PCBP2, KH1 is responsible for interaction with the PV encoded proteinase $3CD^{pro}$, and for binding to RNA (105). A truncated PCBP2 protein containing KH1 domain shows dominant negative effects on PV mRNA translation (105). Together, these data strongly suggest that PCBP2 plays an important role in PV IRES-dependent translation. PCBP2 also binds, albeit weakly, to a second site in PV 5'NTR, the 5'-terminal cloverleaf (CL) structure (domain I in Fig. 2A). At the same time, PCBP2 interacts with PV $3CD^{pro}$ forming a CL/PCBP2/$3CD^{pro}$ complex (106). Based on these molecular interactions and considering the increase of RNA-binding affinity of protein complex $3CD^{pro}$/PCBP2 to domain I, Gamarnik and Andino proposed a molecular switch mechanism from translation to replication of PV genomic RNA (107,108). According to this model, PCBP2 that is bound to domain IV of PV IRES, drives translation of PV mRNA. Upon production of PV proteins, $3CD^{pro}$ then forms the CL/PCBP2/$3CD^{pro}$ complex thereby lowering the affinity of PCBP2 to domain IV. As a consequence, translation of PV mRNA decreases and, instead, replication of PV RNA commences. However, further investigations are required to explain the concomitant occurrence of replication and translation in the PV-infected cells (109).

PCBP2 also binds to the HAV 5'NTR and enhances translation (110). Depletion and repletion experiment supports a positive role of PCBP2 in HAV RNA translation (110). PCBP1 and PCBP2 bind to the 5'NTR of HCV (111), even though the effect of this association in translation remains obscure.

ITAF45 and attenuation of neurovirulence: The cellular transacting factor 45 (ITAF45), also called murine proliferation-associated protein (Mpp1) that is expressed in a cell cycle and proliferation-dependent manner (112,113), binds specifically to the IRES of FMDV, an aphthovirus, and the IRES of Theiler's murine encephalomyelitis virus (TMEV strain: GDVII), a highly neurotropic mouse cardiovirus (114). Both of these viruses carry type II IRESs. For 48S complex formation with the TMEV IRES, eIF2, eIF3, eIF4A, eIF4B, eIF4F, and PTB are required (114). In addition to all these factors, ITAF45 is also mandatory for the equivalent complex formation on the FMDV IRES (114). These observations explain why FMDV is not neurotropic in mice: ITAF45, is not present in the brain tissue of mice (114). On the other hand, the TMEV IRES does not require ITAF45 for function and, thus, TMEV proliferates well in the mouse CNS. This is a good example of intracellular restriction of viral proliferation in a tissue specific manner. A special case of IRES-mediated tissue tropism has been observed in PV/HRV2 chimera in which the PV IRES was exchanged with that of HRV. The chimera, although proliferating in the HeLa cells like PV, is severely restricted in proliferation in human neuroblastoma cells (37,115).

hnRNP L: Heterogeneous nuclear ribonucleoprotein L (hnRNP L) which is homologous to and interacts with PTB, binds to the 5'-terminal nucleotides of the core-coding sequence of HCV (116). The nucleotide sequence downstream of the initiating AUG342 is known to be important for efficient translation of HCV mRNA (36,117). Translational efficiency and affinity to hnRNP L increases gradually as the length of the core-coding sequence expands up to nucleotides 400 of the HCV genome. The role of hnRNP L in translation remains to be elucidated.

What is the Role(s) of RNA-Binding Proteins in IRES-Controlled Translation?

As outlined above, there is compelling evidence that viral IRES-dependent translation requires cellular RNA-binding proteins that are not required for translation of cap-dependent cellular mRNAs. However, the role(s) of these RNA-binding proteins in translation is poorly understood. So far, only relatively few cellular RNA-binding proteins have been discovered that bind to IRESs and enhance their function. This is surprising considering the diversity and abundance of RNA-binding proteins in cells and the enormous size of the IRES elements. Moreover, these cellular RNA-binding proteins seem to be involved in translation of a variety of IRES elements that do not share apparent similarities in primary sequences and secondary structures. However, it is possible that all viral IRESs share a common motif (higher order structure), perhaps of tertiary structure, that is exposed to the IRES-specific proteins for specific complex formation.

We can envision several possible roles of RNA-binding proteins in IRES-dependent translation. The following possibilities are not mutually exclusive (Fig. 3). First, an RNA-binding protein(s) may recruit the translational machinery through a protein-protein interaction or protein-RNA interaction (Fig. 3A). The putative RNA-binding protein may bind directly to the ribosomal subunit, to a canonical translation factor, or even to a putative mediator protein that links RNA-binding proteins with the basal translational machinery. In this respect, the interaction of La protein with 40S ribosomal subunit, perhaps by direct association with 18S rRNA, may give a clue to this kind of mechanism (118). Interactions of other RNA-binding proteins with canonical translational factors and/or components of the ribosomal subunit remain to be investigated.

A. An RNA-binding protein directly interacts with a canonical translation factor or ribosome.

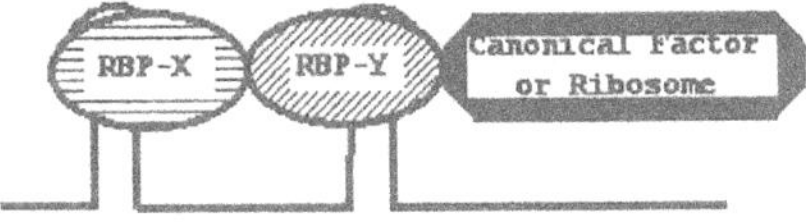

B. RNA-binding proteins maintain a proper RNA conformation.

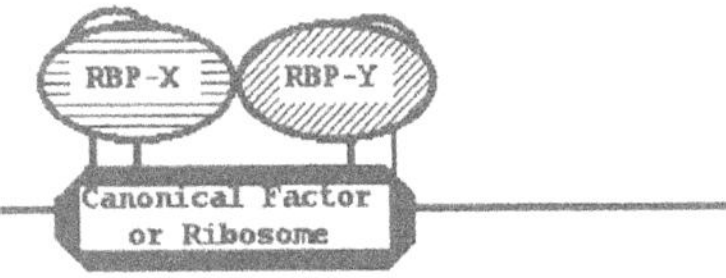

C. RNA-binding proteins function as matchmakers.

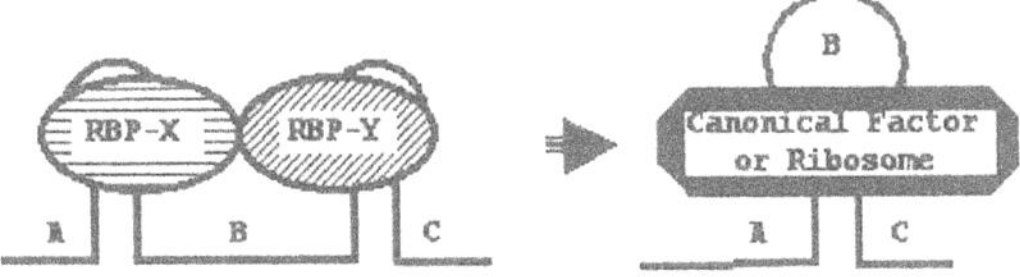

Figure 3. Possible roles of RNA-binding proteins in translation. *RNA-binding proteins are depicted by ovals. Canonical translation factors with or without the basal translational machinery are represented by octagons. Lines depict RNAs.*

Second, RNA-binding proteins may serve as 'clamping proteins' that hold different parts of RNA in a certain configuration (Fig. 3B). Components of the translational machinery may bind exclusively to the RNA portion of the RNA-protein complex maintained by the clamping proteins. In this respect, the requirement of PTB for IRES-dependent translation of EMCV mRNA may serve as an example. The wild type EMCV IRES directs efficient translation without PTB in a reconstituted translation system *in vitro* (88). On the other hand, a mutant EMCV IRES containing an additional adenosine residue at the A-bulge at the stem-loops J-K junction (Fig. 2B) absolutely requires PTB for its function (22,87). It is conceivable that PTB augments translation from the mutant EMCV IRES *in vitro* by maintaining a proper conformation of the RNA required for interaction with translational machinery.

It is worth to note that many RNA-binding proteins engage in homo- or hetero-oligomeric interactions. For instance, PTB (hnRNP I), PCBP 2 (hnRNP E2), hnRNP L, and hnRNP K which all function in IRES-controlled translation, are capable of homomeric interactions with themselves and in heteromeric interactions with others (119). The domains in an hnRNP responsible for protein-protein interactions with different hnRNPs are not the same, even though some parts are overlapping (119). The existence of hnRNP complexes has been well documented by Krecic and Swanson (120). Moreover, RNA-binding proteins often contain more than one RNA-binding domain. For example, PTB and hnRNP L contain four RNA recognition motifs (RRMs), and PCBP and hnRNP K contain three K homology (KH) domains. Considering the presence of multiple RNA-binding domains together with the protein-protein interactions among hnRNPs, it is conceivable that hnRNPs form protein complexes with many "hands" that exert specificity to many different RNA species. It is also possible that a variety of hnRNP complexes with different components of hnRNPs exist in the cell, and that they play different roles in different subcellular compartments (120,121). Since each individual hnRNP possesses a distinct - binding specificity and structure, different hnRNP complexes should have different affinity to different RNA molecules. Therefore, different hnRNP complexes, possibly with other RNA-binding proteins, may interact with different types of IRESs. Alternatively, a putative hnRNP complex with multiple RNA-binding sites may interact with different IRESs by facing the RNAs with a different part of the complex.

An interesting *Xenopus oocyte* assay system has been developed to identify cellular factors necessary for IRES-dependent translation (122). Co-injection of a HeLa cell cytoplasmic extract (or injection of HeLa cell mRNAs prior to the IRES assay) was required for efficient PV IRES activity in *Xenopus oocytes* (122). Intriguingly, a protein or a complex of proteins of approximately 300 kDa was needed for translation of PV mRNA in *Xenopus oocytes* (122). This entity is a good candidate for a putative hnRNP complex even though the identity of it remains obscure. Toyoda and her colleagues showed that a 240 kDa protein complex isolated from HeLa cells was able to stimulate PV IRES activity in RRL. Not surprisingly, the large complex contained PTB and La (123).

Third, RNA-binding proteins may play a role as matchmakers (Fig. 3C). RNA-binding proteins may facilitate inducing a certain configuration of IRES RNA, or maintain a proper structure of IRES RNA, for recognition by the translational machinery. The RNA-binding proteins would function as molecular chaperones and be released from the RNA-protein complex after their action. The activity of La protein in nuclear RNA processing may shed light on this aspect of an RNA-binding protein. La protein binds to the 3' end of nascent RNA polymerase III transcripts and facilitate tRNA processing. La protein is required for the processing of mutant tRNA with unpaired anticodon stem (124). La protein also stabilizes newly synthesized U6 RNA and facilitates assembly of U6 RNA into the U6 snRNP (125). In addition, La protein facilitates efficient Sm protein binding, thus assists formation of the U4/U6 snRNP (126). From these observations it was concluded that La functions as a molecular chaperone facilitating RNA-protein interactions.

A COMPARISON OF TRANSCRIPTION AND TRANSLATION

Both transcription and translation are decoding processes of information stored in nucleic acids. An analogy between transcription and translation has been well elaborated by Sachs and Buratowski (14). Here we will discuss common themes in translation and transcription processes considering the phenomena described above.

Direct Binding of Ribosome and RNA Polymerase to Nucleic Acids: Paradigm of Prokaryotic Gene Expression.

Prokaryotic RNA polymerases directly recognize and bind to promoters via specific sequences immediately upstream of the initiation site (127) (Fig. 4A). Transcription activators, which interact specifically with specific sequences in promoters, stimulate transcription by directly interacting with RNA polymerase. Importantly, activators are not universally required for transcription *in vivo* and, hence, are utilized only at a subset of prokaryotic promoters (128). Similarly, prokaryotic ribosomes directly bind near the initiation codon through RNA-RNA interaction between the purine-rich Shine-Dalgarno (SD) sequence residing 5-7 nucleotides upstream of the initiator AUG, and a complementary pyrimidine-rich sequence at the 3' terminus of 16S rRNA (129) (Fig. 4B). Ribosomal protein S1, which binds to U-rich elements, augments translation by binding to a specific region of mRNA (130).

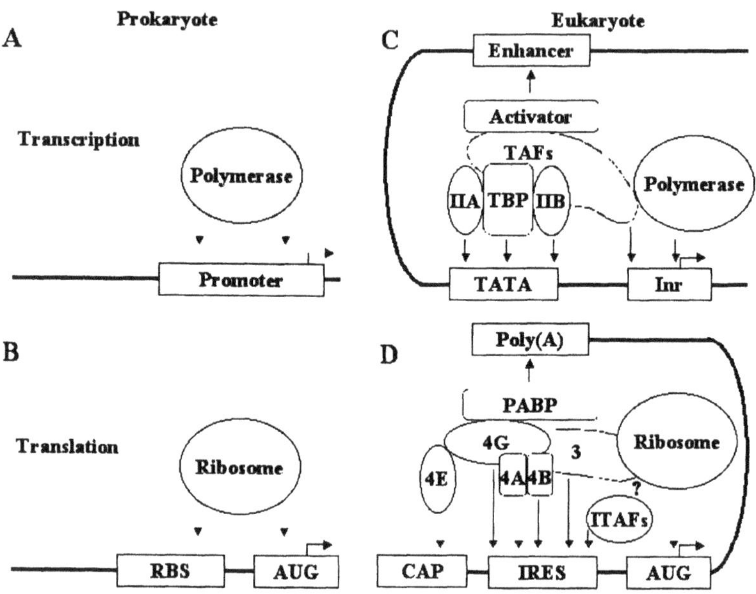

Figure 4. Schematic diagram of transcriptional and translational initiation of gene expression. *Transcription and translation are arbitrarily grouped into prokaryotic and eukaryotic paradigms. In nature, eukaryotes sometimes use a prokaryotic paradigm in transcription and translation, and vice versa. Interactions among proteins are indicated by direct contacts. Arrows indicate interactions between proteins and nucleotides. Thick lines represent either double stranded DNAs in transcription or single stranded RNAs in translation. Abbreviations are as follows: polymerase, RNA polymerase; activator, transcriptional activator; SD, Shine-Dalgarno sequence; TBP, TATA-binding protein; TAF, TATA-binding protein associated factor; IIA, TFIIA; IIB, TFIIB; Inr, initiator region; PABP, poly(A)-binding protein; 4E, eIF4E; 4G, eIF4G, 3, eIF3; 4A, eIF4A; 4B, eIF4B; ITAF, IRES-specific cellular transacting factor such as PTB, La, PCBP, $ITAF_{45}$ and other RNA-binding proteins required for IRES activity.*

It is an intrinsic property of prokaryotic ribosomes and RNA polymerases that they can bind directly to RNA and DNA, respectively. Therefore, regulation of prokaryotic gene transcription occurs mostly by repressors that prevent binding of RNA polymerase to the promoter (127). Translational regulation, in turn, occurs by the accessibility of an SD sequence to ribosomes and rapid turnover of mRNA.

Protein-Assisted Binding of Ribosome and RNA Polymerase: Paradigm of Eukaryotic Gene Expression.

Eukaryotic RNA polymerases and ribosomes do not bind to nucleic acids with high specificity. The binding specificity of these basic transcriptional and translational machineries are provided by a variety of proteins and complexes of proteins that serve as links between nucleic acids and the machineries (14). These mediator proteins render the basic machineries to interact with a variety of nucleic acid elements and to respond to complex signals.

In transcription, the TATA-binding protein (TBP) and its associated factors (TAFs) play a key role in bringing RNA polymerase to the pyrimidine-rich initiator region (Inr) (131) (Fig. 4C). TBP is a DNA-binding protein that binds to the TATA box close to an Inr. TBP binds activators, TAFs, repressors, and general transcription factors. TBP resides not only within a variety of complexes functioning at pol-II transcribed promoters such as TFIID, PCAF, and TFTC but also within SL1 and TFIIIB functioning at pol-I and pol-III promoters, respectively (131). $SNAP_C$ that activate human snRNA promoters, also contains TBP (132). Among the components of TFIID, $dTAF_{II}150$ (d stands for Drosophila) binds to Inr (133), and $dTAF_{II}60/dTAF_{II}40$ contacts a DPE element located downstream of Inr (134).

It is generally believed that sequence specific transcriptional activators facilitate recruitment of TFIID to the promoter. A number of $TAF_{II}s$ interact with transcriptional activators. A variety of activators of the 'acidic' class, including p53 and VP16, interact with $dTAF_{II}40$ (135,136). Spl and NTF activators interact with $dTAF_{II}110$ and $dTAF_{II}150$, respectively (137,138). Hormone receptors function through a number of TAFs including $hTAF_{II}28$ (h stands for human), $hTAF_{II}30$, and $hTAF_{II}135$ (139,140,141), and a variety of activators target $hTAF_{II}55$ (142). These observations underscore a remarkably complex interplay between proteins that may control specific expression of genes.

In translation, we can envision eIF3 as the equivalent of the TAF complexes in transcription (Fig. 4D). Just as the TAF complex interacts with RNA polymerase, eIF3 interacts with the 40S ribosomal subunit. Similarly to the TAF complex, mammalian eIF3 is a big complex of about 600 kDa, composed of at least 11 polypeptides. Numerous translational initiation factors interact with eIF3. eIF1 binds to p110 of eIF3 (143), and eIF5 copurifies with eIF3 (144). eIF4B interacts with p44 of eIF3. The central domain of eIF4G interacts with eIF3 (145), but the eIF3 subunit(s) responsible for the interaction is yet to be identified. Through protein-protein interactions, eIF3 is likely to organize initiation complexes on the 40S ribosomal surface.

eIF3 plays also a pivotal role in translation by interacting with mRNAs. At least four of its subunits, p170, p116 or p110, p66, and p44, directly bind to RNA. It also interacts with mRNA indirectly through protein-protein interaction with RNA binding proteins such as eIF4G and eIF4B. Interaction with eIF4G particularly attracts attention, since eIF4G, in turn, interacts with other RNA-binding proteins eIF4E and poly(A)-binding protein (PABP). eIF4E and PABP are functional analogs of transcriptional activators. eIF4E, the cap-binding protein, facilitates ribosome-binding near the 5' end of mRNA and facilitates 'ribosomal scanning' through the 5'NTR (see above). PABP, on the other hand, binds to the poly(A) tail at the 3' end of mRNA, and enhances translation (146). Simultaneous association of eIF4E and PABP to an mRNA in the presence of eIF4G results in circularization of mRNA (147,148). The cap-structure and the poly(A) tail stimulate translation synergistically through the protein complex of eIF4E, PABP, and eIF4G (146,149,150). Similarly, synergistic stimulation by transcriptional activators occurs in transcription (151). Please note, however, that a certain eukaryotic cellular mRNA (histone mRNA) and numerous viral mRNAs (particularly of plant viruses) lack poly(A) and, hence, evolved exceptional mechanisms to overcome the deficiency.

IRES-Dependent Translation Occurs by Both Direct- and Protein-Assisted Binding of the Ribosome

Whereas the 40S ribosomal subunit is loaded onto the 5' end of mRNA in cap-dependent translation, it is loaded directly onto the IRES element in IRES-dependent translation (see above). Direct binding of eukaryotic ribosomes to the initiation site of mRNA was demonstrated with the cricket paralysis virus (CrPV) IRES, residing in the intergenic region (IGR) (6). Strikingly, the 80S ribosome-mRNA complex is formed with purified 40S and 60S ribosomal subunits without initiator Met-tRNA, eIF2, eIF3, eIF5, eIF5B or GTP hydrolysis (6). A CCU triplet, which is a part of the IRES element, occupies the P-site of 80S ribosomal complex. And translocation of mRNA occurs by the addition of tRNA at the A-site of 80S ribosome-mRNA without peptide bond formation as indicated by the shift in toe print patterns in the presence of cycloheximide. This indicates that the eukaryotic ribosome itself intrinsically can bind to a specific region in the mRNA and commences translation.

The HCV IRES is another example of an mRNA directly interacting with the ribosome. The salt-washed 40S ribosomal subunit can bind to HCV mRNA near the initiation site in the absence of canonical initiation factors (62). 48S ribosome-mRNA complex assembles accurately at the initiation codon on the HCV IRES on inclusion of eIF2, initiator tRNA, and GTP with 40S ribosomal subunit. Intriguingly, eIF3, which is usually required for 48S complex formation, possibly with other factors in 50-70% ammonium sulfate fraction is absolutely necessary for assembly of the 80S complex on HCV mRNA (62). There is no requirement for eIF4A, eIF4B, or eIF4F in formation of the 80S complex on the HCV IRES.

Some IRESs require RNA-binding proteins as well as canonical translation factors for 48S complex formation. eIF2, eIF3, and eIF4F are required for 48S complex formation of the EMCV IRES. eIF4B and PTB stimulate this process (58). 48S complex formation on the TMEV IRES requires eIF2, eIF3, eIF4A, eIF4B, eIF4F and an RNA-binding protein PTB. For the same process on the FMDV IRES, the additional RNA-binding protein ITAF45 is required in addition to the factors required for TMEV IRES function (114).

The roles of cellular factors on IRES-dependent translation are not clear yet. A recent report, however, provides an insight on the role of RNA-binding proteins in translation. Artificial IRESs, composed of 1, 2, or 3 copies of iron-responsive elements (IRE), were inserted into the intergenic region of mRNAs containing two reporter genes (152). These artificial IRESs directed translation of the second cistron in the presence of a fusion protein composed of IRE-binding protein IRP-1 and different parts of eIF4G. The central domain of eIF4G, containing eIF3-binding site and one of the eIF4A-binding sites, was sufficient for the stimulation activity of the artificial IRES (152). This clearly demonstrates that the recruitment of the 40S ribosomal subunit near the initiation codon via an RNA-binding protein leads to translation. Considering this, it is plausible that some of the RNA-binding proteins stimulating IRES-dependent translation function by recruiting ribosomes via protein-protein interactions.

Future Perspectives

An explosion of recent reports has described the existence of cellular IRESs (153-158). Moreover, new RNA-binding proteins interacting with the IRES elements have been reported (101,114,159). Some of the factors are expressed in a cell cycle and development-specific manner (160-162). It can be anticipated that more cellular IRESs will be discovered, since about 3% of mRNAs analyzed by microarrays remain associated in polysomal fraction when most of eIF4G proteins are cleaved by poliovirus infection (163). In addition many of the cellular mRNAs, which require intact eIF4G for function similarly to HRV IRES (60), are likely to contain IRES elements. Such mRNAs would have escaped from the screen described by Johannes et al. (163). Studies on *cis*-acting elements, *trans*-acting factors related to the IRES elements, molecular mechanism of translation, and the modulation of translational efficiency by environmental changes will shed light on understanding the detailed processes in translation.

Summary of key concepts

- ❖ *Translation of eukaryotic mRNAs occurs by either a cap-dependent or an IRES-dependent mechanism.*
- ❖ *The basal translational machinery, including canonical initiation factors and a ribosome, is recruited to the translational initiation codon through protein-protein interactions with RNA binding proteins.*
- ❖ *Regulation of gene expression at the level of translation occurs by changing the availability, or modifying translational factors and RNA binding proteins.*

Study Guide Questions

1) What are the advantages of the utilization of RNA binding proteins in translation over the direct interaction of a ribosome with the initiation codon?
2) Why do so many cellular mRNAs, which play key roles in cell proliferation and apoptosis, contain IRES elements?
3) What are the common features of transcription and translation at the molecular level?

Acknowledgements

The authors are grateful for numerous discussions and suggestions by our colleagues. This work was supported in part by grants of the Molecular Medicine Research Group Programs of MOST, and by HMP-98-B-3-0020 to S.K.J.; and by NIH grants AI15122, AI32100, and AI39485 to E.W.

REFERENCES

1. Jang, S. K., Kraussl ich, H. G., Nicklin, M. J., Duke, G. M., Palmenberg, A. C., and Wimmer, E. (1988). A segment of the 5' nontranslated region of encephalomyocarditis virus RNA directs internal entry of ribosomes during in vitro translation. J Virol *62*, 2636-43.
2. Pelletier, J., and Sonenberg, N. (1988). Internal initiation of translation of eukaryotic mRNA directed by a sequence derived from poliovirus RNA. Nature *334*, 320-5.
3. Tsukiyama-Kohara, K., Iizuka, N., Kohara, M., and Nomoto, A. (1992). Internal ribosome entry site within hepatitis C virus RNA. J Virol *66*, 1476-83.
4. Poole, T. L., C. Wang, R. A. Popp, L. N. Potgieter, A. Siddiqui, and M. S. Collett. (1995). Pestivirus translation initiation occurs by internal ribosome entry. Virology. *206*, 750-4.
5. Sasaki, J., and Nakashima, N. (1999). Translation initiation at the CUU codon is mediated by the internal ribosome entry site of an insect picorna-like virus in vitro. J Virol *73*, 1219-26.
6. Wilson, J. E., Powell, M. J., Hoover, S. E., and Sarnow, P. (2000b). Naturally occurring dicistronic cricket paralysis virus RNA is regulated by two internal ribosome entry sites. Mol Cell Biol *20*, 4990-9.

7. Molla, A., Jang, S. K., Paul, A. V., Reuer, Q., and Wimmer, E. (1992). Cardioviral internal ribosomal entry site is functional in a genetically engineered dicistronic poliovirus. Nature *356*, 255-7.
8. Sasaki, J., and Nakashima, N. 2000. Methionine-independent initiation of translation in the capsid protein of an insect virus. Proc. Natl Acad. Sci. USA *97,* 1512-1515.
9. Wilson, J. E., Pestova, T. V., Hellen, C. U., and Sarnow, P. (2000a). Initiation of protein synthesis from the A site of the ribosome. Cell *102*, 511-20.
10. Muthukrishnan, S., Both, G. W., Furuichi, Y., and Shatkin, A. J. (1975). 5'-Terminal 7-methylguanosine in eukaryotic mRNA is required for translation. Nature *255*, 33-7.
11. Furuichi, Y., and Shatkin, A. J. (2000). Viral and cellular mRNA capping: past and prospects. Adv Virus Res *55*, 135-84.
12. Gale, M., Tan, S-L., and Katze, M.G. 2000. Translational control of viral gene expression in eukaryotes. Microb. Mol. Biol. Rev. *64*, 239-280.
13. Kozak, M. (1986). Point mutations define a sequence flanking the AUG initiator codon that modulates translation by eukaryotic ribosomes. Cell *44*, 283-92.
14. Sachs, A. B., and Buratowski, S. (1997). Common themes in translational and transcriptional regulation. Trends Biochem Sci *22*, 189-92.
15. Pelham, H. R. B., and Jackson, R. J. 1976. An efficient mRNA-dependent translation system from rabbit reticulocyte lysates. Eur. J. Biochem. *67*, 247-256
16. Wimmer, E., and Reichmann, M. E. (1968). Pyrophosphate in the 5' terminal position of a viral ribonucleic acid. Science *160*, 1452-4.
17. Nomoto, A., Kitamura, N., Golini, F., and Wimmer, E. 1977. The 5'-terminal structures of poliovirion RNA and poliovirus mRNA differ only in the genome-linked protein. Proc. Natl. Acad. Sci. USA *74*, 5345-5349.
18. Semler, B. L., Anderson, C. W., Hanecak, R., Dorner, L. F., and Wimmer, E. (1982). A membrane-associated precursor to poliovirus VPg identified by immunoprecipitation with antibodies directed against a synthetic heptapeptide. Cell *28*, 405-12.
19. Wimmer, E. 1982. Genome-linked proteins of viruses. Cell *28*, 199-201.
20. Pilipenko, E. V., Blinov, V. M., Chernov, B. K., Dmitrieva, T. M., and Agol, V. I. (1989a). Conservation of the secondary structure elements of the 5'-untranslated region of cardio- and aphthovirus RNAs. Nucleic Acids Res *17*, 5701-11.
21. Pilipenko, E. V., Blinov, V. M., Romanova, L. I., Sinyakov, A. N., Maslova, S. V., and Agol, V. I. (1989b). Conserved structural domains in the 5'-untranslated region of picornaviral genomes: an analysis of the segment controlling translation and neurovirulence. Virology *168*, 201-9.
22. Jang, S. K., and Wimmer, E. (1990). Cap-independent translation of encephalomyocarditis virus RNA: structural elements of the internal ribosomal entry site and involvement of a cellular 57-kD RNA-binding protein. Genes Dev *4*, 1560-72.
23. Wimmer, E., Hellen, C. U., and Cao, X. (1993). Genetics of poliovirus. Annu Rev Genet *27*, 353-436.
24. Chen, C. Y., and Sarnow, P. (1995). Initiation of protein synthesis by the eukaryotic translational apparatus on circular RNAs. Science *268*, 415-7.
25. Etchison, D., Milburn, S. C., Edery, I., Sonenberg, N., and Hershey, J. W. (1982). Inhibition of HeLa cell protein synthesis following poliovirus infection correlates with the proteolysis of a 220,000-dalton polypeptide associated with eucaryotic initiation factor 3 and a cap binding protein complex. J Biol Chem *257*, 14806-10.
26. Bernstein, H. D., Sonenberg, N., and Baltimore, D. (1985). Poliovirus mutant that does not selectively inhibit host cell protein synthesis. Mol Cell Biol *5*, 2913-23.
27. Krausslich, H. G., Nicklin, M. J., Toyoda, H., Etchison, D., and Wimmer, E. (1987). Poliovirus proteinase 2A induces cleavage of eucaryotic initiation factor 4F polypeptide p220. J Virol *61*, 2711-8.
28. Lamphear, B. J., Yan, R., Yang, F., Waters, D., Liebig, H. D., Klump, H., Kuechler, E., Skern, T., and Rhoads, R. E. (1993). Mapping the cleavage site in protein synthesis

initiation factor eIF-4 gamma of the 2A proteases from human Coxsackievirus and rhinovirus. J Biol Chem *268*, 19200-3.

29. Devaney, M. A., Vakharia, V. N., Lloyd, R. E., Ehrenfeld, E., and Grubman, M. J. (1988). Leader protein of foot-and-mouth disease virus is required for cleavage of the p220 component of the cap-binding protein complex. J Virol *62*, 4407-9.
30. Gingras, A. C., Svitkin, Y., Belsham, G. J., Pause, A., and Sonenberg, N. (1996). Activation of the translational suppressor 4E-BP1 following infection with encephalomyocarditis virus and poliovirus. Proc Natl Acad Sci U S A *93*, 5578-83.
31. Sarnow, P. (1989). Translation of glucose-regulated protein 78/immunoglobulin heavy-chain binding protein mRNA is increased in poliovirus-infected cells at a time when cap-dependent translation of cellular mRNAs is inhibited. Proc Natl Acad Sci U S A *86*, 5795-9.
32. Macejak, D. G., and Sarnow, P. (1991). Internal initiation of translation mediated by the 5' leader of a cellular mRNA [see comments]. Nature *353*, 90-4.
33. Gingras, A. C., Raught, B., and Sonenberg, N. (1999). eIF4 initiation factors: effectors of mRNA recruitment to ribosomes and regulators of translation. Annu Rev Biochem *68*, 913-63.
34. Witherell, G. W., Gil, A., and Wimmer, E. (1993). Interaction of polypyrimidine tract binding protein with the encephalomyocarditis virus mRNA internal ribosomal entry site. Biochemistry *32*, 8268-75.
35. Alexander, L., Lu, H. H., and Wimmer, E. (1994). Polioviruses containing picornavirus type 1 and/or type 2 internal ribosomal entry site elements: genetic hybrids and the expression of a foreign gene. Proc Natl Acad Sci U S A *91*, 1406-10.
36. Lu, H. H., and Wimmer, E. (1996). Poliovirus chimeras replicating under the translational control of genetic elements of hepatitis C virus reveal unusual properties of the internal ribosomal entry site of hepatitis C virus. Proc Natl Acad Sci U S A *93*, 1412-7.
37. Gromeier, M., Alexander, L., and Wimmer, E. (1996). Internal ribosomal entry site substitution eliminates neurovirulence in intergeneric poliovirus recombinants. Proc Natl Acad Sci U S A *93*, 2370-5.
38. Meerovitch, K., Nicholson, R., and Sonenberg, N. (1991). In vitro mutational analysis of cis-acting RNA translational elements within the poliovirus type 2 5' untranslated region. J Virol *65*, 5895-901.
39. Skinner, M. A., Racaniello, V. R., Dunn, G., Cooper, J., Minor, P. D., and Almond, J. W. (1989). New model for the secondary structure of the 5' non-coding RNA of poliovirus is supported by biochemical and genetic data that also show that RNA secondary structure is important in neurovirulence. J Mol Biol *207*, 379-92.
40. Stewart, S. R., and Semler, B. L. (1998). RNA structure adjacent to the attenuation determinant in the 5'-non-coding region influences poliovirus viability. Nucleic Acids Res *26*, 5318-26.
41. Stone, D. M., Almond, J. W., Brangwyn, J. K., and Belsham, G. J. (1993). trans complementation of cap-independent translation directed by poliovirus 5' noncoding region deletion mutants: evidence for RNA-RNA interactions. J Virol *67*, 6215-23.
42. Drew, J., and Belsham, G. J. (1994). trans complementation by RNA of defective foot-and-mouth disease virus internal ribosome entry site elements. J Virol *68*, 697-703.
43. Van Der Velden, A., Kaminski, A., Jackson, R. J., and Belsham, G. J. (1995). Defective point mutants of the encephalomyocarditis virus internal ribosome entry site can be complemented in trans. Virology *214*, 82-90.
44. Jang, S. K., Pestova, T. V., Hellen, C. U., Witherell, G. W., and Wimmer, E. (1990). Cap-independent translation of picornavirus RNAs: structure and function of the internal ribosomal entry site. Enzyme *44*, 292-309.
45. Nicholson, R., Pelletier, J., Le, S. Y., and Sonenberg, N. (1991). Structural and functional analysis of the ribosome landing pad of poliovirus type 2: in vivo translation studies. J Virol *65*, 5886-94.

46. Pilipenko, E. V., Gmyl, A. P., Maslova, S. V., Svitkin, Y. V., Sinyakov, A. N., and Agol, V. I. (1992). Prokaryotic-like cis elements in the cap-independent internal initiation of translation on picornavirus RNA. Cell *68*, 119-31.
47. Robertson, M. E., Seamons, R. A., and Belsham, G. J. (1999). A selection system for functional internal ribosome entry site (IRES) elements: analysis of the requirement for a conserved GNRA tetraloop in the encephalomyocarditis virus IRES. RNA *5*, 1167-79.
48. Zhao and Wimmer
49. Kaminski, A., Howell, M. T., and Jackson, R. J. (1990). Initiation of encephalomyocarditis virus RNA translation: the authentic initiation site is not selected by a scanning mechanism. Embo J *9*, 3753-9.
50. Kaminski, A., Belsham, G. J., and Jackson, R. J. (1994). Translation of encephalomyocarditis virus RNA: parameters influencing the selection of the internal initiation site. Embo J *13*, 1673-81.
51. Pestova, T. V., Hellen, C. U., and Wimmer, E. (1994). A conserved AUG triplet in the 5' nontranslated region of poliovirus can function as an initiation codon in vitro and in vivo. Virology *204*, 729-37.
52. Ohlmann, T., and Jackson, R. J. (1999). The properties of chimeric picornavirus IRESes show that discrimination between internal translation initiation sites is influenced by the identity of the IRES and not just the context of the AUG codon. RNA *5*, 764-78.
53. Pelletier, J., Flynn, M. E., Kaplan, G., Racaniello, V., and Sonenberg, N. (1988). Mutational analysis of upstream AUG codons of poliovirus RNA. J Virol *62*, 4486-92.
54. Kuge, S., Kawamura, N., and Nomoto, A. (1989). Strong inclination toward transition mutation in nucleotide substitutions by poliovirus replicase. J Mol Biol *207*, 175-82.
55. Hellen, C. U., Pestova, T. V., and Wimmer, E. (1994). Effect of mutations downstream of the internal ribosome entry site on initiation of poliovirus protein synthesis. J Virol *68*, 6312-22.
56. Futterer, J., Kiss-Laszlo, Z., and Hohn, T. (1993). Nonlinear ribosome migration on cauliflower mosaic virus 35S RNA. Cell *73*, 789-802.
57. Kuge, S., and Nomoto, A. (1987). Construction of viable deletion and insertion mutants of the Sabin strain of type 1 poliovirus: function of the 5' noncoding sequence in viral replication. J Virol *61*, 1478-87.
58. Pestova, T. V., Shatsky, I. N., and Hellen, C. U. (1996). Functional dissection of eukaryotic initiation factor 4F: the 4A subunit and the central domain of the 4G subunit are sufficient to mediate internal entry of 43S preinitiation complexes. Mol Cell Biol *16*, 6870-8.
59. Borman, A. M., Bailly, J. L., Girard, M., and Kean, K. M. (1995). Picornavirus internal ribosome entry segments: comparison of translation efficiency and the requirements for optimal internal initiation of translation in vitro. Nucleic Acids Res *23*, 3656-63.
60. Borman, A. M., and Kean, K. M. (1997). Intact eukaryotic initiation factor 4G is required for hepatitis A virus internal initiation of translation. Virology *237*, 129-36.
61. Borman, A. M., Le Mercier, P., Girard, M., and Kean, K. M. (1997). Comparison of picornaviral IRES-driven internal initiation of translation in cultured cells of different origins. Nucleic Acids Res *25*, 925-32.
62. Pestova, T. V., Shatsky, I. N., Fletcher, S. P., Jackson, R. J., and Hellen, C. U. (1998). A prokaryotic-like mode of cytoplasmic eukaryotic ribosome binding to the initiation codon during internal translation initiation of hepatitis C and classical swine fever virus RNAs. Genes Dev *12*, 67-83.
63. Pestova, T. V., and Hellen, C. U. (1999). Internal initiation of translation of bovine viral diarrhea virus RNA. Virology *258*, 249-56.

64. Kolupaeva, V. G., Pestova, T. V., and Hellen, C. U. (2000). An enzymatic footprinting analysis of the interaction of 40S ribosomal subunits with the internal ribosomal entry site of hepatitis C virus. J Virol *74*, 6242-50.
65. Buratti, E., Tisminetzky, S., Zotti, M., and Baralle, F. E. (1998). Functional analysis of the interaction between HCV 5'UTR and putative subunits of eukaryotic translation initiation factor eIF3. Nucleic Acids Res *26*, 3179-87.
66. Sizova, D. V., Kolupaeva, V. G., Pestova, T. V., Shatsky, I. N., and Hellen, C. U. (1998). Specific interaction of eukaryotic translation initiation factor 3 with the 5' nontranslated regions of hepatitis C virus and classical swine fever virus RNAs. J Virol *72*, 4775-82.
67. Pestova, T. V., Hellen, C. U., and Shatsky, I. N. (1996). Canonical eukaryotic initiation factors determine initiation of translation by internal ribosomal entry. Mol Cell Biol *16*, 6859-69.
68. Kolupaeva, V. G., Pestova, T. V., Hellen, C. U., and Shatsky, I. N. (1998). Translation eukaryotic initiation factor 4G recognizes a specific structural element within the internal ribosome entry site of encephalomyocarditis virus RNA. J Biol Chem *273*, 18599-604.
69. Lomakin, I. B., Hellen, C. U., and Pestova, T. V. (2000). Physical association of eukaryotic initiation factor 4G (eIF4G) with eIF4A strongly enhances binding of eIF4G to the internal ribosomal entry site of encephalomyocarditis virus and is required for internal initiation of translation. Mol Cell Biol *20*, 6019-29.
70. Lopez de Quinto, S., and Martinez-Salas, E. (2000). Interaction of the eIF4G initiation factor with the aphthovirus IRES is essential for internal translation initiation in vivo [In Process Citation]. RNA *6*, 1380-92.
71. Meyer, K., Petersen, A., Niepmann, M., and Beck, E. (1995). Interaction of eukaryotic initiation factor eIF-4B with a picornavirus internal translation initiation site. J Virol *69*, 2819-24.
72. Rust, R. C., Ochs, K., Meyer, K., Beck, E., and Niepmann, M. (1999). Interaction of eukaryotic initiation factor eIF4B with the internal ribosome entry site of foot-and-mouth disease virus is independent of the polypyrimidine tract-binding protein. J Virol *73*, 6111-3.
73. Brown, B. A., and Ehrenfeld, E. (1979). Translation of poliovirus RNA in vitro: changes in cleavage pattern and initiation sites by ribosomal salt wash. Virology *97*, 396-405.
74. Dorner, A. J., Semler, B. L., Jackson, R. J., Hanecak, R., Duprey, E., and Wimmer, E. (1984). In vitro translation of poliovirus RNA: utilization of internal initiation sites in reticulocyte lysate. J Virol *50*, 507-14.
75. Phillips, B. A., and Emmert, A. (1986). Modulation of the expression of poliovirus proteins in reticulocyte lysates. Virology *148*, 255-67.
76. Borman, A., Howell, M. T., Patton, J. G., and Jackson, R. J. (1993). The involvement of a spliceosome component in internal initiation of human rhinovirus RNA translation. J Gen Virol *74*, 1775-88.
77. Roberts, L. O., Seamons, R. A., and Belsham, G. J. (1998). Recognition of picornavirus internal ribosome entry sites within cells; influence of cellular and viral proteins. RNA *4*, 520-9.
78. Meerovitch, K., Pelletier, J., and Sonenberg, N. (1989). A cellular protein that binds to the 5'-noncoding region of poliovirus RNA: implications for internal translation initiation. Genes Dev *3*, 1026-34.
79. Meerovitch, K., Svitkin, Y. V., Lee, H. S., Lejbkowicz, F., Kenan, D. J., Chan, E. K., Agol, V. I., Keene, J. D., and Sonenberg, N. (1993). La autoantigen enhances and corrects aberrant translation of poliovirus RNA in reticulocyte lysate. J Virol *67*, 3798-807.

80. Shiroki, K., Isoyama, T., Kuge, S., Ishii, T., Ohmi, S., Hata, S., Suzuki, K., Takasaki, Y., and Nomoto, A. (1999). Intracellular redistribution of truncated La protein produced by poliovirus 3Cpro-mediated cleavage. J Virol *73*, 2193-200.
81. Ali, N., and Siddiqui, A. (1997). The La antigen binds 5' noncoding region of the hepatitis C virus RNA in the context of the initiator AUG codon and stimulates internal ribosome entry site-mediated translation. Proc Natl Acad Sci U S A *94*, 2249-54.
82. Ali, N., Pruijn, G. J., Kenan, D. J., Keene, J. D., and Siddiqui, A. (2000). Human La antigen is required for the hepatitis C virus internal ribosome entry site-mediated translation. J Biol Chem *275*, 27531-40.
83. Holcik, M., and Korneluk, R. G. (2000). Functional characterization of the X-linked inhibitor of apoptosis (XIAP) internal ribosome entry site element: role of La autoantigen in XIAP translation. Mol Cell Biol *20*, 4648-57.
84. Holcik, M., Yeh, C., Korneluk, R. G., and Chow, T. (2000). Translational upregulation of X-linked inhibitor of apoptosis (XIAP) increases resistance to radiation induced cell death. Oncogene *19*, 4174-7.
85. Carter, M. S., and Sarnow, P. (2000). Distinct mRNAs that encode La autoantigen are differentially expressed and contain internal ribosome entry sites. J Biol Chem *275*, 28301-7.
86. Hellen, C. U., Witherell, G. W., Schmid, M., Shin, S. H., Pestova, T. V., Gil, A., and Wimmer, E. (1993). A cytoplasmic 57-kDa protein that is required for translation of picornavirus RNA by internal ribosomal entry is identical to the nuclear pyrimidine tract-binding protein. Proc Natl Acad Sci U S A *90*, 7642-6.
87. Kaminski, A., Hunt, S. L., Patton, J. G., and Jackson, R. J. (1995). Direct evidence that polypyrimidine tract binding protein (PTB) is essential for internal initiation of translation of encephalomyocarditis virus RNA. RNA *1*, 924-38.
88. Kaminski, A., and Jackson, R. J. (1998). The polypyrimidine tract binding protein (PTB) requirement for internal initiation of translation of cardiovirus RNAs is conditional rather than absolute. RNA *4*, 626-38.
89. Niepmann, M., Petersen, A., Meyer, K., and Beck, E. (1997). Functional involvement of polypyrimidine tract-binding protein in translation initiation complexes with the internal ribosome entry site of foot-and-mouth disease virus. J Virol *71*, 8330-9.
90. Hunt, S. L., and Jackson, R. J. (1999). Polypyrimidine-tract binding protein (PTB) is necessary, but not sufficient, for efficient internal initiation of translation of human rhinovirus-2 RNA. RNA *5*, 344-59.
91. Hunt, S. L., Hsuan, J. J., Totty, N., and Jackson, R. J. (1999). unr, a cellular cytoplasmic RNA-binding protein with five cold-shock domains, is required for internal initiation of translation of human rhinovirus RNA. Genes Dev *13*, 437-48.
92. Boussadia, O., Jacquemin-Sablon, H., and Dautry, F. (1993). Exon skipping in the expression of the gene immediately upstream of N-ras (unr/NRU). Biochim Biophys Acta *1172*, 64-72.
93. Jacquemin-Sablon, H., Triqueneaux, G., Deschamps, S., le Maire, M., Doniger, J., and Dautry, F. (1994). Nucleic acid binding and intracellular localization of unr, a protein with five cold shock domains. Nucleic Acids Res *22*, 2643-50.
94. Boussadia, O., Amiot, F., Cases, S., Triqueneaux, G., Jacquemin-Sablon, H., and Dautry, F. (1997). Transcription of unr (upstream of N-ras) down-modulates N-ras expression in vivo. FEBS Lett *420*, 20-4.
95. Ali, N., and Siddiqui, A. (1995). Interaction of polypyrimidine tract-binding protein with the 5' noncoding region of the hepatitis C virus RNA genome and its functional requirement in internal initiation of translation. J Virol *69*, 6367-75.
96. Anwar, A., Ali, N., Tanveer, R., and Siddiqui, A. (2000). Demonstration of functional requirement of polypyrimidine tract-binding protein by SELEX RNA during hepatitis C virus internal ribosome entry site-mediated translation initiation [In Process Citation]. J Biol Chem *275*, 34231-5.

97. Gosert, R., Chang, K. H., Rijnbrand, R., Yi, M., Sangar, D. V., and Lemon, S. M. (2000). Transient expression of cellular polypyrimidine-tract binding protein stimulates cap-independent translation directed by both picornaviral and flaviviral internal ribosome entry sites In vivo. Mol Cell Biol *20*, 1583-95.
98. Schultz, D. E., Hardin, C. C., and Lemon, S. M. (1996). Specific interaction of glyceraldehyde 3-phosphate dehydrogenase with the 5'-nontranslated RNA of hepatitis A virus. J Biol Chem *271*, 14134-42.
99. Yi, M., Schultz, D. E., and Lemon, S. M. (2000). Functional significance of the interaction of hepatitis A virus RNA with glyceraldehyde 3-phosphate dehydrogenase (GAPDH): opposing effects of GAPDH and polypyrimidine tract binding protein on internal ribosome entry site function. J Virol 74, 6459-68.
100. Kim, Y. K., Hahm, B., and Jang, S. K. (2000). Polypyrimidine tract-binding protein inhibits translation of bip mRNA [In Process Citation]. J Mol Biol *304*, 119-33.
101. Huez, I., Creancier, L., Audigier, S., Gensac, M. C., Prats, A. C., and Prats, H. (1998). Two independent internal ribosome entry sites are involved in translation initiation of vascular endothelial growth factor mRNA. Mol Cell Biol *18*, 6178-90.
102. Blyn, L. B., Swiderek, K. M., Richards, O., Stahl, D. C., Semler, B. L., and Ehrenfeld, E. (1996). Poly(rC) binding protein 2 binds to stem-loop IV of the poliovirus RNA 5' noncoding region: identification by automated liquid chromatography-tandem mass spectrometry. Proc Natl Acad Sci U S A *93*, 11115-20.
103. Blyn, L. B., Towner, J. S., Semler, B. L., and Ehrenfeld, E. (1997). Requirement of poly(rC) binding protein 2 for translation of poliovirus RNA. J Virol *71*, 6243-6.
104. Walter, B. L., Nguyen, J. H., Ehrenfeld, E., and Semler, B. L. (1999). Differential utilization of poly(rC) binding protein 2 in translation directed by picornavirus IRES elements. RNA *5*, 1570-85.
105. Silvera, D., Gamarnik, A. V., and Andino, R. (1999). The N-terminal K homology domain of the poly(rC)-binding protein is a major determinant for binding to the poliovirus 5'-untranslated region and acts as an inhibitor of viral translation. J Biol Chem *274*, 38163-70.
106. Parsley, T. B., Towner, J. S., Blyn, L. B., Ehrenfeld, E., and Semler, B. L. (1997). Poly (rC) binding protein 2 forms a ternary complex with the 5'-terminal sequences of poliovirus RNA and the viral 3CD proteinase. RNA *3*, 1124-34.
107. Gamarnik, A. V., and Andino, R. (1998). Switch from translation to RNA replication in a positive-stranded RNA virus. Genes Dev *12*, 2293-304.
108. Gamarnik, A. V., and Andino, R. (2000). Interactions of viral protein 3CD and Poly(rC) binding protein with the 5' untranslated region of the poliovirus genome [In Process Citation]. J Virol *74*, 2219-26.
109. Agol, V. I., Paul, A. V., and Wimmer, E. (1999). Paradoxes of the replication of picornaviral genomes. Virus Res *62*, 129-47.
110. Graff, J., Cha, J., Blyn, L. B., and Ehrenfeld, E. (1998). Interaction of poly(rC) binding protein 2 with the 5' noncoding region of hepatitis A virus RNA and its effects on translation. J Virol *72*, 9668-75.
111. Spangberg, K., Goobar-Larsson, L., Wahren-Herlenius, M., and Schwartz, S. (1999). The La protein from human liver cells interacts specifically with the U-rich region in the hepatitis C virus 3' untranslated region. J Hum Virol *2*, 296-307.
112. Radomski, N., and Jost, E. (1995). Molecular cloning of a murine cDNA encoding a novel protein, p38-2G4, which varies with the cell cycle. Exp Cell Res *220*, 434-45.
113. Nakagawa, Y., Watanabe, S., Akiyama, K., Sarker, A. H., Tsutsui, K., Inoue, H., and Seki, S. (1997). cDNA cloning, sequence analysis and expression of a mouse 44-kDa nuclear protein copurified with DNA repair factors for acid-depurinated DNA. Acta Med Okayama *51*, 195-206.
114. Pilipenko, E. V., Pestova, T. V., Kolupaeva, V. G., Khitrina, E. V., Poperechnaya, A. N., Agol, V. I., and Hellen, C. U. (2000). A cell cycle-dependent protein serves as a template-specific translation initiation factor. Genes Dev *14*, 2028-45.

115. Gromeier, M., Lachmann, S., Rosenfeld, M. R., Gutin, P. H., and Wimmer, E. (2000). Intergeneric poliovirus recombinants for the treatment of malignant glioma. Proc Natl Acad Sci U S A *97*, 6803-8.
116. Hahm, B., Kim, Y. K., Kim, J. H., Kim, T. Y., and Jang, S. K. (1998). Heterogeneous nuclear ribonucleoprotein L interacts with the 3' border of the internal ribosomal entry site of hepatitis C virus. J Virol *72*, 8782-8.
117. Reynolds, J. E., Kaminski, A., Kettinen, H. J., Grace, K., Clarke, B. E., Carroll, A. R., Rowlands, D. J., and Jackson, R. J. (1995). Unique features of internal initiation of hepatitis C virus RNA translation. Embo J *14*, 6010-20.
118. Peek, R., Pruijn, G. J., and Van Venrooij, W. J. (1996). Interaction of the La (SS-B) autoantigen with small ribosomal subunits. Eur J Biochem *236*, 649-55.
119. Kim, J. H., Hahm, B., Kim, Y. K., Choi, M., and Jang, S. K. (2000). Protein-protein interaction among hnRNPs shuttling between nucleus and cytoplasm. J Mol Biol *298*, 395-405.
120. Krecic, A. M., and Swanson, M. S. (1999). hnRNP complexes: composition, structure, and function. Curr Opin Cell Biol *11*, 363-71.
121. Shyu, A. B., and Wilkinson, M. F. (2000). The double lives of shuttling mRNA binding proteins. Cell *102*, 135-8.
122. Gamarnik, A. V., and Andino, R. (1996). Replication of poliovirus in Xenopus oocytes requires two human factors. Embo J *15*, 5988-98.
123. Toyoda, H., Koide, N., Kamiyama, M., Tobita, K., Mizumoto, K., and Imura, N. (1994). Host factors required for internal initiation of translation on poliovirus RNA. Arch Virol *138*, 1-15.
124. Yoo, C. J., and Wolin, S. L. (1997). The yeast La protein is required for the 3' endonucleolytic cleavage that matures tRNA precursors. Cell 89, 393-402.
125. Pannone, B. K., Xue, D., and Wolin, S. L. (1998). A role for the yeast La protein in U6 snRNP assembly: evidence that the La protein is a molecular chaperone for RNA polymerase III transcripts. Embo J *17*, 7442-53.
126. Xue, D., Rubinson, D. A., Pannone, B. K., Yoo, C. J., and Wolin, S. L. (2000). U snRNP assembly in yeast involves the La protein [published erratum appears in EMBO J 2000 Jun 1;19(11):2763]. Embo J *19*, 1650-60.
127. McClure, W. R. (1985). Mechanism and control of transcription initiation in prokaryotes. Annu Rev Biochem *54*, 171-204.
128. Struhl, K. (1999). Fundamentally different logic of gene regulation in eukaryotes and prokaryotes. Cell *98*, 1-4.
129. Steitz, J. A., and Jakes, K. (1975). How ribosomes select initiator regions in mRNA: base pair formation between the 3' terminus of 16S rRNA and the mRNA during initiation of protein synthesis in Escherichia coli. Proc Natl Acad Sci U S A *72*, 4734-8.
130. Boni, I. V., Isaeva, D. M., Musychenko, M. L., and Tzareva, N. V. (1991). Ribosome-messenger recognition: mRNA target sites for ribosomal protein S1. Nucleic Acids Res *19*, 155-62.
131. Pugh, B. F. (2000). Control of gene expression through regulation of the TATA-binding protein [In Process Citation]. Gene *255*, 1-14.
132. Henry, R. W., Sadowski, C. L., Kobayashi, R., and Hernandez, N. (1995). A TBP-TAF complex required for transcription of human snRNA genes by RNA polymerase II and III. Nature *374*, 653-6.
133. Verrijzer, C. P., Yokomori, K., Chen, J. L., and Tjian, R. (1994). Drosophila TAFII150: similarity to yeast gene TSM-1 and specific binding to core promoter DNA. Science *264*, 933-41.
134. Burke, T. W., and Kadonaga, J. T. (1997). The downstream core promoter element, DPE, is conserved from Drosophila to humans and is recognized by TAFII60 of Drosophila. Genes Dev *11*, 3020-31.

135. Goodrich, J. A., Hoey, T., Thut, C. J., Admon, A., and Tjian, R. (1993). Drosophila TAFII40 interacts with both a VP16 activation domain and the basal transcription factor TFIIB. Cell *75*, 519-30.

136. Farmer, G., Colgan, J., Nakatani, Y., Manley, J. L., and Prives, C. (1996). Functional interaction between p53, the TATA-binding protein (TBP), andTBP-associated factors in vivo. Mol Cell Biol *16*, 4295-304.

137. Hoey, T., Weinzierl, R. O., Gill, G., Chen, J. L., Dynlacht, B. D., and Tjian, R. (1993). Molecular cloning and functional analysis of Drosophila TAF110 reveal properties expected of coactivators. Cell *72*, 247-60.

138. Chen, J. L., Attardi, L. D., Verrijzer, C. P., Yokomori, K., and Tjian, R. (1994). Assembly of recombinant TFIID reveals differential coactivator requirements for distinct transcriptional activators. Cell *79*, 93-105.

139. May, M., Mengus, G., Lavigne, A. C., Chambon, P., and Davidson, I. (1996). Human TAF(II28) promotes transcriptional stimulation by activation function 2 of the retinoid X receptors. Embo J *15*, 3093-104.

140. Jacq, X., Brou, C., Lutz, Y., Davidson, I., Chambon, P., and Tora, L. (1994). Human TAFII30 is present in a distinct TFIID complex and is required for transcriptional activation by the estrogen receptor. Cell *79*, 107-17.

141. Mengus, G., May, M., Carre, L., Chambon, P., and Davidson, I. (1997). Human TAF(II)135 potentiates transcriptional activation by the AF-2s of the retinoic acid, vitamin D3, and thyroid hormone receptors in mammalian cells. Genes Dev *11*, 1381-95.

142. Chiang, C. M., and Roeder, R. G. (1995). Cloning of an intrinsic human TFIID subunit that interacts with multiple transcriptional activators. Science *267*, 531-6.

143. Fletcher, C. M., Pestova, T. V., Hellen, C. U., and Wagner, G. (1999). Structure and interactions of the translation initiation factor eIF1. Embo J *18*, 2631-7.

144. Bandyopadhyay, A., and Maitra, U. (1999). Cloning and characterization of the p42 subunit of mammalian translation initiation factor 3 (eIF3): demonstration that eIF3 interacts with eIF5 in mammalian cells. Nucleic Acids Res *27*, 1331-7.

145. Lamphear, B. J., Kirchweger, R., Skern, T., and Rhoads, R. E. (1995). Mapping of functional domains in eukaryotic protein synthesis initiation factor 4G (eIF4G) with picornaviral proteases. Implications for cap-dependent and cap-independent translational initiation. J Biol Chem *270*, 21975-83.

146. Tarun, S. Z., Jr., and Sachs, A. B. (1997). Binding of eukaryotic translation initiation factor 4E (eIF4E) to eIF4G represses translation of uncapped mRNA. Mol Cell Biol *17*, 6876-86.

147. Christensen, A. K., Kahn, L. E., and Bourne, C. M. (1987). Circular polysomes predominate on the rough endoplasmic reticulum of somatotropes and mammotropes in the rat anterior pituitary. Am J Anat *178*, 1-10.

148. Wells, S. E., Hillner, P. E., Vale, R. D., and Sachs, A. B. (1998). Circularization of mRNA by eukaryotic translation initiation factors. Mol Cell *2*, 135-40.

149. Tarun, S. Z., Jr., and Sachs, A. B. (1995). A common function for mRNA 5' and 3' ends in translation initiation in yeast. Genes Dev *9*, 2997-3007.

150. Tarun, S. Z., Jr., Wells, S. E., Deardorff, J. A., and Sachs, A. B. (1997). Translation initiation factor eIF4G mediates in vitro poly(A) tail-dependent translation. Proc Natl Acad Sci U S A *94*, 9046-51.

151. Carey, M., Lin, Y. S., Green, M. R., and Ptashne, M. (1990). A mechanism for synergistic activation of a mammalian gene by GAL4 derivatives. Nature *345*, 361-4.

152. De Gregorio, E., Preiss, T., and Hentze, M. W. (1999). Translation driven by an eIF4G core domain in vivo. Embo J *18*, 4865-74.

153. Chappell, S. A., Edelman, G. M., and Mauro, V. P. (2000). A 9-nt segment of a cellular mRNA can function as an internal ribosome entry site (IRES) and when present in linked multiple copies greatly enhances IRES activity. Proc Natl Acad Sci U S A *97*, 1536-41.

154. Coldwell, M. J., Mitchell, S. A., Stoneley, M., MacFarlane, M., and Willis, A. E. (2000). Initiation of Apaf-1 translation by internal ribosome entry. Oncogene *19*, 899-905.
155. Cornelis, S., Bruynooghe, Y., Denecker, G., Van Huffel, S., Tinton, S., and Beyaert, R. (2000). Identification and characterization of a novel cell cycle-regulated internal ribosome entry site. Mol Cell *5*, 597-605.
156. Henis-Korenblit, S., Strumpf, N. L., Goldstaub, D., and Kimchi, A. (2000). A novel form of DAP5 protein accumulates in apoptotic cells as a result of caspase cleavage and internal ribosome entry site-mediated translation. Mol Cell Biol *20*, 496-506.
157. Lauring, S. A., and Overbaugh, J. (2000). Evidence that an IRES within the Notch2 coding region can direct expression of a nuclear form of the protein.[In Process Citation]. Mol Cell *6*, 939-45.
158. Pyronnet, S., Pradayrol, L., and Sonenberg, N. (2000). A cell cycle-dependent internal ribosome entry site. Mol Cell *5*, 607-16.
159. Sella, O., Gerlitz, G., Le, S. Y., and Elroy-Stein, O. (1999). Differentiation-induced internal translation of c-sis mRNA: analysis of the cis elements and their differentiation-linked binding to the hnRNP C protein. Mol Cell Biol *19*, 5429-40.
160. Galy, B., Maret, A., Prats, A. C., and Prats, H. (1999). Cell transformation results in the loss of the density-dependent translational regulation of the expression of fibroblast growth factor 2 isoforms. Cancer Res *59*, 165-71.
161. van der Velden, A. W., and Thomas, A. A. (1999). The role of the 5' untranslated region of an mRNA in translation regulation during development. Int J Biochem Cell Biol *31*, 87-106.
162. Fernandez, J., Yaman, I., Mishra, R., Merrick, W. C., Snider, M. D., Lamers, W. H., and Hatzoglou, M. (2001). IRES-mediated translation of a mammalian mRNA is regulated by amino acid availability. J Biol Chem, in press.
163. Johannes, G., Carter, M. S., Eisen, M. B., Brown, P. O., and Sarnow, P. (1999). Identification of eukaryotic mRNAs that are translated at reduced cap binding complex eIF4F concentrations using a cDNA microarray. Proc Natl Acad Sci U S A *96*, 13118-23.

2

TRANSLATIONAL REGULATION OF MASKED MATERNAL mRNAs IN EARLY DEVELOPMENT

Nancy Standart

Department of Biochemistry, University of Cambridge, Cambridge, UK

Gene expression in early development, at a time when transcription is silent, is essentially regulated at the level of protein synthesis in a wide variety of organisms. Overall, there is modest activation of the translational machinery at the time when the oocytes or eggs resume meiosis. More importantly, in every case examined in detail, specific sub-sets of mRNA are recruited onto polysomes from a masked form associated with proteins (mRNP). In contrast to 'house-keeping' mRNAs such as actin, tubulin and ribosomal protein mRNAs, which are actively translated in immature oocytes, mRNAs encoding proteins required for entry and progression through the cell cycle (including cyclins, c-mos and ribonucleotide reductase) are translationally inert until oocytes are induced to undergo meiotic maturation or fertilization, when their products are required (1,2). The control of mRNAs encoding cell cycle regulatory proteins in early development has been extensively characterized in lower and higher eukaryotes in the last decade; this research has uncovered one of the best-understood mRNA-specific translational regulators, cytoplasmic polyadenylation element binding protein (CPEB), the major subject of this chapter.

BACKGROUND

Early Studies of Masked mRNAs

Masked mRNA was the term coined by A.S. Spirin in the mid 1960s to describe the state of messenger RNA isolated from early fish embryos and sea urchin eggs. mRNA was proposed to be associated with proteins in ribonucleoprotein particles (mRNP) (99). Unless steps were taken to de-proteinize the mRNP, they were inactive in a translation assay *in vitro*.

However, perfectly proper and active template was obtained following phenol extraction or trypsinization; implying that these treatments removed inhibitors (repressors) of translation that normally hold maternal mRNA in a masked state.

Masked maternal mRNAs also regulate sexual fates in the *C. elegans* hermaphrodite germ line (3) and the specification of pattern along the anteroposterior body axis in *Drosophila* by generation of protein gradients from localized mRNAs (4). The question of how the initial state of repression is imposed and how it is relieved to allow expression is intriguing at several levels. First, early development, and this is true for all organisms examined – ranging from marine invertebrates, worms, and flies, to frog, mouse and man – is conspicuously a period when gene expression is essentially governed by translational control, rather than transcriptional control. Secondly, the control mechanisms target particular RNAs, implying the recognition of specific sequences and/or specific RNA-binding proteins. Lastly, the processes regulated by the translational repressors and/activators are of fundamental physiological importance. Lessons that are being learnt from this wealth of examples complement studies of the somewhat rarer cases of translational control of somatic mRNAs such as ferritin (5), 15-lipoxygenase (6) and ribosomal protein mRNAs (7) as well as those that mediate synaptic plasticity (8).

Several underlying principles have emerged from genetic and biochemical studies of masked mRNAs. The Y-box family of nucleic acid-binding proteins, with relatively low RNA-sequence specificity, participate in the general packaging of mRNA as it emerges from the nucleus. Regulatory elements specifying repression lie in the 3' untranslated region (UTR). They are generally short, apparently unstructured motifs, often present in more than one copy, that mediate the binding of specific trans-acting factors, the masking repressors. Full control is sometimes achieved in conjuction with 5' UTR elements; mRNAs that are controlled by both localization and translation tend to have more complex, structured motifs. Strikingly, there is no single pathway by which mRNAs are regulated in early development – control may be exerted by interfering with the function of the 5' cap structure or the 3' poly(A) tail, a mixture of both or by as yet unknown means.

In *Xenopus* oocytes, the principal vertebrate model system in this field, mRNA is associated with two abundant phosphoproteins (known as mRNP3 and mRNP4) and several minor components (9). mRNP3 and mRNP4 are respectively, very similar and identical to FRGY2, a protein independently characterized as a transcription factor in the oocyte and subsequently classified as the prototype of a family of related proteins, the Y-box proteins (10). The Y-box family of nucleic acid-binding proteins, of relatively low RNA-sequence specificity, appear to participate in the general packaging and translational masking of mRNA as it emerges from the nucleus in *Xenopus* oocytes, mammalian germ cells and somatic cells ((11-14); reviewed in 15-17). However, the FRGY2 proteins, while contributors to the masked state, cannot be responsible for the sequence-specific changes seen on oocyte maturation or at fertilization.

Our early studies focused on the regulatory mechanisms responsible for translational activation of surf clam (*Spisula solidissima*) maternal mRNAs. The advantages of using this system were the ability of cell-free lysates from oocytes and activated eggs/embryos to support characteristic and stage-specific patterns of protein synthesis (18) and the ability to activate oocyte lysates *in vitro* by raising Ca^{2+} and pH (19, 20). Exploitation of these unique properties of clam lysates allowed us to characterize 3'UTR-mediated masking and more recently, to clone and characterize the specific RNA-binding protein p82/CPEB. Upon fertilization, clam oocytes, arrested in prophase I, complete meiosis and proceed directly into the mitotic cell division cycles. During meiotic maturation, three abundant maternal mRNAs are recruited into polysomes encoding cyclins A and B and the small subunit of ribonucleotide reductase (RR), whose products enable cell cycle progression and DNA synthesis. Cyclin A and RR mRNAs, masked in oocyte lysates, can be specifically translationally activated *in vitro* by oligonucleotide and RNAse H-mediated scission or by antisense RNAs directed to the 'masking elements' located approximately in the centre of their 3'UTRs. The antisense RNAs, by forming double-stranded structures, are postulated to prevent the binding of repressor proteins thus leading to translational activation (18). Using UV-crosslinking, we characterized an 82 kDa oocyte protein (p82) which selectively binds the U-rich RR and cyclin A masking elements. p82 is associated with masked mRNAs in low salt gel filtration columns, but is removed from the RNP peak by 0.5 M KCl, conditions known to activate these mRNAs (19). Thus p82 has several characteristics expected of a translational masking/repressor protein. However, sequence-specific masking was not well understood till more recently, following on from studies of translational activation of maternal mRNAs in *Xenopus*, mouse, clam and *Drosophila* by cytoplasmic elongation of their poly(A) tails, and the cloning and characterization of the regulatory factors.

Cytoplasmic Polyadenylation, CPEs and CPEB

Cytoplasmic polyadenylation is a highly regulated and conserved mechanism that dramatically increases translation during meitotic maturation and after fertilization (reviewed in (21)). It was in fact in *Spisula* oocytes that Rosenthal and colleagues made the seminal observation that translationally activated mRNAs undergo poly(A) extension, and conversely, that deadenylated mRNAs are silenced (22). This correlation was observed in many organisms (1). Studies in *Xenopus* revealed that individual mRNAs undergo poly(A) extension at varying times during meiosis and after fertilization, and from different initial to different final lengths (23-25). The 3'UTR sequences that promote polyadenylation, when microinjected into maturing eggs or in egg cell-free lysates, include one or more copies of a U-rich cytoplasmic polyadenylation element (CPE), consensus $U_{4\text{-}6}A_{1\text{-}3}U$ in fairly close proximity to the ubiquitous nuclear polyadenylation signal, AAUAAA. Both elements are needed to support cytoplasmic poly(A) extension and stimulate translation during oocyte maturation. Their relative position can influence the timing and degree of polyadenylation (24,26), but systematic mutational analyses of either the CPE motifs or their location have not been performed. mRNAs lacking a CPE-motif in their 3'UTR mRNAs such as ribosomal protein mRNAs, lose their poly(A) tail by default at GVBD, and are concomitantly released from polysomes, providing further evidence of the tight connection between poly(A) tail length and translational efficiency (27,28). Since the original pioneering studies (29,30), mRNAs in a wide range of organisms have been found to be regulated by CPEs. Recent examples of biologically important mRNAs include those encoding *Xenopus* cyclin B1 (26,31-33), lamin B1 (34), wee-1 (35), mouse tissue-type plasminogen activator (tPA) and cyclin B1 (36,37), clam ribonucleotide reductase and cyclin A (38) and *Drosophila bicoid* (39). A particularly striking example is provided by *c-mos* mRNA whose polyadenylation and consequent translation is a pivotal regulatory step in meiotic maturation of *Xenopus* and mouse oocytes (40, 41).

CPEB was first cloned and characterized as a specific 62 kDa *Xenopus* B4 RNA CPE-binding protein (42, 43). It has 2 RRMs (RNA recognition motif) and an unusual zinc finger in its C-terminus; all three contribute to RNA recognition (43). Evidence obtained with CPEB antibodies support the positive role of CPEB in cytoplasmic polyadenylation, initially of B4 mRNA (42) and subsequently of *c-mos*, cdk-2, cyclins and G10 mRNAs (44). CPEB binds the CPEs of all these mRNAs, and is necessary to support their polyadenylation in egg lysates. Injection of CPEB antibody into oocytes not only prevents polyadenylation, but also blocks progesterone-induced maturation, suggesting that CPEB is critical for early development (44). *Xenopus* CPEB is the founder member of a growing family of proteins in both vertebrates and invertebrates – *Drosophila orb* (45, 46), clam p82 (38, 47), *C. elegans* CPB-1-4 (48) as well as the more closely related mouse, zebrafish and human homologs (48-50, 50a). All CPEBs share the C-terminal RNA-binding domains, the N-termini are far more varied in sequence (38, 48). Intriguingly, it seems that the role of the different members is not confined to cytoplasmic polyadenylation or oogenesis. Moreover, *Drosophila*, *C.elegans*, zebrafish and man possess 2 or more isoforms; in the nematode these perform distinct roles in spermatogenesis (48).

Dual Role of CPEB in Translational Repression and Cytoplasmic Polyadenylation

The pioneering work on mouse oocyte tPA mRNA (36, 51) formed the framework for the notion that one 3' UTR element may perform dual roles in regulating maternal mRNA translation. In this study, the 3'UTR adenylation control element (ACE), that causes polyadenylation upon resumption of meiosis, was shown to also mediate translational masking in primary oocytes by interaction with a repressor protein. The ACE, which supports deadenylation in growing oocytes as well as the subsequent meiotic readenylation, is very similar to CPEs in sequence (52). However, despite the apparent relatedness of function and sequence between CPEs and the ACE, the ACE-binding protein, ~80 kDa, does not appear to correspond in size to the known isoform of mouse CPEB, and remains to be characterized (36). The identification of the clam putative translational repressor p82 (19) as a CPEB homologue (47) provided one of the first hints that CPEB itself may have dual roles in regulating translation in early development. As expected from its homology to *Xenopus* CPEB, clam p82 binding sites in the RR 3'UTR which resemble the U-rich CPEs are required to support its polyadenylation in egg lysates and anti-p82 antibodies prevent polyadenylation. In line with our previous observations, p82/CPEB also acts as a translational repressor in the oocyte; anti-p82 antibodies specifically activate translation of masked mRNAs in oocyte lysates (38). These data suggested that p82/CPEB acts first as a repressor of translation in immature oocytes and subsequently participates in the activation of translation by cytoplasmic polyadenylation (38).

In parallel studies, the role of vertebrate CPEB as a repressor was extensively documented. *Xenopus* cyclin B1, lamin and wee-1, and mouse cyclin B1 3' UTRs repress reporter mRNA translation in oocytes, providing low levels of reporter mRNA are injected. Higher levels of mRNA arc not repressed, implicating the action of a saturable masking factor. Deletion and mutational analyses pointed to the CPEs as the primary repression elements (32-35, 37). Since these were shown to bind CPEB (with the exception of lamin B1), the inescapable conclusion was that CPEs and CPEB cause translational repression in oocytes, in additional to their role in polyadenylation in eggs. In contrast, 3'UTR sequences derived from *Xenopus* cyclin A1, B2 and *c-mos* only function in polyadenylation, not in repression (33, 44). An interesting difference between CPEs promoting polyadenylation and repression is their number – one copy is sufficient for polyadenylation, repeated copies appear necessary for repression; possibly implying the requirement for the tighter binding of multiple CPEBs to prevent translation (see Table 1). These studies also highlight another interesting contrast – some mRNAs, such as cyclin B1 mRNA strictly require polyadenylation in maturing oocytes for derepression (unmasking) (32, 33, 37); others such as *Xenopus* wee-1 and mouse tPA can be unmasked without extensive polyadenylation (34, 35). Similarly, polyadenylation and unmasking are uncoupled in the clam lysate (18). It seems that individual mRNAs regulated by CPEs (and other motifs, see (53)) are differentially subject to control by repression and polyadenylation – some mRNAs require a secondary level of activation (polyadenylation) over and above simple relief of repression (unmasking); others do not. Whether this reflects differences in the interaction of regulatory factors, or the temporal context of activation is not yet clear (for discussion see Wickens ref 2).

Xenopus	
cyclin B1	...GUUUUUAAUGUUUUACUGGUUUU**AAUAAA**GCUCAUUUUAACAUG
cyclin A1	...UAACUUGUGAUGGUGUUAAGUGUUUUU**AAUAAA**CUGACUUUACUCAA
cyclin B2	...AUUUUUAUU-57nt-**AAUAAA**ACUUCACAUUUUUUAUUU
c-mos	...660nt-UUUUAUAUGUAUGUGUUGUUUUAU-1240nt-UUUUAU**AAUAAA**GAAAUUGAUUUGUCU
lamin B1	...GUUUUAU-44nt-UUUUUUUUAUU-48nt-**AAUAAA**AGGGGAUUUAAAUAC
wee-1	...UUUAUUGACUUUUUUUUUUUUAUUAUCUUAUUGUCUUUUA**AAUAAA**AAUUUUAAUGUGUA
mouse	
t-PA	...AUUUUAAUCUAUUUUAGAUUUUAC-37nt-**AAUAAA**UUCAGAGGUAUUUUUCACACUUU
cyclin B1	...CUUUUUUUA-154nt-AUUUUAU-301nt-GUUUUAAU-22nt-**AAUAAA**AUUUAUUGGUGGAAAGCUUUCACAAUU
clam	
RR	..UUUUAAU-23nt-UUUUUUAUU-89nt-UUUUAAU-13nt-UUUUAU-113nt-UUUUAUCAGGUUUUAAAU-35nt-**AAUAAA**UUUAAUGUGUG

Table 1: 3' UTR cytoplasmic polyadenylation elements. *Nuclear polyadenylation elements are bold, CPEs are underlined.*

How do events at the 3' end of mRNA affect ribosome binding at the 5' end? Considerable evidence supports the so-called 'closed loop model' of eukaryotic mRNA in which communication between the 5' cap structure and the 3' poly(A) tail synergistically stimulates translation. In yeast and mammals, the cap and poly(A) tail-binding proteins, eIF4E and PABP, mediate this effect through their interactions with different domains of eIF4G. Strikingly, atomic force microscopy experiments provide physical data that capped, polyadenylated mRNA circularizes in the presence of these 3 factors. mRNA circularization elegantly provides a basis for permitting ribosomes to recycle promptly following termination of protein synthesis and for stabilization of full-length mRNA (reviewed in (54)). In *Xenopus*, methylation of the 5' cap at N-7 (which enhances eIF4E-binding) and polyadenylation stimulate translation synergistically during oocyte maturation (55). Cap ribose methylation has also been proposed to enhance translation during maturation, in a polyadenylation-dependent manner, but such cap modification is not observed in all activated mRNAs (55, 56). Since PABP, and in particular, the binding of eIF4G with PABP, stimulates translation (57, 58), and is critical for *Xenopus* oocyte maturation (58), it seems fairly certain that the same 5'-3' contacts are also made in early development. 3'UTR-mediated repressors may interfere with the closed loop form of mRNA, either directly or indirectly by preventing or disrupting the positive 5'-3' contacts. Activation of translation may result from simple relief of repression, resulting from repressor modification such as phosphorylation and degradation. More importantly, polyA tail lengthening would serve to recruit PABP (57). A recently solved puzzle concerns the levels of *Xenopus* PABP in oocytes. The understanding for many years was that their content of PABP was unusually low (compared to somatic cells) (59), certainly far lower than poly(A)+RNA-binding sites. It turns out that these oocytes contain perfectly reasonable levels of a novel, embryonic form of PABP, called ePAB (60). ePAB, like the canonical PABP, protects polyadenylated RNA from degradation (60, 61); whether it has additional roles is not yet known.

CPEB-Interacting Proteins

In *Xenopus*, CPEB binds a 150 kDa protein called maskin, similar to human TACC3 (transforming acidic coiled-coil), as revealed in co-immunoprecipitation and yeast two hybrid assays. Before maturation, CPEB sequesters the cap-binding factor eIF4E, indirectly through maskin, and prevents productive eIF4F complex formation and hence ribosome recruitment. Upon progesterone-treatment, maskin releases eIF4E, and active translation can ensue (62). Thus maskin, which contains an eIF4E-binding motif similar to those in eIF4G and eIF4E-binding proteins, forms an unproductive bridge between CPE-containing mRNAs and eIF4E in the oocyte (62). Interestingly, maskin levels themselves are regulated during oogenesis; significant levels are not observed till stage VI, suggesting that in earlier stages CPEB represses in a maskin-independent manner (63). In *Spisula*, anti-p82/CPEB co-immunoprecipitations identified the clam homologue of a DEAD-box RNA helicase as a protein that indirectly (through RNA) binds CPEB in oocytes, but not in mature eggs (64). The *Xenopus* member of this RCK/p54 helicase family Xp54 is, significantly, an abundant and integral component of stored mRNP in oocytes, with *bona fida* helicase (unwinding) activity (65). Though the role of this helicase in masking is not yet known, a recent study suggesting that an RNA helicase can actively disrupt an RNA-protein interaction (66) means that its mechanism need not necessarily be confined to unwinding stretches of double-stranded RNA. In maturing oocytes, *Xenopus* CPEB acts as a recruitment factor to CPE-containing mRNAs of the cytoplasmic forms of cleavage and polyadenylation specificity factors (CPSF) (67) required for robust polyadenylation during meiotic maturation (68).

Regulation of CPEB by Phosphorylation and Degradation

It is doubtless noteworthy that the two functions of CPEB are temporally and cell-cycle stage distinct, and that the protein is modified between the two stages. CPEB is phosphorylated and subsequently degraded in maturing oocytes of *Xenopus* (42), zebrafish (50), mouse (37) and clam (47). Hyperphosphorylation, leading to retardation of CPEB in denaturing gels (42, 47) is due to cdc2 kinase, activated at GVBD (19, 20, 47, 69). Cdc2 phosphorylation of CPEB may mediate its proteolysis. Mutations of putative cdc2 S/P and T/P sites in the clam CPEB N-terminus stabilize the protein during oocyte maturation. Inspection of CPEB sequences identified two short N-terminal motifs conserved between clam and vertebrate CPEB (47). The deletion of the second island, rich in PEST residues, resulted in a stable and phosphorylated species. Thus, while phosphorylation appears necessary for degradation, it is not sufficient (*George Thom and N.S., unpublished*). The role of the PEST region may be to target phosphorylated CPEB to the ubiquitin-proteasome proteolysis machinery (70, 70a). In addition to cdc2 kinase, kinases activated early during meiotic maturation have been implicated in phosphorylating CPEB, including MAP kinase in *Spisula* (20, 71) and Eg2 kinase (72, 73) in *Xenopus*. The mitogen-activated protein kinase signaling pathway stimulates *mos* mRNA cytoplasmic polyadenylation during *Xenopus* oocyte maturation (74), but the precise role of CPEB phosphorylation by MAP kinase is not known. Eg2 phosphorylates *Xenopus* CPEB on one or two consecutive LDSR motifs, also found in other vertebrate CPEBs, but absent from invertebrates (75). This modification increases the affinity of the 160 kDa CPSF subunit for CPEB, neatly explaining the connection between progesterone-triggered kinase signaling, CPEB and cytoplasmic polyadenylation of CPE-containing mRNAs (68). Our current understanding then is that early during maturation, phosphorylation enhances CPEB's role in polyadenylation, and subsequent cdc2 phoshorylation targets the protein for degradation. Perhaps surprisingly, in considering the rapidity with which cyclins are degraded during meiotic and early mitotic cell divisions, CPEB's proteolysis, at least in *Xenopus*, is gradual (42). Intriguingly, the small proportion of protein that remains intact in early embryogenesis is found on centrosomes, and may be responsible for regulating localized synthesis of cyclin B1 (63). The observations discussed in this and the preceding section, are represented diagramatically in Fig. 1.

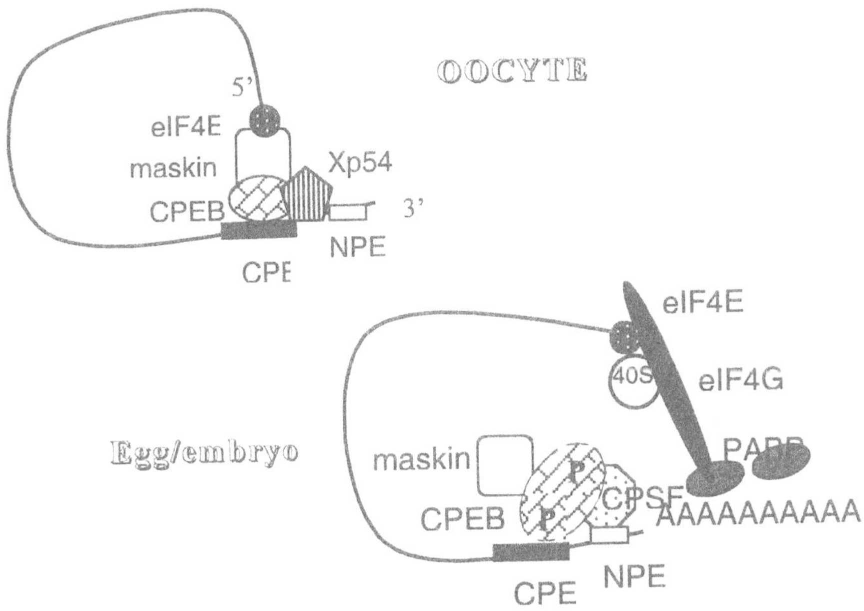

Figure 1. Model of CPE/CPEB control of translation in the oocyte and egg.

Parallels, Contrasts and Remaining Questions

To what extent are the mechanistic principles underlying CPE- and CPEB-mediated translational regulation mirrored in other systems? In other words, is polyA elongation a ubiquitous activating process? Do all 3' UTRs repressors inhibit the function of eIF4E? The simple answer is no in both cases, pointing to the diversity of regulatory controls.

Thus, *Drosophila, bicoid, torso* and *toll* RNAs are regulated by polyadenylation. These RNAs are present in oocytes, but are not translated till after egg activation or fertilization, and poly(A) tail elongation is required for translational activation (39, 76). In contrast, translational activation of *oskar* (77) and *nanos* (78) mRNA is independent of polyadenylation. Unlocalized *oskar* mRNA is repressed through multiple BRE elements (bruno-response elements) in its 3'UTR by bruno, a 68 kDa ELAV-type RNA-binding protein with 3 RRMs (79, 80). Activation of *oskar* mRNA is only observed at the posterior pole, and requires a 5' UTR derepressor element, and its specific binding proteins, p68 and p50. p50 also binds the 3' UTR BREs, and acts as a co-repressor of *oskar* translation through its interaction with the BRE (77). Derepression does not appear to involve repressor degradation or modification or cytoplasmic polyadenylation, but rather an interaction between proteins linking the 5' and 3' ends of the mRNA (77), possibly mediated by Staufen's dsRBD5 domain (81). Translational activation of *nanos* mRNA is also independent of polyadenylation changes and requires recognition by localization factors of *nanos* 3' UTR motifs distinct but overlapping those that mediate translational repression by Smaug (82-84). The vasa DEAD-box helicase (similar to eIF4A) is required to promote *nanos* and *oskar* translation (78, 85), possibly through its interaction with a *Drosophila* yIF2 homologue (86).

Some 3'UTR repressors, including *C. elegans* GLD-1, interfere with their target RNA (in this case, *tra-2*) poly(A)'s role in translation, and/or promote deadenylation (87). *Drosophila* bicoid represses *caudal* translation in a cap-dependent manner (88), by recognizing and binding the 3'UTR BBR (Bcd-binding region) (89). Bicoid may inhibit translation by its interaction with eIF4E through a putative eIF4E-binding site (*Rivera-Pomar, personal communication*). Control by Bicoid is thus reminiscent of CPEB, which binds eIF4E indirectly through maskin. Other repressors (such as *Drosophila* Nanos and Pumilio, and rabbit lipoxygenase DICE binding protein) however, exert their effects in a cap-and /or polyA-independent manner, suggesting targets downstream of cap recognition and scanning to the initiator AUG, e.g. ribosome assembly at the AUG (90-92).

CPE/CPEB-mediated regulation of translation remains one of the best-understood examples of control of masked maternal mRNA. Recent investigations also highlight remaining questions and unexpected roles in neurons and spermatogenesis. While CPEB is undoubtedly one of the principal CPE-binding proteins, there is also compelling evidence for other regulatory proteins that bind CPEs or whose binding sites overlap with CPEs (32-35). In the same vein, 3'UTR motifs other than CPEs also regulate translation, such as Vg1 mRNA during oogenesis (93, *and Otero, L.J., Devaux, A., and Standart, N. A 250nt UA-rich element in the 3'-untranslated region of Xenopus laevis Vg1 mRNA represses translation both in vivo and in vitro. Submitted to RNA, 2001*) or wee-1 mRNA late in oocyte maturation (35). And lastly, it is interesting to note that two of the four CPEB isoforms in *C. elegans* perform critical functions in spermatogenesis, not oogenesis (48). CPB-1 binds FBF, a *fem-3* RNA-binding protein, which together with NANOS-3 controls the sperm-oocyte switch (48, 94, 95), while the second homolog, FOG-1, controls germ cell fates by regulating the translation of specific messenger RNAs (96). The *Drosophila* FBF homolog, Pumilio, also interacts with Nanos to regulate *hunchback* mRNA (90, 97, 98). Whether vertebrate CPEB similarly mediates its role by association with, for example, *Xenopus* FBF/pum (100) and nanos homologs remains to be seen.

Summary of key concepts

- *Changes in gene expression during early development are governed by regulating the translation of maternal mRNAs, such as those encoding cyclins and c-mos.*
- *Translational control in early development is mediated bu 3' untranslated region (3'UTR) cis-elements and their specific trans-acting binding factors.*
- *Activation of translation during meiotic maturation results from regulated cytoplasmic polyadenylation, a process that requires U-rich cytoplasmic polyadenylation elements proximal to the ubiquitous nuclear polyadenylation signal AAUAAA.*
- *mRNAs are also repressed in oocytes; some mRNAs are subject to both repression in oocytes and subsequent activation in eggs and early embryos.*
- *CPEB (cytoplasmic polyadenylation element binding protein) is an RRM-containing protein that specifically recognizes CPEs, and has dual functions in early development, namely, repression of translation in oocytes and activation of translation in eggs and early embryos.*
- *CPEBs are a conserved family of RNA-binding proteins, subject to phosphorylation and regulated proteolysis during oocyte maturation.*

Study Guide Questions

1) Gene expression is usually regulated at the transcriptioional level – why is control of translation the principal way to regulate gene expression in early development? In what other cell types might this also be the case?
2) Translational control is achieved by RNA motifs residing in the 3'UTR of mRNAs. How do we envisage ribosome binding to the 5' cap structure being regulated from the 3' end?
3) Consider how extending the poly(A) tail of an mRNA could enhance its translation.

Acknowledgements

The author thanks the past and current members of the lab, Jim Walker, Yoshinao Katsu, Nicola Minshall and George Thom for their contributions, experimental and discoursive, to the CPEB work in this lab. Work was funded initially by the MRC, and more recently by the Wellcome Trust.

REFERENCES

1. Standart, N. (1992). Masking and unmasking of maternal mRNA. Semin. Dev. Biol. **3,** 367-379.
2. Wickens, M., Goodwin, E., Kimble, J., Strickland, S., and Hentze, M., *Translational control of developmental decisions*, in *Translational control of gene expression*, N. Sonenberg, J. Hershey, and M. Mathews, Editors. 2000, Cold Spring Harbor Laboratory Press: Cold Spring Harbor, New York. p. 295-370.
3. Goodwin, E.R. and Evans, T.C. (1997). Translational control of development in *C. elegans*. Seminars in Cell and Dev. Biol. **8,** 551-559.
4. St Johnston, D. (1995). The intracellular localization of messenger RNAs. Cell. **81,** 161-170.
5. Rouault, T.A. and Harford, J.B., *Translational control of ferritin synthesis.*, in *Translatonal control of gene expression*, N. Sonenberg, J. Hershey, and M. Mathews, Editors. 2000, Cold Spring Harbor Laboratory Press: Cold Spring Harbor, New York. p. 655-670.
6. Ostareck-Lederer, A., Ostareck, D.H., and Hentze, M.W. (1998). Cytoplasmic regulatory functions of the KH-domain proteins hnRNP K and E1/E2. Trends Biochem. Sci. **23,** 409-411.
7. Meyuhas, O. and Hornstein, E., *Translational control of TOP mRNAs.*, in *Translational control of gene expression*, N. Sonenberg, J. Hershey, and M. Mathews, Editors. 2000, Cold Spring Harbor Laboratory Press: Cold Spring Harbor, New York. p. 671-694.
8. Wu, L., Wells, D., Tay, J., Mendis, D., Abbott, M.-A., Barnitt, A., Quinlan, E., Heynen, A., Fallon, J.R., and Richter, J.D. (1998). CPEB-mediated cytoplasmic polyadenylation and the regulation of experience-dependent translation of a-CaMKII mRNA at synapses. Neuron. **21,** 1129-1139.
9. Murray, M.T., Krohne, G., and Franke, W.W. (1991). Different forms of soluble cytoplasmic mRNA binding proteins and particles in *Xenopus laevis* oocytes and embryos. J. Cell Biol. **112,** 1-11.
10. Murray, M.T., Schiller, D.L., and Franke, W.W. (1992). Sequence analysis of cytoplasmic mRNA-binding proteins of *Xenopus* oocytes identifies a family of RNA-binding proteins. Proc. Natl. Acad. Sci. USA. **89,** 11-15.
11. Bouvet, P. and Wolffe, A.P. (1994). A role for transcription and FRGY2 in masking maternal mRNA within *Xenopus* oocytes. Cell. **77,** 931-941.
12. Braddock, M., Muckenthaler, M., White, M.R.H., Thorburn, A.M., Sommerville, J., Kingsman, A.J., and Kingsman, S.M. (1994). Intron-less RNA injected into the nucleus of *Xenopus* oocytes accesses a regulated translation control pathway. Nucl. Acids Res. **22,** 5255-5264.
13. Davydova, E.K., Evdokimova, V.M., Ovchinnikov, L.P., and Hershey, J.W. (1997). Overexpression in COS cells of p50, the major core protein associated with mRNA, results in translation inhibition. Nucl. Acids Res. **25,** 2911-2916.
14. Davies, H.G., Giorgini, F., Fajardo, M.A., and Braun, R.E. (2000). A sequence-specific RNA binding complex expressed in murine germ cells contains MSY2 and MSY4. Dev. Biol. **221,** 87-100.
15. Standart, N. and Jackson, R. (1994). Y the message is masked? Curr. Biol. **4,** 939-941.
16. Matsumoto, K. and Wolffe, A.P. (1998). Gene regulation by Y-box proteins: coupling control of transcription and translation. Trends Cell Biol. **8,** 318-323.
17. Sommerville, J. and Ladomery, M. (1996). Masking of mRNA by Y-box proteins. FASEB J. **10,** 435-43.
18. Standart, N., Dale, M., Stewart, E., and Hunt, T. (1990). Maternal mRNA from clam oocytes can be specifically unmasked *in vitro* by antisense RNA complementary to the 3'-untranslated region. Genes Dev. **4,** 2157-2168.

19. Walker, J., Dale, M., and Standart, N. (1996). Unmasking mRNA in clam oocytes: Role of phosphorylation of a 3' UTR masking element-binding protein at fertilization. Dev. Biol. **173,** 292-305.
20. Katsu, Y., Minshall, N., Nagahama, Y., and Standart, N. (1999). Ca^{2+} is required for phosphorylation of clam p82/CPEB *in vitro*: Implications for dual and independent roles of MAP and cdc2 kinases. Dev. Biol. **209,** 186-199.
21. Richter, J.D., *Influence of polyadenylation-induced translation on metazoan development and neuronal synaptic function.*, in *Translational control of gene expression*, N. Sonenberg, J. Hershey, and M. Mathews, Editors. 2000, Cold Spring Harbor Laboratory Press: Cold Spring Harbor, New York. p. 785-806.
22. Rosenthal, E.T., Tansey, T.R., and Ruderman, J.V. (1983). Sequence-specific adenylations and deadenylations accompany changes in the translation of maternal messenger RNA after fertilization of *Spisula* oocytes. J. Mol. Biol. **166,** 309-327.
23. Sheets, M., Fox, C., Hunt, T., Vande Woude, G., and Wickens, M. (1994). The 3'-untranslated regions of c-*mos* and cyclin mRNAs stimulate translation by regulating cytoplasmic polyadenylation. Genes Dev. **8,** 926-938.
24. Simon, R., Tassan, J.-P., and Richter, J.D. (1992). Translational control by poly(A) elongation during *Xenopus* development: Differential repression and enhancement by a novel cytoplasmic polyadenylation element. Genes Dev. **6,** 2580-2591.
25. Simon, R. and Richter, J. (1994). Further analysis of cytoplasmic polyadenylation in *Xenopus* embryos and identification of embryonic cytoplasmic polyadenylation element-binding proteins. Mol. Cell. Biol. **14,** 7867-7875.
26. de Moor, C.H. and Richter, J., D. (1997). The *mos* pathway regulates cytoplasmic polyadenylation in *Xenopus* oocytes. Mol. Cell Biol. **17,** 6419-6426.
27. Varnum, S.M. and Wormington, W.M. (1990). Deadenylation of maternal mRNAs during *Xenopus* oocyte maturation does not require specific *cis*-sequences: a default mechanism for translational control. Genes Dev. **4,** 2278-2286.
28. Fox, C.A. and Wickens, M. (1990). Poly(A) removal during oocyte maturation: a default reaction selectively prevented by specific sequences in the 3'-UTR of certain maternal mRNAs. Genes Dev. **4,** 2287-2298.
29. McGrew, L.L., Dworkin-Rastl, E., Dworkin, M.B., and Richter, J.D. (1989). Poly(A) elongation during *Xenopus* oocyte maturation is required for translational recruitment and is mediated by a short sequence element. Genes Dev. **3,** 803-815.
30. Fox, C.A., Sheets, M.D., and Wickens, M.P. (1989). Poly(A) addition during maturation of frog oocytes: distinct nuclear and cytoplasmic activities and regulation by the sequence UUUUUAU. Genes Dev. **3,** 2151-2162.
31. Ballantyne, S., Daniel, J., D. L., and Wickens, M. (1997). A dependent pathway of cytoplasmic polyadenylation reactions linked to cell cycle control by c-*mos* and CDK1 activation. Mol. Biol. Cell. **8,** 1633-1648.
32. de Moor, C. and Richter, J.D. (1999). Cytoplasmic polyadenylation elements mediate masking and unmasking of cyclin B1 mRNA. EMBO J. **18,** 2294-2303.
33. Barkoff, A.F., Dickson, K.S., Gray, N.K., and Wickens, M. (2000). Translational control of cyclin B1 mRNA during meiotic maturation: coordinated repression and cytoplasmic polyadenylation. Dev. Biol. **220,** 97-109.
34. Ralle, T., Gremmels, D., and Stick, R. (1999). Translational control of nuclear lamin B1 mRNA during oogenesis and early development of *Xenopus*. Mech. Dev. **84,** 89-101.
35. Charlesworth, A., Welk, J., and MacNicol, A.M. (2000). The temporal control of wee1 mRNA translation during *Xenopus* oocyte maturation is regulated by cytoplasmic polyadenylation elements within the 3'-untranslated region. Dev. Biol. **227,** 706-719.
36. Stutz, A., Conne, B., Huarte, J., Gubler, P., Völkel, V., Flandin, P., and Vassalli, J.D. (1998). Masking, unmasking, and regulated polyadenylation cooperate in the translational control of a dormant mRNA in mouse oocytes. Genes Dev. **12,** 2535-2548.

37. Tay, J., Hodgman, R., and Richter, J. (2000). The control of cyclin B1 mRNA translation during mouse oocyte maturation. Dev. Biol. **221,** 1-9.
38. Minshall, N., Walker, J., Dale, M., and Standart, N. (1999). Dual roles of p82, the clam CPEB homolog, in cytoplasmic polyadenylation and translational masking. *RNA*. **5,** 27-38.
39. Sallés, F.J., Lieberfarb, M.E., Wreden, C., Gergen, J.P., and Strickland, S. (1994). Coordinate initiation of *Drosophila* development by regulated polyadenylation of maternal messenger RNAs. Science. **266,** 1996-1999.
40. Sheets, M.D., Wu, M., and Wickens, M. (1995). Polyadenylation of *c-mos* mRNA as a control point in *Xenopus* meiotic maturation. Nature. **374,** 511-516.
41. Gebauer, F., Xu, W., Cooper, G., and Richter, J. (1994). Translational control by cytoplasmic polyadenylation of c-*mos* mRNA is necessary for oocyte maturation in the mouse. EMBO J. **13,** 5712-5720.
42. Hake, L.E. and Richter, J.D. (1994). CPEB is a specificity factor that mediates cytoplasmic polyadenylation during *Xenopus* oocyte maturation. Cell. **79,** 617-627.
43. Hake, L.E., Mendez, R., and Richter, J.D. (1998). Specificity of RNA binding by CPEB: Requirement for RNA recognition motifs and a novel zinc finger. Mol. Cell. Biol. **18,** 685-693.
44. Stebbins-Boaz, B., Hake, L.E., and Richter, J.D. (1996). CPEB controls the cytoplasmic polyadenylation of cyclin, Cdk2 and c-*mos* mRNAs and is necessary for oocyte maturation in *Xenopus*. EMBO J. **15,** 2582-2592.
45. Lantz, V., Chang, J., Horabin, J., Bopp, D., and Schedl, P. (1994). The *Drosophila orb* RNA-binding protein is required for the formation of the egg chamber and establishment of polarity. Genes Dev. **8,** 598-613.
46. Chang, J., Tan, L., and Schedl, P. (1999). The *Drosophila* CPEB homolog, *orb,* is required for oskar protein expression in oocytes. Dev. Biol. **215,** 91-106.
47. Walker, J., Minshall, C., Hake, L., Richter, J., and Standart, N. (1999). The clam 3'UTR masking element-binding protein p82 is a member of the CPEB family. *RNA*. **5,** 14-26.
48. Luitjens, C., Gallegos, M., Kraemer, B., Kimble, J., and Wickens, M. (2000). CPEB proteins control two key steps in spermatogenesis in *C. elegans*. Genes Dev. **14,** 2596-2609.
49. Gebauer, F. and Richter, J. (1996). Mouse cytoplasmic polyadenylylation element binding protein: An evolutionary conserved protein that interacts with the cytoplasmic polyadenylylation elements of *c-mos* mRNA. Proc. Natl. Acad. Sci. USA. **93,** 14602-14607.
50. Bally-Cuif, L., Schatz, W.J., and Ho, R.K. (1998). Characterization of the zebrafish Orb/CPEB-related RNA-binding protein and localization of maternal components in the zebrafish oocyte. Mech. Dev. **77,** 31-47.
50a Welk, J.F., Charlesworth, A., Smith, G.D, and MacNichols, A.M. (2001). Identification and characterization of the gene encoding human cytoplasmic polyadenylation element binding protein. Gene 263:113-120.
51. Stutz, A., Huarte, J., Gubler, P., Conne, B., Belin, D., and Vassalli, J.-D. (1997). *In vivo* antisense oligodeoxynucleotide mapping reveals masked regulatory elements in an mRNA dormant in mouse oocytes. Mol. Cell. Biol. **17,** 1759-1767.
52. Huarte, J., Stutz, A., O'Connell, M.L., Gubler, P., Belin, D., Darrow, A.L., Strickland, S., and Vassali, J.-D. (1992). Transient translational silencing by reversible mRNA deadenylation. Cell. **69,** 1021-1030.
53. Culp, P.A. and Musci, T.J. (1998). Translational activation and cytoplasmic polyadenylation of FGF receptor-1 are independently regulated during *Xenopus* oocyte maturation. Dev. Biol. **193,** 63-76.
54. Sachs, A., Physical and functional interactions between the mRNA cap structure and the poly(A) tail, in Translational control of gene expression, N. Sonenberg, J. Hershey,

and M. Mathews, Editors. 2000, Cold Spring Harbor Laboratory Press: Cold Spring Harbor, New York.

55. Gillian-Daniel, D.L., Gray, N.K., Astrom, J., Barkoff, A., and Wickens, M. (1998). Modifications of the 5' cap of mRNAs during Xenopus oocyte maturation: independence from changes in poly(A) length and impact on translation. Mol. Cell Biol. **18,** 6152-6153.
56. Kuge, H. and Richter, J. (1995). Cytoplasmic 3' poly(A) addition induces 5' cap ribose methylation: implications for translational control of maternal mRNA. EMBO J. **14,** 6301-6310.
57. Gray, N., Coller, J., Dickson, K., and Wickens, M. (2000). Multiple portions of poly(A)-binding protein stimulate translation *in vivo*. EMBO J. **19,** 4723-4733.
58. Wakiyama, M., Imataka, H., and Sonenberg, N. (2000). Interaction of eIF4G with poly(A)-binding protein stimulates translation and is critical for *Xenopus* oocyte maturation. Curr Biol. **10,** 1147-1150.
59. Zelus, B.D., Giebelhaus, D.H., Eib, D.W., Kenner, K.A., and Moon, R.T. (1989). Expression of the poly(A)-binding protein during development of *Xenopus laevis*. Mol. Cell. Biol. **9,** 2756-2760.
60. Voeltz, G.K., Ongkasuwan, J., Standart, N., and Steitz, J.A. (2001). A novel embryonic poly(A) binding protein, ePAB, regulates mRNA deadenylation in *Xenopus* egg extracts. Genes and Dev. 15, 774-778.
61. Wormington, M., Searfoss, A., and Hurney, C. (1996). Overexpression of poly(A) binding protein prevents maturation-specific deadenylation and translational inactivation in *Xenopus* oocytes. EMBO J. **15,** 900-909.
62. Stebbins-Boaz, B., Cao, Q., de Moor, C.H., Mendez, R., and Richter, J.D. (1999). Maskin is a CPEB-associated factor that transiently interacts with eIF-4E. Mol. Cell. **4,** 1017-1027.
63. Groisman, I., Huang, Y.-S., Mendez, R., Cao, Q., Therkauf, W., and Richter, J. (2000). CPEB, maskin, and cyclin B1 mRNA at the mitotic apparatus: Implications for local translational control of cell division. Cell. **103,** 435-447.
64. Minshall, N., Thom, G., and Standart, N. Conserved role of a DEAD-box helicase in mRNA masking. Submitted to RNA, 2001.
65. Ladomery, M., Wade, E., and Sommerville, J. (1997). Xp54, the *Xenopus* homologue of human RNA helicase p54, is an integral component of stored mRNP particles in oocytes. Nucl. Acids Res. **25,** 965-973.
66. Jankowsky, E., Gross, C.H., Shuman, S., and Pyle, A.M. (2001). Active disruption of an RNA-protein interaction by a DExH/D RNA helicase. Science. **291,** 121-125.
67. Dickson, K.S., Bilger, A., Ballantyne, S., and Wickens, M.P. (1999). The cleavage and polyadenylation specificity factor in *Xenopus laevis* oocytes is a cytoplasmic factor involved in regulated polyadenylation. Mol. Cell. Biol. **19,** 5707-5717.
68. Mendez, R., Murthy, K.G.K., Ryan, K., Manley, J.L., and Richter, J.D. (2000).Phosphorylation of CPEB by Eg2 mediates the recruitment of CPSF into an active cytoplasmic polyadenylation complex. Mol. Cell. **6,** 1253-1259.
69. Paris, J., Swenson, K., Piwnica-Worms, H., and Richter, J.D. (1991). Maturation-specific polyadenylation: *in vitro* activation by $p34^{cdc2}$ and phosphorylation of a 58-kD CPE-binding protein. Genes Dev. **5,** 1697-1708.
70. Rechsteiner, M. and Rogers, S. (1996). PEST sequences and regulation by proteolysis. Trends Biochem. Sci. **21,** 267-271.

70a. Reverte, C.G., Ahearn, M.D., and Hake, L.E. (2001). CPEB degradation during *Xenopus* oocyte maturation rewuires a PEST domain and the 26S proteasome. Dev. Biol. 231:447-458.

71. Shibuya, E.K., Boulton, T.G., Cobb, M.H., and Ruderman, J.V. (1992). Activation of p42 MAP kinase and the release of oocytes from cell cycle arrest. **11,** 3963-3975.
72. Andresson, T. and Ruderman, J.V. (1998). The kinase Eg2 is a component of the Xenopus oocyte progesterone-activated signaling pathway. EMBO J. **17,** 5627-5637.

73. Frank-Vaillant, M., Haccard, O., Thibier, C., Ozon, R., Arlot-Bonnemains, Y., Prigent, C., and Jessus, C. (2000). Progesterone regulates the accumulation and the activation of Eg2 kinase in *Xenopus* oocytes. J. Cell Sci. **113,** 1127-1138.
74. Howard, E.L., Charlesworth, A., Welk, J., and MacNicol, A.M. (1999). The mitogen-activated protein kinase signaling pathway stimulates *mos* mRNA cytoplasmic polyadenylation during *Xenopus* oocyte maturation. Mol. Cell. Biol. **19,** 1990-1999.
75. Mendez, R., Hake, L.E., Andresson, T., Littlepage, L.E., Ruderman, J.V., and Richter, J.D. (2000). Phosphorylation of CPE binding factor by Eg2 regulates translation of c-*mos* mRNA. Nature. **404,** 302-307.
76. Verrotti, A., Thompson, S., Wreden, C., Strickland, S., and Wickens, M. (1996). Evolutionary conservation of sequence elements controlling cytoplasmic polyadenylation. Proc. Natl. Acad. Sci. USA. **93,** 9027-9032.
77. Gunkel, N., Yano, T., Markussen, F.-H., Olsen, L.C., and Ephrussi, A. (1998). Localization-dependent translation requires a functional interaction between the 5' and 3' ends of *oskar* mRNA. Genes. Dev. **12,** 1652-1664.
78. Gavis, E.R., Lunsford, L., Bergsten, S.E., and Lehmann, R. (1996). A conserved 90 nucleotide element mediates translational repression of nanos RNA. Development. **122,** 2791-2800.
79. Kim-Ha, J., Kerr, K., and Macdonald, P. (1995). Translational regulation of *oskar* mRNA by Bruno, an ovarian RNA-binding protein, is essential. Cell. **81,** 403-412.
80. Webster, P.J., Liang, L., Berg, C.A., Lasko, P., and Macdonald, P.M. (1997). Translational repressor *bruno* plays multiple roles in development and is widely conserved. Genes Dev. **11,** 2510-2521.
81. Micklem, D.R., Adams, J., Grunert, S., and St. Johnston, D. (2000). Distinct roles of two conserved Staufen domains in *oskar* mRNA localization and translation. EMBO J. **19,** 1366-1377.
82. Smibert, C., Lie, Y., Shillinglaw, W., Henzel, W., and Macdonald, P. (1999). Smaug, a novel and conserved protein, contributes to repression of nanos mRNA translation in vitro. *RNA.* **5,** 1535-1547.
83. Dahanukar, A., Walker, J.A., and Wharton, R.P. (1999). Smaug, a novel RNA-binding protein that operates a translational switch in *Drosophila*. Mol. Cell. **4,** 209-218.
84. Crucs, S., Chatterjee, S., and Gavis, E. (2000). Overlapping but distinct RNA elements control repression and activation of nanos translation. Mol Cell. **5,** 457-467.
85. Markussen, F.-H., Michon, A.-M., Breitwieser, W., and Ephrussi, A. (1995). Translational control of *oskar* generates Short OSK, the isoform that induces pole plasm assembly. Development. **121,** 3723-3732.
86. Carrera, P., Johnstone, O., Nakamura, A., Casanova, J., Jackle, H., and Lasko, P. (2000). VASA mediates translation through interaction with a *Drosophila* yIF2 homolog. Mol Cell. **5,** 181-187.
87. Thompson, S., Goodwin, E., and Wickens, M. (2000). Rapid deadenylation and poly(A)-dependent translational repression mediated by the *Caenorhabditis elegans* tra-2 3' untranslated region in the *Xenopus* embryos. Mol. Cell. Biol. **20,** 2129-2137.
88. Niessing, D., Dostatni, N., Jackle, H., and Rivera-Pomar, R. (1999). Sequence interval within the PEST motif of bicoid is important for translational repression of *caudal* mRNA in the anterior region of the *Drosophila* embryo. EMBO J. **18,** 1966-1973.
89. Rivera-Pomar, R., Niessing, D., Schmidt-Ott, U., Gehring, W.J., and Jackle, H. (1996). RNA binding and translational suppression by *bicoid.* Nature. **379,** 746-749.
90. Wharton, R.P., J., S., Lee, T., Patterson, M., and Murata, Y. (1998). The pumilio RNA-binding domain is also a translational repressor. Mol. Cell. **1,** 863-872.
91. Ostareck-Lederer, A., Ostareck, D.H., Standart, N., and Thiele, B.J. (1994). Translation of 15-lipoxygenase mRNA is controlled by a protein that binds to a repeated sequence in the 3' untranslated region. EMBO J. **13,** 1476-1481.

92. Ostareck, D.H., Ostareck-Lederer, A., Shatsky, I.N., and Hentze, M.W. (2001). Lipoxygenase mRNA silencing in erythroid differentiation: The 3'UTR regulatory complex controls 60S ribosomal subunit joining. Cell 104, 281-290
93. Wilhelm, J., Vale, R., and Hegde, R. (2000). Coordinate control of translation and localization of Vg1 mRNA in *Xenopus* oocytes. Proc Natl Acad Sci U S A. **97,** 13132-13137.
94. Zhang, B., Gallegos, M., Puoti, A., Durkin, E., Fields, S., Kimble, J., and Wickens, M.P. (1997). A conserved RNA-binding protein that regulates sexual fates in the *C. elegans* hermaphrodite germ line. Nature. **390,** 477-484.
95. Kraemer, B., Crittenden, S., Gallegos, M., Moulder, G., Barstead, R., Kimble, J., and Wickens, M. (1999). NANOS-3 and FBF proteins physically interact to control the sperm-oocyte switch in *Caenorhabditis elegans.* Curr. Biol. **9,** 1009-1018.
96. Jin, S.W., Kimble, J., and Ellis, R.E. (2001). Regulation of cell fate in *Caenorhabditis elegans* by a novel cytoplasmic polyadenylation element binding protein. Dev. Biol. **229,** 537-553.
97. Zamore, P., Williamson, J., and Lehmann, R. (1997). The pumilio protein binds RNA through a conserved domain that defines a new class of RNA-binding proteins. *RNA.* **3,** 1421-1433.
98. Sonoda, J. and Wharton, R.P. (1999). Recruitment of Nanos to *hunchback* mRNA by Pumilio. Genes Dev. **13,** 2704-2712.
99. Spirin, A.S., *On 'masked' forms of messenger RNA in early embryogenesis and in other differentiating systems.* Current Topics in Developmetal Biology, ed. A.A. Moscona and A. Monroy. Vol. I. 1966, New York: Academic Press. 1-38.
100. Nakahata, S., Katsu, Y., Mita, K., Inoue, K., Nagakama, Y., and Yamashita, Y. (2001). Biochemical identification of Xenopus pumilioasa sequence-specific cyclin B1 mRNA-binding protein that physically interacts with a Nanos homolog, Xcat-2, and a cytoplasmic polyadenylation element-binding protein. J. Biol. Chem. 276:20945-20953.

3

POLY-C BINDING PROTEINS: CELLULAR REGULATORS OF mRNA FATE AND FUNCTION

Andrea V. Gamarnik and Raul Andino*
*Instituto de Investigaciones Bioquimicas, Buenos Aires, Argentina, and *University of California, San Francisco, CA*

Cellular mRNAs are subjected to multiple levels of regulation, and an increasing number of RNA-binding proteins that control mRNA function have been identified. Recently, several laboratories showed that a family of proteins called poly(rC) binding proteins, PCBPs, play important roles in translation regulation and mRNA stabilization. PCBPs are RNA binding proteins that contain three copies of a conserved RNA binding domain called KH also found in other RNA binding proteins such as hnRNP K (1). PCBPs are involved in both normal cellular mRNA metabolism and in the regulation of viral RNA utilization. Here we will review the recent advances in PCBP function in cellular and viral RNA metabolism as well as structure determinants for RNA recognition. We will discuss possible mechanisms by which PCBPs may control different aspects of mRNA fate, including subcellular localization, regulation of translation, RNA stability as well as the role of PCBP in viral RNA replication.

ROLE OF PCBP IN CELLULAR mRNA METABOLISM: REGULATION OF STABILITY AND TRANSLATION

The activity of mRNA in the cell can be regulated through several mechanisms, such as subcellular localization, stability, and translation activation or inhibition. The translation rates of mRNA can be controlled by binding of regulatory proteins to the 5' or 3' untranslated regions (5' or 3'UTRs). The cellular proteins PCBP1 and PCBP2 (also referred to as hnRNP E1 and 2, or αCP1 and 2) play an important role in the regulation of mRNA translation and stability. The specific binding of PCBPs to sequences located within the 3'UTR of certain mRNAs can determine the fate of the targeted RNA, enhancing or silencing translation as well as stabilizing the RNA molecule. PCBPs are widely expressed in many different tissues with higher levels of expression in skeletal muscle (1). These proteins are mostly cytoplasmic, however, a nuclear localization has also been reported (1, 2).

Turnover of mRNA is a critical step in the regulation of gene expression. In eukaryotic cells, the decay rates of individual mRNAs vary by more than two orders of magnitude (3,4). A growing number of examples indicate that the rates of degradation of specific mRNAs can regulate gene expression in response to environmental cues or the developmental program (for review see 5).

PCBP binding to the 3'UTR of a number of cellular mRNAs increases their stability. The best-studied example is α-globin mRNA, which accumulates during terminal erythroid differentiation. The unusual stability of this mRNA is conferred by a pyrimidine rich region in the 3'UTR that forms an RNP complex (called the α-complex) with three closely related isoforms of PCBP: PCBP1, PCBP2, and sPCBP2 (6). Formation of this complex is necessary for stabilization, as single mutations that disrupt the α-complex destabilize the mRNA. Such is the case in the common thalassemia known as Constant Spring mutation, which results in ribosome readthrough into the 3'UTR (7,8). Biochemical analysis of the α-complex using *in vitro* binding assays and two hybrid systems revealed that several proteins, including AUF 1 and PCBP, are necessary for α-complex formation (9). Although these proteins co-assemble in a synergistic manner, PCBPs bind directly to the α-globin mRNA and can form a minimal α-complex with a defined 20 nucleotides sequence characterized by short stretches of cytidines interrupted by one or two uridines (10). Disruption of α-complex formation by specific mutations in the context of the full length mRNA results in a direct, translation-independent destabilization of the mRNA in transfected cells (8). Even though formation of the α-complex is essential to maintain α-globin mRNA stability, the precise mechanisms of stabilization (and degradation) are unclear.

Database analysis revealed that several highly stable mRNAs have sequence similarities to the 20 nucleotide long pyrimidine-rich motif found in the α-globin 3'UTR (11). These mRNAs, which include collagen-α1(I), tyrosine hydroxylase (TH), erythropoietin (EPO), and the 15-lipoxygenase (LOX), also assemble an RNP complex containing PCBPs at this pyrimidine rich region (Fig. 1). Interestingly, the stability of these mRNAs is regulated in a temporal manner in response to different cellular conditions. Since these rapid changes in mRNA stability allow cells of different types or at different developmental stages to quickly fine-tune their gene expression, it will be important to determine how PCBP activity changes under different conditions.

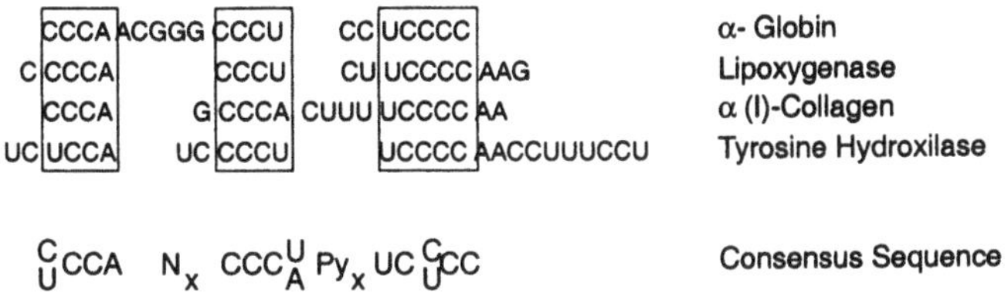

	CCCA	ACGGG		CCCU	CC	UCCCC		α- Globin
C	CCCA			CCCU	CU	UCCCC	AAG	Lipoxygenase
	CCCA		G	CCCA	CUUU	UCCCC	AA	α (I)-Collagen
UC	UCCA		UC	CCCU		UCCCC	AACCUUUCCU	Tyrosine Hydroxilase

$^{C}_{U}$CCA N_x CCC$^{U}_{A}$ Py_x UC $^{C}_{U}$CC — Consensus Sequence

Figure 1. Sequence alignment of pyrimidine-rich segments within the 3' UTR of human α - globin, rabbit 15-lipoxygenase, human α1(I)-collagen, and rat tyrosine hydroxylase*. The sequences shared by all four UTR's are boxed. A consensus sequence is shown under the sequence alignment (11).*

Interestingly, PCBP participates in the regulation of three mRNAs in response to very different conditions. Changes in the synthesis of collagen-α1(I) are associated with both normal growth or repair processes and with several pathological conditions. For instance, in cirrhotic livers, collagen-a1(I) is expressed mainly by hepatic stellate (HS) cells. In a normal liver, quiescent HS cells express only trace amounts of collagen; however upon a fibrogenic stimulus there is a 60- to 70-fold increase in the amount of collagen-α1(I) mRNA. The transcription rate of this mRNA increases only two-fold but the half-life dramatically increases from 1.5 h to 24 h. PCBP binds to a C-rich sequence localized 24 nucleotides downstream to the stop codon (12). Mutation of this C-rich sequence abolishes stabilization of the collagen-α1(I) mRNA. Interestingly, PCBP is present in both quiescent and activated HS cells, but formation of an RNP complex is observed using cytoplasmic extracts from activated but not with quiescent HS cells. This suggests that PCBP is required but not sufficient for complex formation. In contrast, an RNP complex is formed at the 3'UTR of the α-globin mRNA using both quiescent and activated HS cell cytoplasmic extracts (12). Thus, RNP complexes formed in the 3' UTR of collagen-α1(I) and α-globin mRNAs must be qualitatively different. Additional proteins appear to bind the 3'UTR of the collagen-α1(I) mRNA, one of them is a 68kDa protein present in both nuclear and cytoplasmic extracts (13), but the role of this protein in maintaining mRNA stability is still unclear.

The stability of tyrosine hydroxylase mRNA is regulated by oxygen tension in PC12 cells. In lower oxygen tension tyrosine hydroxylase mRNA has a 3 fold longer half-life (14). A pyrimidine-rich sequence within the 3'UTR was shown to form an RNP complex in a hypoxia-inducible manner (15). This 27 nucleotide long sequence (called hypoxia inducible protein binding site, HIPBS) interacts with PCBP1 and PCBP2. It has been shown that specific mutations that abolish complex formation destabilize tyrosine hydroxylase mRNA. Cytoplasmic extracts from PC12 cells exposed to hypoxia conditions contain an increased amount of proteins that specifically bind to HIPBS. PCBP1 concentration increases two-fold when the oxygen concentration decreases, but PCBP2 does not change (16). A short fragment of the tyrosine hydroxylase 3'UTR containing the HIPBS element is sufficient to confer augmented mRNA stability on a heterologous mRNA; however it is insensitive to hypoxia. This observation indicates that the RNP complex formed at HIPBS element is necessary but other *cis*-acting sequences and/or trans-acting factors are also required for hypoxia regulated mRNA stability. In addition, it has been recently reported that the erythropoietin mRNA stability is also regulated in hypoxic conditions and that an RNP complex that includes PCBP proteins is involved in this process (17).

Translation of mRNA in eukaryotic cells is also highly regulated. Current evidence indicates that translation of several cellular and viral mRNAs is determined by the specific and regulated interaction of certain proteins with RNA elements in the 5' and 3' UTRs (for reviews see 18, 19). Although many of these *cis*-acting RNA elements have been defined, only a few *trans*-acting regulatory proteins are known, and the mechanisms by which they regulate translation are only now beginning to unfold.

In the case of the very stable LOX mRNA, the binding of PCBP1 and another member of the KH-domain family, hnRNP K, to the 3'UTR results in translational silencing (20). LOX is a key enzyme in erythroid cell differentiation, and its mRNA, which is the most abundant after the globin mRNA in erythroid precursor cells, is not translated until the enucleated reticulocytes reach the final stage of maturation in the peripheral blood. This temporal regulation of translation depends on a pyrimidine-rich repetitive sequence known as differentiation control element (DICE) located in the 3' UTR of the LOX mRNA (21). DICE consists of multiple tandem copies (10 in rabbit and 4 in human) of the pyrimidine-rich motif. LOX mRNA translational silencing was reconstituted in a cell free translation extract using recombinant PCBP and hnRNP K and a minimal functional DICE containing two repeats. In addition, DICE is fully functional in heterologous mRNAs. The presence of PCBP and/or hnRNP K is capable of silencing cap-dependent and IRES-dependent translation of a DICE-containing reporter gene. Even thought a single repeat is sufficient for PCBP binding *in vitro*, at least two repeats are necessary for translational silencing, suggesting that multiple proteins and/or multiple RNA-protein contacts are required for function.

ROLE OF PCBP IN VIRAL TRANSLATION AND RNA REPLICATION

The role of PCPB in viral replication is a good example of how host cell factors participate in the life cycle of eukaryotic viruses. In addition to controlling the fate and activity of cellular mRNAs, PCBPs also form specific RNP complexes with viral RNAs. PCBP1 and PCBP2 facilitate poliovirus translation through the interaction with two of the six domains of the viral 5'UTR (Fig. 2). While the majority of cellular mRNAs depend on the 5'cap structure to initiate translation, poliovirus initiates translation internally via a cap-independent mechanism from an RNA element termed the IRES (internal ribosomal entry site). The mechanism by which the translation apparatus recognizes IRES sequences is still unknown, but it has been proposed that several initiation factors as well as other cellular proteins, such as PCBP, participate in this process. PCBP1 and PCBP2 specifically bind C-rich sequences present in both the first stem-loop of the viral 5'UTR, which folds into a cloverleaf-like structure, and the stem-loop IV of the IRES (22, 2, 23, 24). The binding of PBCP to the cloverleaf RNA was mapped using chemical and enzymatic probing and confirmed by RNA mobility shift studies using mutated RNAs (25, 26). The presence of three cytosines in stem-loop B of the cloverleaf is essential for protein recognition as well as for viral viability. In addition, mutations that abolish PCBP binding to the stem-loop IV RNA impair poliovirus translation (22, 2, 23). A functional role of PCBPs in poliovirus translation was also suggested by depletion of PCBPs from HeLa cell translation extracts and reconstitution of the system by readdition of the recombinant proteins (2, 23). The KH1 domain of PCBP is sufficient to bind specifically both the cloverleaf and the stem loop IV RNAs. The presence of KH1 inhibits translation from a poliovirus IRES in HeLa cell extracts as well as in microinjected oocytes, presumably by acting as a dominant negative molecule that competes with the full length protein (27). Furthermore, initiation of poliovirus translation in *Xenopus laevis* oocytes is strongly inhibited by microinjection of antibodies against PCBP or decoy RNAs that contain PCBP binding sequences (2, 24,28). These results indicate that PCBP is required for poliovirus IRES mediated translation.

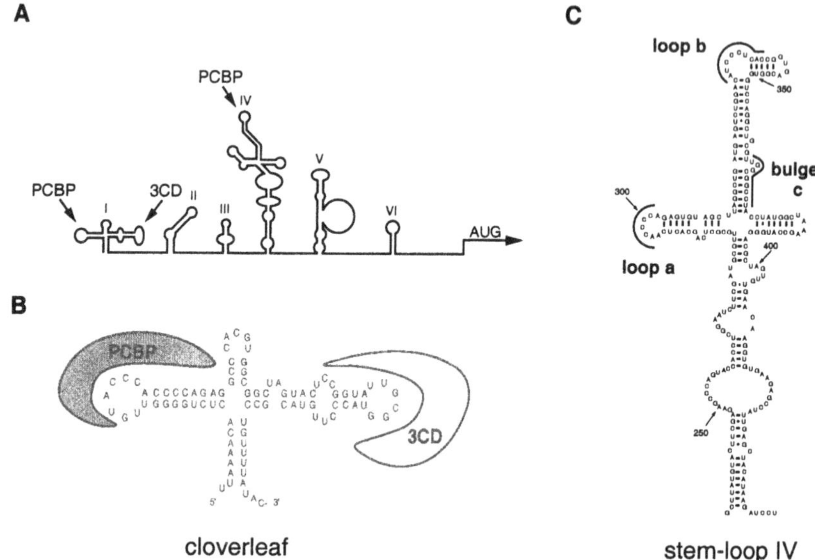

Figure 2. A. Schematic representation of the secondary structure of the poliovirus 5' UTR. *Predicted stem loops I to VI are indicated. AUG represents the translation initiation codon. The specific sites of PCBP binding are indicated. The binding site for the viral protein 3CD is also shown.* ***B****. Representation of the ribonucleoprotein complex formed by the poliovirus cloverleaf structure present at the 5' end of the genomic RNA. The viral protein 3CD and the cellular protein PCBP are shown interacting with the specific target sequences.* ***C****. Predicted secondary structure of domain IV of the poliovirus 5'UTR. PCBP protected regions determined by footprinting analysis are indicated by black lines (loop a, loop b, and bulge c).*

Both PCBP1 and PCBP2 form a low affinity complex with the cloverleaf RNA (Kd~95nM); but together with the viral protein 3CD, precursor of the viral polymerase 3D and the viral protease 3C, they are incorporated into a high affinity ternary ribonucleoprotein complex (Kd~1nM) (25, 26, 29) (Fig. 2B). Ternary complex formation is required for RNA synthesis (25, 28). Mutations that impair ternary complex formation abolish RNA synthesis. Since PCBP greatly enhances the binding of 3CD to its target RNA, an important role of PCBP in replication may be to facilitate the binding of 3CD to the cloverleaf RNA. However, because the poliovirus genome is genetically diverse, it seems that 3CD could have co-evolved with the viral RNA to interact efficiently in the absence of cellular factors. The fact that PCBPs are required for binding of 3CD suggests both a structural and a functional role of PCBP during RNA replication. In addition, the interactions of the PCBP proteins and 3CD with the cloverleaf RNA seem to determine whether the genomic RNA is used as a template for protein synthesis or RNA replication. Binding of PCBPs to the cloverleaf stimulates viral translation, while binding of 3CD downregulates translation and promotes negative strand RNA synthesis (28). Thus, it appears that PCBP is involved in different aspects of viral mRNA function. At early stages of infection, PCBP is required for viral mRNA translation; as the viral cycle progresses, PCBP facilitates the recruitment of the viral encoded polymerase precursor, 3CD. This complex, in turn, downregulates translation and promotes the use of the viral RNA as a template for viral RNA replication. Interestingly, it has recently been shown that PCBP and 3CD both interact with an RNP complex at the 3' poly(A) tail to promote RNA circularization (see below).

PCBPs have also been implicated in the activation of translation of other viral RNAs. PCBP2 is required for translation of coxsackievirus and human rhinovirus but not by encephalomyocarditis virus and foot-and-mouth disease virus (30). These data indicate that PCBP2 is essential for the internal initiation of translation on picornavirus type I IRES elements but is dispensable for translation directed by the structurally distinct type II elements. Also, PCBP promotes protein synthesis from Hepatitis A Virus (31). In contrast, PCBP1, PCBP2, and hnRNP K induce translational silencing of the Human Papilloma Virus Type 16 L2 mRNA (32). It has been demonstrated that RNA elements present in the 3' UTR of the L2 RNA act in *cis* to reduce mRNA translation without substantially affecting mRNA levels. This element could also function inhibiting translation of heterologous mRNAs. Using RNA gel shift assays and UV cross-linking, it has been shown that several cellular proteins, including PCBPs and hnRNP K, specifically interact with the 3'UTR of L2 RNA. Expression of PCBP1, PCBP2, and hnRNP K in HeLa cells decreases translation of L2 mRNA. *In vitro* studies suggested that each protein (PCBP1, PCBP2, or hnRNP K) could inhibit L2 translation independently. Thus the specific role of each protein in the L2 RNP complex remains to be established. More recently, the interactions of PCBP with hepatitis C and Norwalk virus RNAs have been reported but the functional significance of these interactions remains unclear (33, 34).

Structural Basis of RNA Recognition by PCBP

PCBP1 and PCBP2 contain three copies of the RNA-binding KH motif (K homologous), first described in hnRNP K (35). The arrangement of the KH motifs within several RNA binding proteins, such as hnRNP K, Nova protein, PCBP1, and PCBP2, is similar: two consecutive KH domains at the amino terminus, followed by a region of variable sequence and length before the third KH motif (Fig. 3A) (1, 27). The degree of homology within the corresponding domain in all of these proteins is higher than that shared by KH motifs within the same polypeptide. (1). No other known RNA binding motif is found within PCBPs, and indeed the KH domains are able to function as discreet and independent nucleic acid binding units.

A

PCBP 2

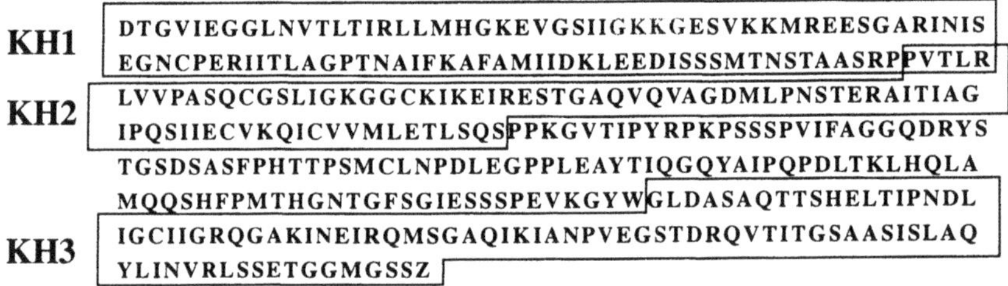

B

KH1

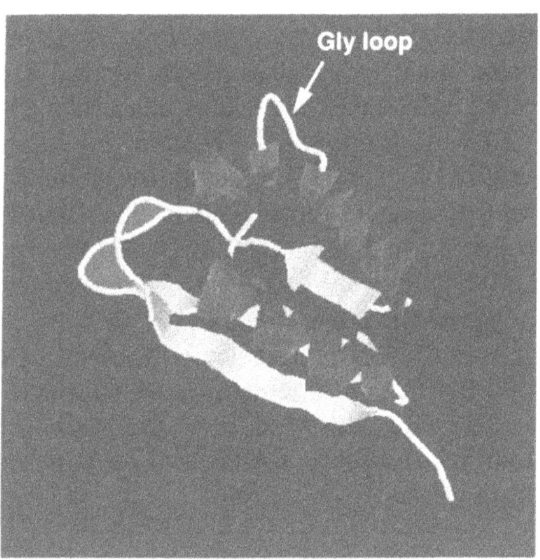

Figure 3. A. Amino acid sequence of PCBP2. *Boxed areas indicate the sequences representing each of the three KH domains.* ***B. Ribbon diagram of the 6th KH-domain from vigilin.***

Expression of each of the three domains demonstrated that the isolated KH1 and KH3 domains of PCBP1 and PCBP2 can specifically bind to poly(rC) homopolymers (36). However, the basis for the association of these KH containing proteins with their specific RNA targets remains to be established.

Musco and co-workers were the first to solve the tertiary structure of a KH-domain. The 6th KH-domain of vigilin consists of a stable three-stranded antiparallel β-sheet, oriented against three helices (37). This stable baabba fold exposes a putative RNA-binding surface: the tetrapeptide Gly-Lys-X-Gly between the first two helices. The central position in this exposed loop is occupied by a semiconserved positively charged residue (Fig. 3B). The conservation of these residues in the flexible loop structure could be an intrinsic feature of the KH motif and might play a functional role.

Interestingly, the structure of the KH domain provides insight into the molecular basis of fragile X syndrome. This inherited disease is caused by a mutation at the isoleucine 304 in the KH-domain of the FMR protein. The homologous mutation in the KH domain of the vigilin protein disrupts the domain structure and its capacity to interact with RNA (38). The RNA target of FMR1 has not been defined.

Although both the KH1 and KH3 domains of PCBPs bind with high affinity to poly(rC) homopolymers (27, 36), only KH1 was capable of specifically interacting with the cloverleaf and the stem loop IV of the poliovirus 5'UTR. These results suggest that the KH1 domain is the major RNA binding determinant for the recognition of the poliovirus-specific RNA targets by PCBPs. However, the KH2 and KH3 domains must play a role in these interactions because mutations in these domains have a detrimental effect on the binding by the full-length protein (27). Similarly, mutation of any of the three KH domains of hnRNP K abolishes RNA binding (38), but only the KH3 fragment interacts with poly(rC) with high affinity (36). It is not clear, however, how the KH2 and KH3 stabilize the interaction of PCBP with the viral RNA. The lack of cooperativity in the interactions of the individually expressed domains with the viral RNA suggest that all three motifs must be tethered within a single polypeptide in order to have optimal affinity for the RNA.

Mechanisms and Regulation of PCBP Function

The formation of PCBP-containing RNP complexes are involved in a wide variety of processes: stabilizing RNA molecules, increasing or silencing cellular or viral translation, or promoting viral replication. Yet, PCBPs bind to relatively simple pyrimidine rich elements present at the 3'UTR of α-globin, collagen-α1(I), TH, LOX mRNAs as well as to different domains within the UTRs of viral RNAs. This raises important mechanistic and regulatory questions: how is the formation of specific complexes regulated under different environmental conditions and how do the RNP complexes communicate with the translation and/or RNA degradation machinery to perform their specific function?

PHOSPHORYLATION AND INTERACTION WITH OTHER PROTEINS

One possible mechanism to control RNP complex formation is by regulating the phosphorylation state of PCBP. It has been shown that the RNA binding ability of PCBP is reduced upon phosphorylation of the protein, providing a mechanism for releasing PCBP from the RNA (1). In addition, the interaction of PCBP with the RNA target is probably modulated by association with other proteins. In the best studied example, the presence of the virally encoded 3CD increases the affinity of PCBP for the poliovirus cloverleaf RNA by two orders of magnitude (29). Thus, the binding of accessory proteins to PCBP and/or to the RNA could determine the specific role of different complexes.

PCBP and mRNA Stability

Although the increased stability of different mRNAs requires PCBP binding, its sole presence is not sufficient, suggesting the requirement of additional proteins. The identity of these factors is unclear at the moment. However, using the yeast two hybrid screen, it was shown that PCBP specifically interacts with the AUF1 proteins also known as hnRNP D (9). AUF1 proteins interact with ARE (A-U rich elements) which mediate mRNA decay. The ARE appears to stimulate deadenylation and subsequent degradation of the mRNA and mRNA degradation. Interestingly, it was shown that disruption of PCBP binding to the α-globin mRNA resulted in deadenylation of the RNA (39,40) (Fig. 4A). In addition, PCBP can interact with the poly(A) binding protein (PABP). Thus, PCBP could promote mRNA stabilization by its interaction with PABP and AUF 1. Formation of such a complex may induce conformational changes in the mRNA structure so that it no longer exposes RNA elements recognized by the degradation machinery.

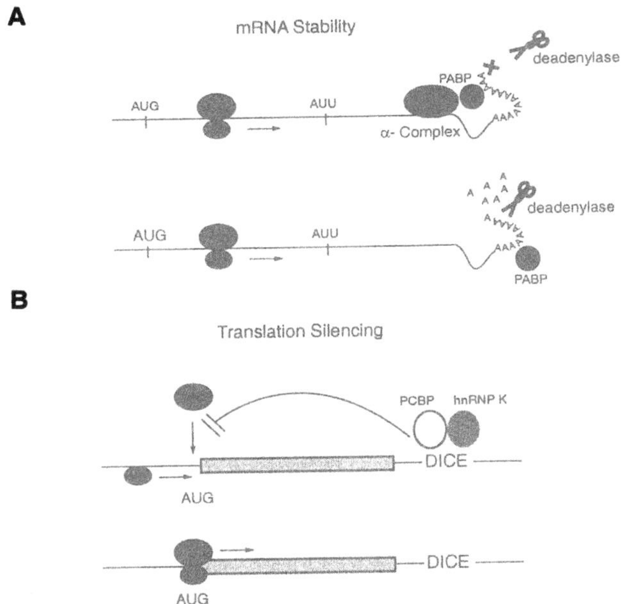

Figure 4. ***Mechanism of PCBP function A. mRNA stability model****. Formation of the a-complex (containing PCBP) protects the mRNA from degradation. The mRNA is denoted by a line, the translation start AUG and stop UAA sites, and the travelling ribosomes are shown. The interaction of the poly(A) binding protein (PABP) with the poly(A) tail and with the a-complex required for mRNA stability are shown at the 3' end.* ***B. Translation silencing model.*** *Formation of the ribonucleoprotein complex between DICE sequences and PCBP/hnRNP K at the 3' UTR of LOX mRNA is indicated. The 43S complex that contains the small ribosomal subunit is shown scanning the 5'UTR. The inhibition of 60S ribosomal subunit joining is indicated. At the bottom efficient translation is achieved in the absence of complex formation.*

PCBP AND TRANSLATION CONTROL

PCBP can either activate or inhibit translation of specific mRNAs. It appears that the positive effect on translation is mediated by interactions of PCBP with the 5'UTR, while the translational silencing involves the interaction of PCBP with 3'UTRs. Also, PCBP seems to stimulate translation of RNAs that initiate the process mediated by IRES elements, but not of the general cap dependent translation. The mechanisms by which PCBP regulates translation are unclear. For instance, PCBPs could be acting as a molecular bridge interacting with both canonical and non-canonical factors implicated in the initiation process (reviewed by (41)). Alternatively, PCBPs could act as RNA chaperones inducing a conformational change in the IRES necessary for the recognition of the RNA by the translation machinery. Furthermore, it has been described that PCBPs could facilitate communication between the 5' and 3' ends of the mRNA by interacting with the PABP bound to the poly(A) tail. However, it is not known whether this interaction can stimulate IRES dependent translation.

The molecular mechanism of translational silencing of the LOX mRNA has been examined in detail in the last few years (20,42). The binding of PCBP and hnRNP K to the RNA sequences of DICE prevent translation initiation (Fig. 4B). This mechanism is independent of the poly(A) tail, since a reconstituted translation system can induce translational silencing from an mRNA lacking a poly(A) tail. In this case, a protein-protein interaction between PCBP and hnRNP K could be important for silencing. Each of these proteins binds the LOX mRNA and silence translation. However, the presence of both proteins simultaneously appears to have a more profound effect on translation. Interestingly, hnRNP K accumulates in the nucleus, and a change in the phosphorylation state determines its cytoplasmic localization (43). Notably, formation of the silencing complex (PCBP/hnRNP K-DICE) does not interfere with the formation of the 48S pre-initiation complex, but prevents formation of a stable 80S complex (42). Also, it was shown that particular modes of translation initiation are not affected by DICE. The cricket paralysis virus IRES utilizes an unusual initiation mechanism that involves direct 80S ribosome formation on the initiation codon without requirement of initiation factors. An artificial mRNA in which initiation is mediated by the cricket paralysis virus IRES is not susceptible to silencing by PCBP/hnRNP K-DICE complex. These studies suggest that inhibition of translation mediated by DICE complex may interfere with joining of the ribosomal subunits.

PCBP and Viral RNA Replication

The interaction of PCBP with the 5' cloverleaf of poliovirus is essential for RNA replication. A high affinity ternary complex that forms at the 5' end of the poliovirus genomic RNA includes the cloverleaf-like RNA structure, PCBPs, and 3CD. After infection, a binary complex is formed between the cloverleaf RNA and PCBP; however after viral protein synthesis, the viral RNA polymerase precursor 3CD is incorporated into the complex (Fig. 5).

One possible function of PCBP in replication is to facilitate 3CD binding. Interestingly, the isolated KH1 domain stimulates the formation of this high affinity ternary complex. How does KH1 facilitate high affinity ternary complex formation to the poliovirus 5'UTR? There are at least two possible mechanisms: First, binding of KH1 may trigger a conformational change in the cloverleaf RNA to generate a target structure that interacts better with 3CD. Alternatively, protein-protein interactions between KH1 and 3CD could stabilize the complex.

Evidence has been presented indicating that the cloverleaf complex is required for initiation of negative strand RNA synthesis (44) (28) (45). It is intriguing that specific binding of the viral polymerase precursor to the 5'-end of the RNA genome is required for negative-strand RNA synthesis, given that initiation of negative-strand RNA synthesis takes place at the opposite end of the genomic RNA, namely at the poly(A)-tail. Interestingly, the presence of a poly(A)-tail is also important for negative strand synthesis. The mechanism by which this 5'RNP complex facilitates initiation of RNA synthesis has recently been elucidated (45). By examining the *cis*- and *trans*-acting elements involved in the initiation of negative-strand RNA, it was more recently shown that a long-range interaction between RNP complexes formed at the ends of the viral genome is necessary for RNA replication.

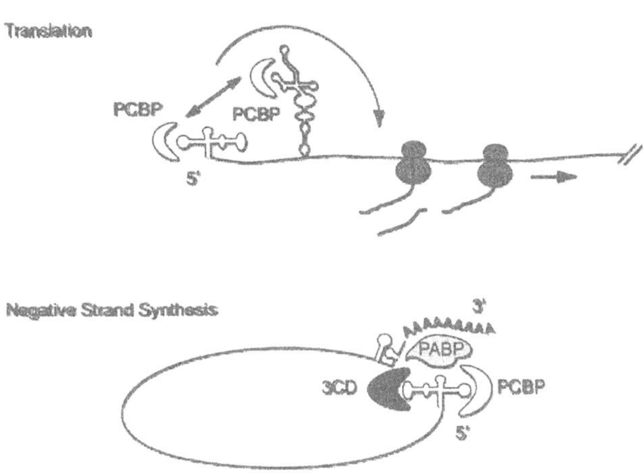

Figure 5. ***Model for translation and replication of poliovirus RNA.*** *The interaction of PCBP with the cloverleaf and domain IV of the poliovirus 5'UTR required for IRES dependent translation is shown. At the bottom, PCBP facilitates the recruitment of 3CD to form the ternary complex with the cloverleaf RNA. This complex decreases viral translation and promotes negative strand synthesis. The interaction between 3CD and PCBP with the PABP that mediates the circularization of the viral RNA is indicated.*

Initiation of negative strand RNA synthesis requires a poly(A) tail at the 3' end and the cloverleaf structure located at the 5' end of the genome. Poly(A)-binding protein 1 (PABP 1) interacts with both the poly(A) RNA-tail and both proteins that bind the cloverleaf structure, 3CD and PCBP (45). Thus, it seems that the viral polymerase precursor 3CD and PCBP bind to the RNA synthesis promoter at the 5'-end of the genomic RNA and reaches its site of action within the poly(A)-tail of the genome via circularization of the genomic RNA using an RNA-protein-protein-RNA-bridge that involves at least two cellular factors, PCBP and PABP 1. However it remains unclear whether PCBP play a direct catalytic or a structural role in the initiation of RNA synthesis.

CONCLUSIONS

PCBPs are key mediators of RNA function. Their multiple roles appear to be regulated by phosphorylation and by interaction with other RNA binding proteins. An attractive possibility is that these complexes function by changing the conformation of the RNA, thereby acting as RNA chaperones.

Summary of key concepts

- *PCBPs are RNA binding proteins involved in both cellular and viral mRNA metabolism.*
- *PCBPs contain three copies of a conserved RNA binding domain called KH, also found in a variety of RNA binding proteins such as hnRNP K.*
- *The specific binding of PCBPs to pyrimidine rich elements present at the 5' and/or the 3' untranslated region of mRNAs controls the stability as well as the translation of the target RNA.*

Study Guide Questions

1) Why are PCBPs important regulators of gene expression?
2) What are the RNA targets of PCBP binding?
3) Describe the mechanism of translational silencing of the LOX mRNA.
4) How do PCBPs regulate viral translation and poliovirus replication?

REFERENCES

1. Leffers H, Dejgaard K, Celis JE. 1995 Characterization of two major cellular poly(rC)-binding human proteins, each containing three K-homologous (KH) domains. European Journal of Biochemistry 230(2):447-53.
2. Gamarnik AV, Andino R. 1997 Two functional complexes formed by KH domain containing proteins with the 5' noncoding region of poliovirus RNA. RNA 3(8):882-92.

3. Singer RH, Penman S. 1973 Messenger RNA in HeLa cells: kinetics of formation and decay. Journal of Molecular Biology 78(2):321-34.
4. Spradling A. 1975 Calculation of mRNA lifetimes using ribosomal RNA labeling as a metabolic probe. Cell 4(2):139-40.
5. Ross J. 1995 mRNA stability in mammalian cells. Microbiological Reviews59(3):423-50.
6. Wang X, Kiledjian M, Weiss IM, Liebhaber SA. 1995 Detection and characterization of a 3' untranslated region ribonucleoprotein complex associated with human alpha-globin mRNA stability [published erratum appears in Mol Cell Biol 1995 Apr;15(4):2331]. Molecular and Cellular Biology 15(3):1769-77.
7. Weiss IM, Liebhaber SA. 1994 Erythroid Cell-Specific Determinants of Alpha-Globin Mrna Stability. Molecular and Cellular Biology 14(12):8123-8132.
8. Weiss IM, Liebhaber SA. 1995 Erythroid cell-specific mRNA stability elements in the alpha 2-globin 3' nontranslated region. Molecular and Cellular Biology 15(5):2457-65.
9. Kiledjian M, DeMaria CT, Brewer G, Novick K. 1997 Identification of AUF1 (heterogeneous nuclear ribonucleoprotein D) as a component of the alpha-globin mRNA stability complex [published erratum appears in Mol Cell Biol 1997 Oct;17(10):6202]. Molecular and Cellular Biology 17(8):4870-6.
10. Chkheidze AN, Lyakhov DL, Makeyev AV, Morales J, Kong J, Liebhaber SA. 1999 Assembly of the alpha-globin mRNA stability complex reflects binary interaction between the pyrimidine-rich 3' untranslated region determinant and poly(C) binding protein alphaCP. Molecular and Cellular Biology 19(7):4572-81.
11. Holcik M, Liebhaber SA. 1997 Four highly stable eukaryotic mRNAs assemble 3' untranslated region RNA-protein complexes sharing cis and trans components. Proceedings of the National Academy of Sciences of the United States of America 94(6):2410-4.
12. Stefanovic B, Hellerbrand C, Holcik M, Briendl M, Liebhaber SA, Brenner DA. 1997 Posttranscriptional regulation of collagen alpha 1(I) mRNA in hepatic stellate cells. Molecular and Cellular Biology 17(9):5201-5209.
13. Maatta A, Penttinen RPK. 1994 Nuclear and Cytoplasmic Alpha-1(I) Collagen Mrna-Binding Proteins. FEBS Letters 340(1-2):71-77.
14. Czyzykkrzeska MF, Furnari BA, Lawson EE, Millhorn DE. 1994 Hypoxia Increases Rate of Transcription and Stability of Tyrosine Hydroxylase Messenger Rna in Pheochromocytoma (Pc12) Cells. Journal of Biological Chemistry 269(1):760-764.
15. Czyzykkrzeska MF, Beresh JE. 1996 Characterization of the Hypoxia-Inducible Protein Binding Site Within the Pyrimidine-Rich Tract in the 3'-Untranslated Region of the Tyrosine Hydroxylase Mrna. Journal of Biological Chemistry 271(6):3293-3299.
16. Paulding WR, Czyzyk-Krzeska MF. 1999 Regulation of tyrosine hydroxylase mRNA stability by protein-binding, pyrimidine-rich sequence in the 3 '-untranslated region. Journal of Biological Chemistry 274(4):2532-2538.
17. Czyzyk-Krzeska MF, Bendixen AC. 1999 Identification of the poly(C) binding protein in the complex associated with the 3' untranslated region of erythropoietin messenger RNA. Blood 93(6):2111-20.
18. Sachs AB, Sarnow P, Hentze MW. 1997 Starting at the beginning, middle, and end: translation initiation in eukaryotes. Cell 89(6):831-8.
19. Jackson RJ, Wickens M. 1997 Translational controls impinging on the 5'-untranslated region and initiation factor proteins. Curr Opin Genet Dev 7(2):233-41.
20. Ostareck DH, Ostareck LA, Wilm M, Thiele BJ, Mann M, Hentze MW. 1997 mRNA silencing in erythroid differentiation: hnRNP K and hnRNP E1 regulate 15-lipoxygenase translation from the 3' end. Cell 89(4):597-606.
21. Ostareckléderer A, Ostareck DH, Standart N, Thiele BJ. 1994 Translation of 15-Lipoxygenase Messenger Rna Is Inhibited By a Protein That Binds to a Repeated Sequence in the 3' Untranslated Region. Embo Journal 13(6):1476-1481.

22. Blyn LB, Swiderek KM, Richards O, Stahl DC, Semler BL, Ehrenfeld E. 1996 Poly(rC) binding protein 2 binds to stem-loop IV of the poliovirus RNA 5' noncoding region: identification by automated liquid chromatography-tandem mass spectrometry. Proc Natl Acad Sci U S A 93(20):11115-20.
23. Parsley TB, Towner JS, Blyn LB, Ehrenfeld E, Semler BL. 1997 Poly (rC) binding protein 2 forms a ternary complex with the 5'-terminal sequences of poliovirus RNA and the viral 3CD proteinase. RNA 3(10):1124-34.
24. Blyn LB, Towner JS, Semler BL, Ehrenfeld E. 1997 Requirement of poly(rC) binding protein 2 for translation of poliovirus RNA. Journal of Virology 71(8):6243-6.
25. Andino R, Rieckhof GE, Baltimore D. 1990 A functional ribonucleoprotein complex forms around the 5' end of poliovirus RNA. Cell 63(2):369-80.
26. Andino R, Rieckhof GE, Achacoso PL, Baltimore D. 1993 Poliovirus RNA synthesis utilizes an RNP complex formed around the 5'- end of viral RNA. EMBO J 12(9):3587-98.
27. Silvera D, Gamarnik AV, Andino R. 1999 The N-terminal K homology domain of the poly(rC)-binding protein is a major determinant for binding to the poliovirus 5'-untranslated region and acts as an inhibitor of viral translation. Journal of Biological Chemistry 274(53):38163-70.
28. Gamarnik AV, Andino R. 1998 Switch from translation to RNA replication in a positive-stranded RNA virus. Genes and Development 12(15):2293-304.
29. Gamarnik AV, Andino R. 2000 Interactions of viral protein 3CD and poly(rC) binding protein with the 5' untranslated region of the poliovirus genome. Journal of Virology 74(5):2219-26.
30. Walter BL, Nguyen JH, Ehrenfeld E, Semler BL. 1999 Differential utilization of poly(rC) binding protein 2 in translation directed by picornavirus IRES elements. RNA 5(12):1570-85.
31. Graff J, Cha J, Blyn LB, Ehrenfeld E. 1998 Interaction of poly(rC) binding protein 2 with the 5' noncoding region of hepatitis A virus RNA and its effects on translation. Journal of Virology 72(12):9668-75.
32. Collier B, GoobarLarsson L, Sokolowski M, Schwartz S. 1998 Translational inhibition in vitro of human papillomavirus type 16 L2 mRNA mediated through interaction with heterogenous ribonucleoprotein K and Poly(rC)-binding proteins 1 and 2. Journal of Biological Chemistry 273(35):22648-22656.
33. Spangberg K, Schwartz S. Poly(C)-binding protein interacts with the hepatitis C virus 5 ' untranslated region. Journal of General Virology 1999;80(PT6):1371-1376.
34. Gutierrez-Escolano AL, Brito ZU, del Angel RM, Jiang X. 2000 Interaction of cellular proteins with the 5 ' end of Norwalk virus genomic RNA. Journal of Virology 74(18):8558-8562.
35. Siomi H, Matunis MJ, Michael WM, Dreyfuss G. 1993 The pre-mRNA binding K protein contains a novel evolutionarily conserved motif. Nucleic Acids Res21(5):1193-8.
36. Dejgaard K, Leffers H. 1996 Characterization of the nucleic-acid-binding activity of KH domains. Different properties of different domains. European Journal of Biochemistry 241(2):425-31.
37. Musco G, Stier G, Joseph C, Castiglione MM, Nilges M, Gibson TJ, Pastore A. 1996 Three-dimensional structure and stability of the KH domain: molecular insights into the fragile X syndrome. Cell 85(2):237-45.
38. Siomi H, Choi MY, Siomi MC, Nussbaum RL, Dreyfuss G. 1994 Essential Role For Kh Domains in Rna Binding - Impaired Rna Binding By a Mutation in the Kh Domain of Fmr1 That Causes Fragile X Syndrome. Cell 77(1):33-39.
39. Wang Z, Kiledjian M. 2000 The poly(A)-binding protein and an mRNA stability protein jointly regulate an endoribonuclease activity. Molecular and Cellular Biology 20(17):6334-41.

40. Wang Z, Day N, Trifillis P, Kiledjian M. 1999 An mRNA stability complex functions with poly(A)-binding protein to stabilize mRNA in vitro. Molecular and Cellular Biology 19(7):4552-60.
41. Andino R, Böddeker N, Silvera D, Gamarnik AV. 1999 Intracellular determinants of picornavirus replication. Trends in Microbiology 7(2):76-82.
42. Ostareck DH, Ostareck-Lederer A, Shatsky IN, Hentze MW. 2001 Lipoxygenase mRNA silencing in erythroid differentiation: The 3'UTR regulatory complex controls 60S ribosomal subunit joining. Cell 104(2):281-90.
43. Habelhah H, Shah K, Huang L, Ostareck-Lederer A, Burlingame AL, Shokat KM, Hentze MW, Ronai Z. 2001 ERK phosphorylation drives cytoplasmic accumulation of hnRNP-K and inhibition of mRNA translation. Nature Cell Biology 3(3):325-330.
44. Barton DJ, O'Donnell BJ, Flanegan JB. 2001 5' cloverleaf in poliovirus RNA is a cis-acting replication element required for negative-strand synthesis. EMBO Journal 20(6):1439-48.
45. Herold J, Andino R. 2001 Poliovirus RNA replication requires genome circularization through a protein-protein bridge. Molecular Cell 7(3):581-591.

4

AUTOREGULATION OF PROTEIN SYNTHESIS BY TRANSLATION

Guim Kwon, Guang Xu, Wilhelm S. Cruz, Connie A. Marshall and Michael L. McDaniel

Washington University School of Medicine, St Louis, MO

In mammalian cells mRNA translation is activated by a variety of stimuli including hormones such as insulin and growth factors. The stimulatory effects of nutrients, in particular, glucose and amino acids, on protein synthesis have recently become a focus of new interest. Acute exposure of pancreatic β-cells to glucose, for example, is known to activate insulin biosynthesis by stimulating translation of pre-existing prepro-insulin mRNA. This regulation at the level of translation rather than at the level of insulin gene transcription ensures a rapid replenishment of stored insulin content after exocytosis. Dietary amino acids have also been shown to stimulate muscle protein synthesis after food intake (1, 2). This anabolic effect may be attributed in part to an increase in amino acid supply to muscle, thereby augmenting substrate availability for peptide synthesis. However, many recent studies suggest that amino acids also function independently as nutritional signaling molecules that regulate mRNA translation. In this chapter, the effects of nutrients, glucose and amino acids, on the regulation of protein synthesis by activating key regulatory translation factors will be discussed primarily focusing on pancreatic β-cells.

GLUCOSE STIMULATES INSULIN SECRETION AND BIOSYNTHESIS BY PANCREATIC β-CELLS

Insulin is a key hormone that regulates glucose homeostasis in mammals. The insulin response to food ingestion is determined by direct actions of glucose and certain amino acids on the pancreatic β-cell together with indirect actions of both hormonal and neural stimuli (3). Among these insulin secretagogues, glucose is a major nutrient regulator for insulin secretion by β-cells. The normal secretory pathway induced by glucose starts with its transport into the β-cell by a GLUT-2 transporter (4, 5). Phosphorylation of glucose by glucokinase, and subsequent metabolism leads to the generation of ATP and an increase in the ATP/ADP ratio. This results in closure of ATP-sensitive K^+ channels in the β-cell plasma membrane, cellular depolarization, opening of voltage-dependent Ca^{2+} channels and Ca^{2+} influx. The resulting increase in cytoplasmic Ca^{2+} concentration then triggers the secretory machinery, culminating in the discharge of insulin by exocytosis (6).

Metabolism of glucose regulates a variety of physiological processes in addition to insulin secretion. These processes include insulin biosynthesis, β-cell replication, and the synthesis of a number of β-cell proteins. Glucose exerts a specific stimulatory effect on insulin gene expression over time intervals of hours (7, 8), whereas the biosynthesis of insulin is significantly controlled within minutes at the level of protein translation (9-11). Although the enhancement of insulin synthesis by glucose at the translational level occurs rapidly and does not require synthesis of new mRNA, limited information is available on processes that regulate translation. More recent studies indicate that insulin is only one of a number of β-cell proteins whose synthesis is regulated by glucose at the level of translation (12). Glucose has been shown to increase the biosynthesis of a secretory granule membrane protein, SGM 110 (13), and also chromogranin A (14) to a similar extent as insulin, suggesting that translational control may extend to a variety of other β-cell proteins.

PHAS-I, an Inhibitor of Eukaryotic Initiation Factor (eIF)-4E, and p70^{s6k}, a Serine and Threonine Kinase for Ribosomal Protein S6, Regulate the Initiation of Protein Translation.

The initiation phase of mRNA translation is generally rate-limiting for protein synthesis. Initiation is mediated in part by the eIF4F complex, which is composed of three subunits, eIF-4γ, eIF-4α, and eIF-4E. eIF-4γ is a large subunit (Mr 220,000) that binds eIF-4α (Mr 45,000) and eIF-4E (Mr 25,000). eIF-4α is an ATP-dependent helicase and eIF-4E is the mRNA cap-binding protein. Thus, eIF-4F is involved in both the recognition of the capped mRNA and melting of secondary structure in the 5'-untranslated region. eIF-4E is the least abundant of the eIF-4F subunits, and it is generally believed that the amount of eIF-4E is limiting for translation initiation.

The availability of eIF-4E is regulated by PHAS-I (phosphorylated, heat and acid stable protein that is regulated by insulin) that is identical to eIF-4E binding protein-1(4E-BP1) and first identified in rat adipocytes (15, 16). The nonphosphorylated form of PHAS-I binds tightly to eIF-4E, and prevents eIF-4E from binding to eIF-4γ. When phosphorylated in the appropriate site(s), PHAS-I dissociates from eIF-4E, allowing its participation in translation initiation.

The S6Ks are a family of mitogen-activated, serine/threonine protein kinases that are essential in translation and cell cycle control (17, 18). The S6Ks are currently known to be comprised of four members derived from alternative mRNA processing and translation of distinct gene products (19). These kinases have been shown to be involved in the selective translation of a unique family of mRNAs, mediating the multiple phosphorylation of 40S ribosomal protein S6 (20, 21). The mRNA family members whose expression are controlled at the translational level by S6K are characterized by an oligopyrimidine tract at their 5' transcriptional start site or 5'TOP (oligopyrimidine tract at transcription site) (22). These mRNAs encode for components of the translational apparatus, including ribosomal proteins and translational elongation factors whose increased expression is essential for cell growth and proliferation (23).

PHAS-I is phosphorylated both *in vitro* and *in vivo* by a variety of protein kinases. Casein kinase II, protein kinase C, and mitogen-activated protein kinase have been reported to phosphorylate recombinant PHAS-I (24). Insulin-like growth factor-1 and platelet-derived growth factor in smooth muscle cells and insulin in 3T3 L1 adipocytes increase PHAS-I and $p70^{s6k}$ phosphorylation by a rapamycin-sensitive pathway. This mitogen-induced phosphorylation is also sensitive to the phosphatidylinositol 3-kinase (PI 3-K) inhibitor, wortmannin (25). Recent studies have implicated PHAS-I and $p70^{s6k}$ in a signaling pathway involving the mammalian target of rapamycin (mTOR). mTOR, a serine and threonine protein kinase, stimulates the phosphorylation of PHAS-I and $p70^{s6k}$ in a parallel manner. Thus, the rapamycin-FKBP12 complex that inhibits mTOR results in the dephosphorylation of these translational regulators (26-28). The extent to which rapamycin affects total protein synthesis varies by cell type. These studies suggest that PHAS-I and $p70^{s6k}$ have important roles in regulating translation in a variety of cells by signal transduction pathways that involve protein phosphorylation.

Glucose Stimulates PHAS-I and $p70^{s6k}$ Phosphorylation Via Autocrine Effects of Insulin

Recent studies have identified insulin and mitogenic signaling pathways in pancreatic β-cell similar to those described previously in insulin-sensitive tissues such as liver, muscle, and fat (29-31). These findings have raised the possibility that activation of a functional insulin receptor may target downstream insulin signaling proteins including insulin receptor substrates 1 and 2 (IRS-1 and 2) and provide an important mechanism for the autoregulation of β-cell function. In particular, specific functional parameters that may be regulated by this autocrine mechanism include secretion, protein translation and synthesis, and β-cell proliferation. This area is especially relevant to the development of type 2 diabetes since it is now recognized that this disease is a consequence of defects in insulin signaling in insulin-sensitive tissues in combination with defects in insulin secretion and insulin and growth factor signaling necessary to maintain sufficient β-cell mass. As shown in Fig. 1, we have attempted to define this insulin and mitogenic signaling pathway in β-cells whereby activation of a functional insulin receptor by endogenous or exogenous insulin results in phosphorylation of IRS-1 and 2 which then activates the downstream signaling elements PI 3-K and mTOR. As discussed, PHAS-I and $p70^{s6k}$ are regulated in a parallel manner through phosphorylation by mTOR, and use this mitogenic signaling pathway.

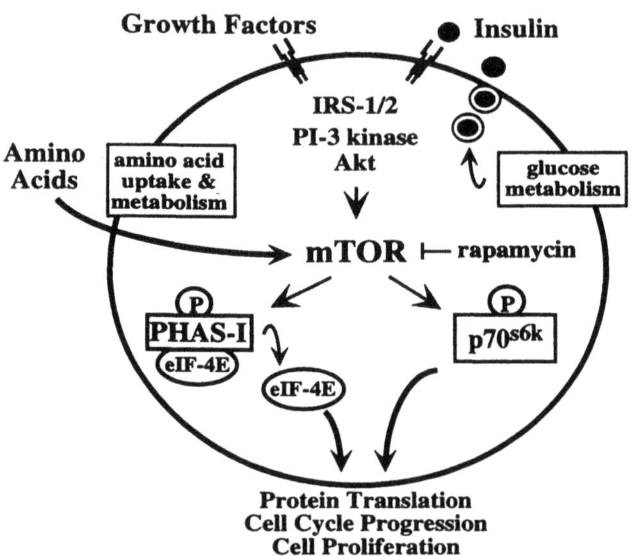

Figure 1. Proposed model for insulin and amino acid-induced mitogenic signaling in β-cells. *Abbreviations are the following: IRS-1 and 2, insulin receptor substrate 1 and 2; PI-3 kinase, phosphoinositide 3-kinase; mTOR, mammalian target of rapamycin; PHAS-I, phosphorylated heat- and acid-stable protein regulated by insulin;* $p70^{s6k}$*, p70 S6 kinase; eIF-4E, eukaryotic initiation factor-4E.*

As shown in Fig. 2, PHAS-I generally appears as 3 bands when visualized by immunoblotting and by gel shift analysis. These bands represent non-phosphorylated α (lane 7), intermediate phosphorylated β, and highly phosphorylated γ isoforms. Exposure of insulin-sensitive cells such as adipocytes to insulin or growth factors results in a shift of β and α bands to the highly phosphorylated and slower migrating PHAS-γ which is associated with the release of eIF-4E and enhanced protein translation (32).

PHASE-I γ β α	1	2	3	4	5	6	7
Minutes	10	10	30	30	180	180	
Glucose (mM)	5.5	20	5.5	20	5.5	20	
PHAS-Iγ (% of total PHAS-I)	52	72	43	86	48	71	

Figure 2. Glucose stimulation of isolated rat islets causes rapid and stable phosphorylation of PHAS-I. *Rat islets (300) were incubated in 1 ml of CMRL (basic media) containing 5.5 or 20 mM glucose for 10, 30, or 180 min at 37°C. The medium was removed, and processed for Western blotting of PHAS-I. Lane 7 contains adipocytes as a positive control at 180 min. Reproduced with permission (32).*

The exposure of islets to glucose (20 mM) enhances PHAS-γ formation by isolated rat islets in a time and concentration dependent manner (Fig. 2). The increased phosphorylation of PHAS-I in response to glucose occurs rapidly within 10 minutes or less and is stable for at least 3 hours.

In order to confirm an autocrine effect of insulin interacting with its receptor on the β-cell surface, the ability of exogenous insulin to stimulate phosphorylation of PHAS-I was evaluated (Fig. 3).

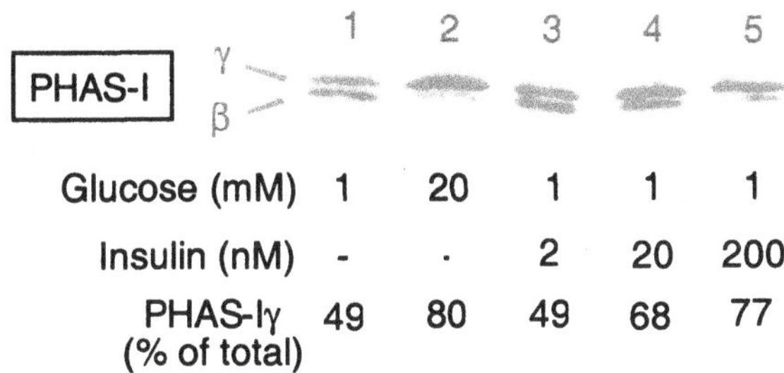

Figure 3. Insulin modulates glucose-induced phosphorylation of PHAS-I in isolated rat islets. *Rat islets (200) were serum- and glucose-depleted in 1 ml of CMRL for 2h at 37°C. Medium was replaced with 1 ml of CMRL + 1 mM glucose, 20 mM glucose, or 1 mM glucose + 2, 20, or 200 nM insulin for 30 min. Islets were processed for immunoblotting of PHAS-I. Reproduced with permission (32).*

Exposure of islets to exogenous insulin at 2, 20 and 200 nM under basal glucose levels of 1 mM (lanes 3-5) markedly stimulated the formation of PHAS-γ in a dose-dependent manner, similar to that observed with glucose

(20 mM) (lanes 1 & 2). Our previous studies have also shown that this enhanced formation of PHAS-γ by β-cells is associated with the release of eIF-4E and inhibited by rapamycin (32). In addition, insulin secreting β-cell lines also demonstrated this enhanced phosphorylation of PHAS-I, suggesting that β-cells of the pancreatic islet are responsible for these effects.

Serendipitously, we discovered that amino acids are required for glucose- or insulin-stimulated PHAS-I phosphorylation (32). Our previous studies had characterized the ability of glucose and insulin to mediate the phosphorylation of PHAS-I by islets incubated in tissue culture media. Similar experiments were subsequently conducted in Krebs-Ringer bicarbonate buffer (KRBB) as shown in Fig. 4.

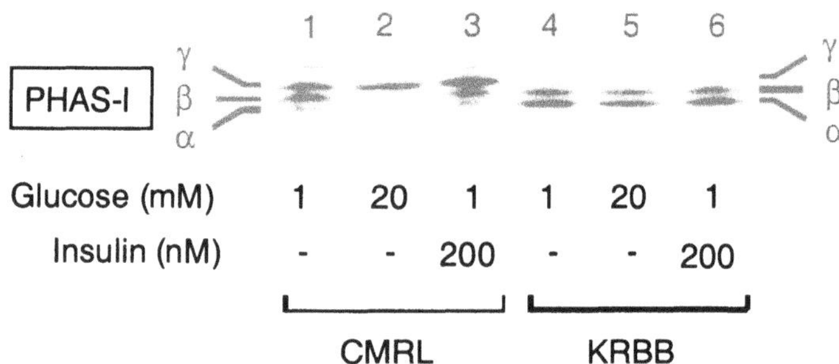

Figure 4. Amino acids are essential for phosphorylation of PHAS-I in rat islets. *Rat islets (200) were serum- and glucose-depleted in 1 ml of CMRL (0.1% BSA) or KRBB (0.1% BSA) for 2 h at 37°C. Islets were then stimulated for 30 min in 1 ml of CMRL or KRBB, 1 mM glucose, 20 mM glucose, or 1 mM glucose + 200 nM insulin. Islets were processed for immunoblotting of PHAS-I. Reproduced with permission (32)*

Unexpectedly, it was found that neither glucose nor insulin was able to enhance the formation of PHAS-γ by islets incubated in KRBB in the absence of amino acids (lanes 5 & 6) compared to islets incubated in tissue culture media (lanes 1 & 2), even though islets secrete insulin normally in KRBB. These studies demonstrated that amino acids are essential for glucose or insulin and IGF-1-stimulated PHAS-γ formation.

Branched-Chain Amino Acids Mediate the Effects of Amino Acids on PHAS-I Phosphorylation

In addition to the requirement of amino acids to facilitate glucose, insulin and IGF-1-mediated phosphorylation of PHAS-I, we also evaluated the ability of amino acids alone to activate this mTOR signaling pathway in RINm5F cells, a β-cell line (33). In these experimental protocols, KRBB was supplemented with amino acids over the range of 0.1-10X following a 2h preincubation in KRBB (amino acid and glucose free) to achieve a basal state of PHAS-I phosphorylation.

In these initial studies, essential amino acids were as effective in mediating the formation of PHAS-γ (69%) compared with a complete complement of amino acids (73%). Although non-essential amino acids and glutamine did cause some PHAS-γ formation, 30 and 40%, respectively, the effect was not nearly as significant as that produced by either essential or a complete complement of amino acids (data not shown). Therefore, essential amino acids appeared to be responsible for amino acid-mediated PHAS-I phosphorylation by RINm5F cells.

Since essential amino acids consist of a mixture of 12 different amino acids, the ability of these 12 amino acids individually to stimulate PHAS-I phosphorylation was determined (33). As shown in Fig. 5, the branched-chain amino acids leucine (72%), isoleucine (65%), and valine (77%) were among the most effective in stimulating the formation of PHAS-γ in comparison to a complete complement of amino acids. It is noted that valine (10 mM) also results in a significant appearance of the hyperphosphorylated form of PHAS-δ. Of the remaining essential amino acids (methionine, phenylalanine, tryptophan, cysteine, threonine, tyrosine, arginine, histidine, and lysine), only tyrosine (65%) was as effective as the branched-chain amino acids (leucine, isoleucine, and valine) to induce the formation of PHAS-γ. We therefore focused on the branched-chain amino acids, although these data do not negate a role for the other amino acids in the regulation of protein translation.

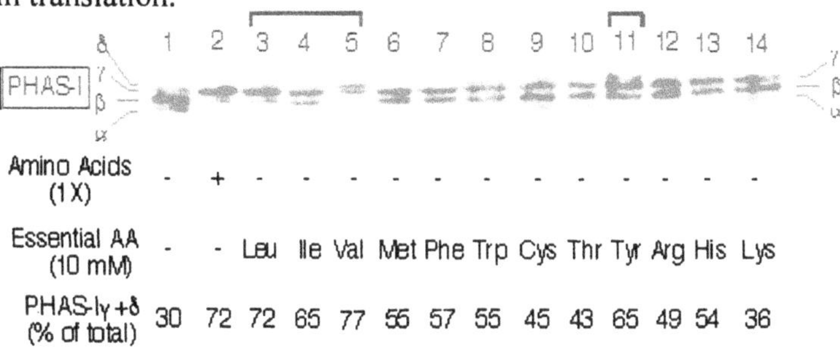

Figure 5. Effects of individual essential amino acids on the phosphorylation of PHAS-I in RINm5F cells. *RINm5F cells were preincubated in KRBB in the absence of glucose and amino acids for 2 h. Following preincubation, buffer was replaced with KRBB containing 10 mM of a single essential amino acid as indicated for 30 min. Cells were processed for immunoblotting of PHAS-I. Reproduced with permission (33).*

Branched-Chain Amino Acids Stimulate the Phosphorylation of PHAS-I and p70^{s6k} Independently of Insulin and Other Growth Factors

Since some of these amino acids are secretagogues for insulin secretion by β-cells, it was determined whether endogenous insulin secreted by

RINm5F cells was involved in amino acid-induced phosphorylation of PHAS-I by an autocrine effect. In this approach, the ability of inhibitors of tyrosine kinase activity, an upstream event in insulin signaling, to interfere with branched-chain amino acid-mediated phosphorylation of PHAS-I was evaluated (33). Genistein (10 mM) and herbimycin A (1 mM), two tyrosine inhibitors with different mechanisms of action (34, 35), had no effect on the ability of branched-chain amino acids to stimulate the phosphorylation of PHAS-I (Fig. 6).

Genistein at concentrations of 1 and 100 μM also had no effect (data not shown). Genistein and herbimycin A are effective tyrosine kinase inhibitors in islets and RINm5F cells. Previously, we have shown that these inhibitors block interleukin-1 stimulated tyrosine kinase activation, resulting in the inhibition of NFκB translocation to the nucleus and blockage of subsequent transcription and translation of the inducible nitric oxide synthase gene by rat islets and RINm5F cells (36, 37). These data and other approaches to block insulin secretion suggested that branched-chain amino acids induce phosphorylation of PHAS-I independently of the upstream insulin signaling pathway by β-cells (32, 33).

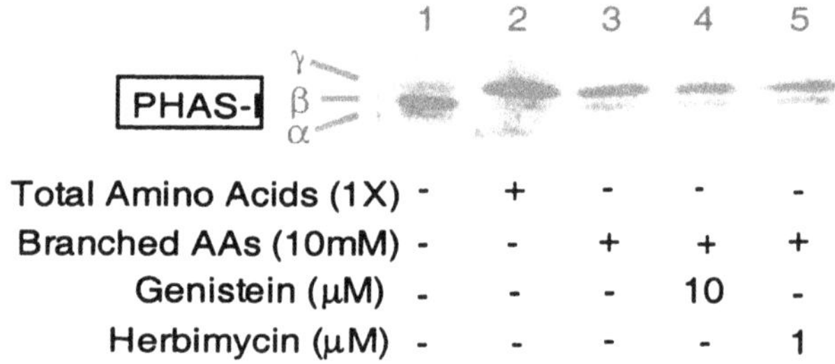

	1	2	3	4	5
Total Amino Acids (1X)	-	+	-	-	-
Branched AAs (10mM)	-	-	+	+	+
Genistein (μM)	-	-	-	10	-
Herbimycin (μM)	-	-	-	-	1

Figure 6. Branched-chain amino acids (leucine, isoleucine, and valine) stimulate phosphorylation of PHAS-I in RINm5F cells independent of endogenous insulin. *RINm5F cells were preincubated in 3 ml of KRBB in the absence of glucose and amino acids for 2 h. During the last hour of preincubation, genistein (10 μM) or herbimycin A (1 μM) were added to the cells as indicated. Following preincubation, buffer was replaced with KRBB containing branched-chain amino acids (leucine, isoleucine, and valine at 3.3 mM for each amino acid) plus inhibitors as indicated for 30 min. RINm5F cells were processed for immunoblotting of PHAS-I. Reproduced with permission (33)*

Leucine is the Key Amino Acid That Stimulates Phosphorylation of PHAS-I and p70^{s6k} Via mTOR

Of the three branched-chain amino acids that activate the mTOR signaling pathway, leucine has generated significant interest due to its unique ability at physiological concentrations to stimulate protein synthesis, inhibit lysosomal autophagy (33, 38-42), and enhance pancreatic β-cell replication in the fetal rodent pancreas (43). Although our previous studies demonstrated that leucine dose-dependently enhanced the phosphorylation of PHAS-I and p70^{s6k} in a rapamycin-dependent manner, this effect was only achieved at leucine concentrations greater than physiological levels (33). Our recent studies indicate this effect was due to the lack of synergy with other amino acids. In our previous experimental design, each amino acid was evaluated individually in KRBB for its ability to mediate the phosphorylation of PHAS-I or p70^{s6k}. An alternative experimental design is to remove individual amino acids from the complete complement of amino acids and assess these same parameters. This latter experimental approach permits an assessment of the synergy of leucine with other amino acids that may enhance its ability to activate the mTOR signaling pathway at physiological concentrations (44).

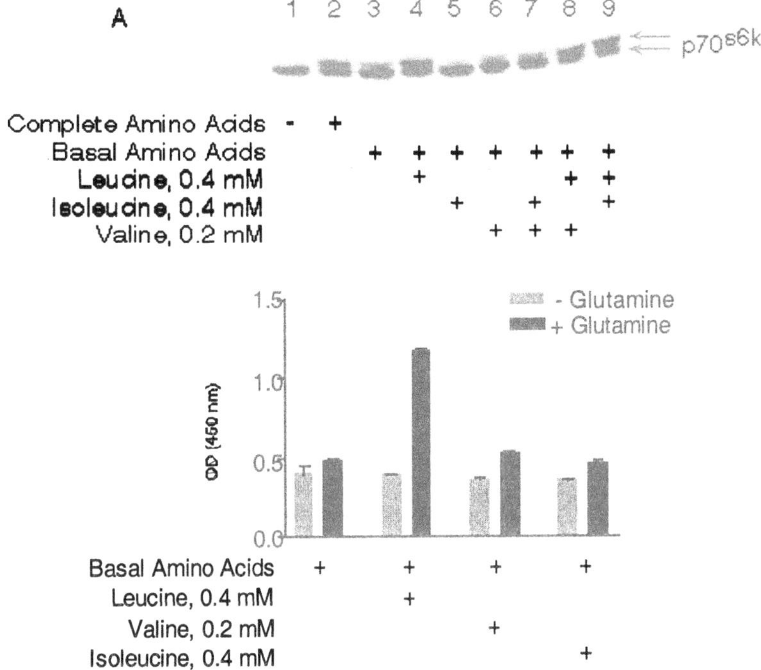

Figure 7. (A) Leucine-induced phosphorylation of p70^{s6k} at a physiological concentration in RINm5F cells. *Cells were preincubated in KRBB in the absence of glucose and amino acids for 2 h. Medium was then replaced with KRBB containing either complete amino acids as a positive control or basal amino acids, which excluded leucine, isoleucine, or valine. Leucine, isoleucine, or valine was then added as indicated for 30 min. Cells were processed for immunoblotting of p70^{s6k}.* ***(B) Effects of glutamine on branched-chain amino acid-mediated metabolism by RINm5F cells.*** *Cells were preincubated in KRBB in the absence of glucose and amino acids for 1 h. Medium was replaced with KRBB containing basal amino acids +/- glutamine (2 mM), leucine, isoleucine, or valine as indicated for 2 h with the MTS reagent. The 450 nm absorbance values were measured at 2 h. Reproduced with permission (44)*

As shown in Fig. 7A, incubation of RINm5F cells in the complete absence of amino acids in KRBB for 2 h results in the dephosphorylation of p70^{s6k} to basal levels (lane 1). The addition of a complete complement of amino acids stimulates p70^{s6k} phosphorylation with the appearance of a slower migrating band (lane 2) and a return to basal levels with the removal of the three branched-chain amino acids, leucine, isoleucine and valine (lane 3). Leucine at 0.4 mM promotes the phosphorylation of p70^{s6k} (lane 4) compared to basal conditions, but isoleucine (0.4 mM), valine (0.2 mM) or the combination of isoleucine and valine (lanes 5-7) does not mimic this effect. However, leucine in the presence of valine or isoleucine (lanes 8 & 9) again restores phosphorylation of p70^{s6k}. To determine if leucine under these conditions also increases cellular metabolism by β-cells, the MTS assay was employed as a measure of NAD(P)H production (Fig. 7B).

Interestingly, only leucine (0.4 mM) in the presence of glutamine (2 mM), as a precursor for glutamate that serves as a substrate for glutamate dehydrogenase (GDH), significantly enhances β-cell metabolism suggesting that mitochondrial metabolism is important for leucine to activate p70^{s6k}.

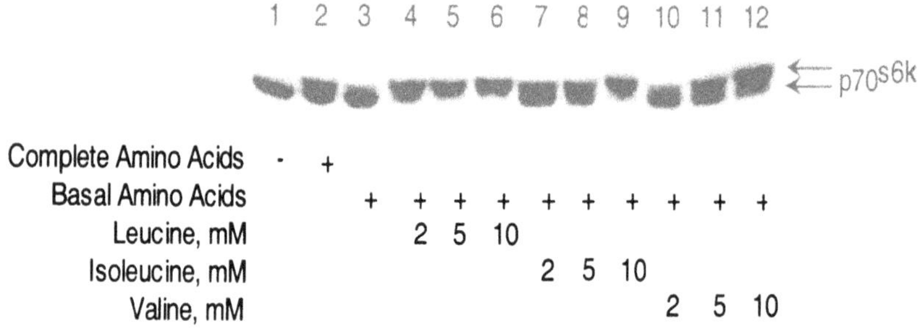

Figure 8. Isoleucine and valine-induced phosphorylation of p70^{s6k} at greater than physiological concentrations in RINm5F cells. *Cells were preincubated in KRBB in the absence of glucose and amino acids for 2 h. Medium was then replaced with KRBB containing either complete amino acids as a positive control or basal amino acids, which excluded leucine, isoleucine, or valine. Leucine, isoleucine, or valine was then added as indicated for 30 min. Cells were processed for immunoblotting of p70^{s6k}. Reproduced with permission (44).*

As shown in Fig. 8, leucine at 2, 5 and 10 mM causes complete activation of p70^{s6k} (lanes 4-6), whereas isoleucine and valine mimic these effects only at higher than physiological conditions (10 mM) (lanes 9 & 12). This may be explained by the fact that GDH is activated by isoleucine and valine only at higher concentrations than leucine.

In addition, the ability of leucine to promote phosphorylation of p70^{s6k} as shown in Fig. 9 is blocked by rapamycin (25 nM), an inhibitor of mTOR and wortmannin (100 μM), an inhibitor of phosphoinositide 3-kinase (PI 3-K).

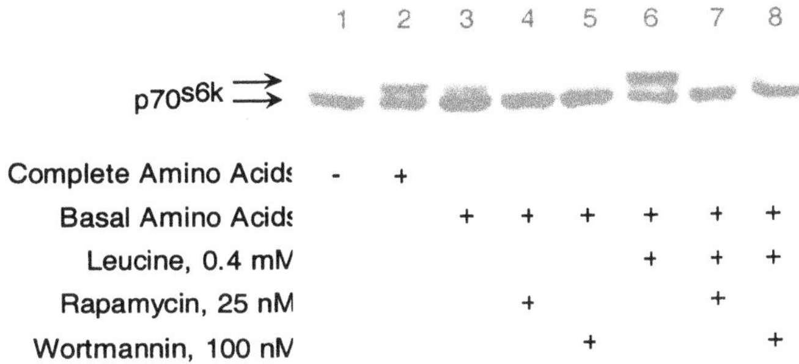

Figure 9. Effects of rapamycin and wortmannin on leucine-induced phosphorylation of p70^{s6k} in RINm5F cells. *Cells were preincubated in KRBB in the absence of glucose and amino acids for 2 h. During the last h of preincubation, rapamycin or wortmannin was added to cells. Medium was then replaced with KRBB containing either complete amino acids as a positive control or basal amino acids, which excluded leucine, isoleucine, or valine. Leucine +/- inhibitors were then added as indicated for 30 min. Cells were processed for immunoblotting of p70^{s6k}. Reproduced with permission (44).*

Our next approach was to determine if we could duplicate this more physiological experimental model whereby leucine activates the mTOR pathway with a minimum of amino acids (44). Our objective was to determine if leucine and glutamine would be sufficient to promote phosphorylation of p70^{s6k} in the absence of all other amino acids.

Our rationale for this minimal model is that leucine is required to provide substrate for its metabolism by the β-cell mitochondria via the oxidative decarboxylation pathway and also serve as an allosteric activator of GDH.

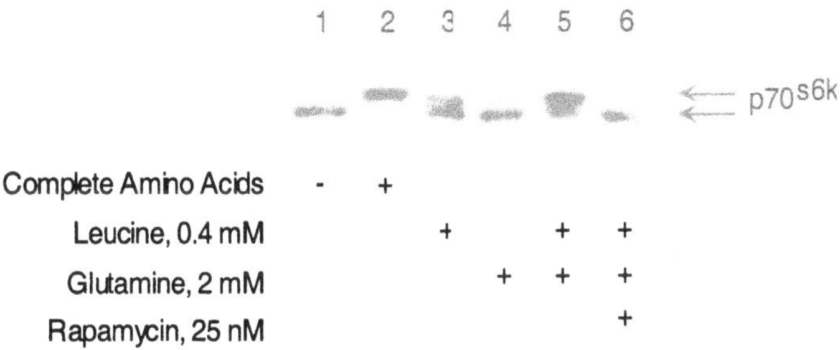

Figure 10. Leucine and glutamine-induced phosphorylation of p70^{s6k} is rapamycin-sensitive. *Cells were preincubated in KRBB in the absence of glucose and amino acids for 2 h. During the last h of preincubation, rapamycin (25 nM) was added to cells. Medium was replaced with KRBB containing leucine alone, glutamine alone or both +/- rapamycin as indicated for 30 min. Cells were processed for immunoblotting of p70^{s6k}. Reproduced with permission (44).*

Glutamine is necessary as a source of glutamate for GDH-mediated production of α-ketoglutarate and its subsequent metabolism by the β-cell mitochondria (see Fig. 11). As shown in Fig. 10, leucine (0.4 mM) or glutamine (2 mM) (lanes 3 & 4) alone produced a small increase in p70^{s6k}

phosphorylation above basal values, whereas a combination of leucine and glutamine markedly activated p70^{s6k} (lane 5). This activation of p70^{s6k} was completely blocked by rapamycin (lane 6).

Proposed model for insulin and leucine-mediated mitogenic signaling in pancreatic β-cells

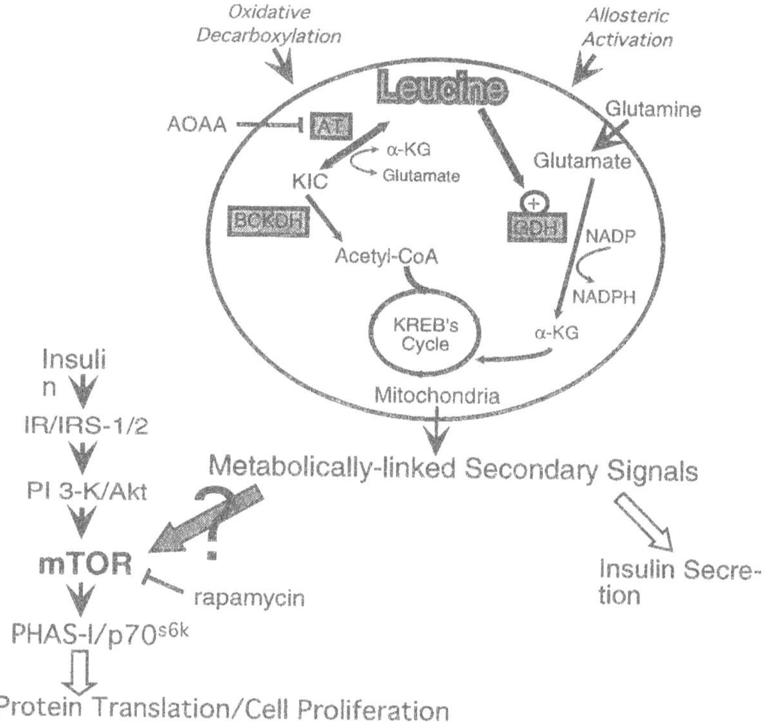

Figure 11. Proposed model for leucine's role in β-cell mitogenic signaling. *Abbreviations are the following: AT, amino transferase; AOAA, aminooxyacetic acid; KIC, α-ketoisocaproic acid; BCKDH, branched-chain α-keto-acid dehydrogenase; GDH, glutamate dehydrogenase; α-KG, α-ketoglutarate; IR, insulin receptor; IRS, insulin receptor substrate; PI-3K, phosphoinositide 3-kinase; mTOR, mammalian target of rapamycin; PHAS-I, phosphorylated heat- and acid-stable protein regulated by insulin; p70^{s6k}, p70 S6 kinase. Reproduced with permission (44).*

As illustrated schematically in Fig. 11, it has been previously established that leucine-induced insulin secretion from β-cells involves increased mitochondrial metabolism by oxidative decarboxylation and allosteric activation of glutamate dehydrogenase (GDH).

Based on our findings, it appears that these same metabolic pathways that generate signals for leucine-induced insulin secretion are necessary to activate the mTOR signaling pathway involved in protein translation. Support for this conclusion is that a minimal model consisting of only leucine and glutamine as substrates for oxidative decarboxylation and an activator of GDH, respectively, confirmed the requirement for these two metabolic components and mimicked closely the synergistic interaction achieved by a complete complement of amino acids to stimulate the phosphorylation of $p70^{s6k}$ and PHAS-I in a rapamycin-sensitive manner. Although leucine utilizes the same metabolic pathways to activate mTOR as insulin exocytosis, the mediators appear different since leucine activation of protein translation is rapamycin sensitive and insulin independent. It is proposed that the metabolic signals produced by glucose, insulin, leucine and glutamine work together in the β-cell to autoregulate translation through mTOR and contribute to growth-related protein synthesis and β-cell proliferation.

Summary of key concepts

- *Insulin released by pancreatic β-cells upon glucose stimulation is responsible, in part by an autocrine mechanism, for the stimulation of insulin receptor signaling by pancreatic β-cells, which regulate insulin secretion, biosynthesis of β-cell proteins, and ultimately β-cell proliferation.*
- *Two key translational regulatory proteins, PHAS-I and $p70^{s6k}$, are activated through mTOR by glucose or exogenous insulin, the latter confirming the autocrine mechanism.*
- *Amino acids are necessary for glucose or insulin-stimulated PHAS-I and $p70^{s6k}$ activation. Furthermore, branched-chain amino acids, leucine, isoleucine, and valine, activate PHAS-I and $p70^{s6k}$ independently of glucose or insulin. Of the three branched-chain amino acids, leucine is most effective.*
- *Metabolism of leucine by the mitochondria appears to be essential for its ability to activate PHAS-I and $p70^{s6k}$. The minimal model consisting of only leucine and glutamine was sufficient to activate PHAS-I and $p70^{s6k}$ in the absence of all other amino acids, further confirming the pivotal role of mitochondrial metabolism in activating PHAS-I and $p70^{s6k}$.*
- *It is proposed that the metabolic signals produced by glucose and amino acids work together in the β-cell to autoregulate translation through mTOR and contribute to growth-related protein synthesis and β-cell proliferation.*

Study Guide Questions

1) Glucose stimulates insulin secretion by pancreatic β-cells. Describe the multiple steps from glucose transport into the β-cells to insulin release.
2) PHAS-I, an inhibitor of eukaryotic initiation factor (eIF)-4E, and $p70^{s6k}$, a serine and threonine kinase for ribosomal protein S6, regulate the initiation of protein translation. Describe the specific roles of PHAS-I and $p70^{s6k}$ in protein translation.
3) Glucose stimulates PHAS-I and $p70^{s6k}$ phosphorylation via autocrine effects of insulin. What are the components of insulin receptor signaling in β-cells?
4) Branched-chain amino acids stimulate phosphorylation of PHAS-I and $p70^{s6k}$ independently of insulin and other growth factors. Describe the experiment that supports this finding.
5) Leucine is the key amino acid that stimulates phosphorylation of PHAS-I and $p70^{s6k}$. Describe the experiment that supports this finding.
6) Both glucose and amino acids stimulate PHAS-I and $p70^{s6k}$ through mTOR. Name the inhibitor of mTOR used for the experiments that supports this finding.

Acknowledgements

This study was supported by an American Diabetes Association Research Grant (to M. L. M.), an American Diabetes association Mentor-based Fellowship (to G. X), and National Institutes of Health Grants DK06181 (to M. L. M.), F32 DK08748 (to G. K.).

REFERENCES

1. Gautsch, T. A., Anthony, J. C., Kimball, S. R., Paul, G. L., Layman, D. K., and Jefferson, L. S. 1998 Eukaryotic initiation factor 4E availability regulates skeletal muscle protein synthesis during recovery from exercise. Am. J. Physiol. 274, C406-C414
2. Yoshizawa, F., Kimball, S. R., Vary, T. C., and Jefferson, L. S. 1998 Effect of dietary protein on translation initiation in rat skeletal muscle and liver. Am. J. Physiol. 275, E814-E820
3. Thorens, B., and Waeber, G. 1993 Glucagon-like peptide-I and the control of insulin secretion in the normal state and in NIDDM. Diabetes 42, 1219-1225
4. Lenzen, S., and Tiedge, M. 1994 Regulation of pancreatic beta-cell glucokinase and GLUT2 glucose transporter gene expression. Biochem. Soc. Trans. 22, 1-6
5. Dunne, M. J., Harding, E. A., Jaggar, J. H., and Squires, P., E. 1994 Ion channels and the molecular control of insulin secretion. Biochem. Soc. Trans. 22, 6-12
6. Flatt, P. R., Barnett, C. R., and Swanston-Flatt, S. K. 1993 Mechanisms of pancreatic b-cell dysfunction and glucose toxicity in non-insulin-dependent diabetes. Biochem. Soc. Trans. 22, 18-23
7. Goodison, S., Kenna, S., and Ashcroft, S. J. 1992 Control of insulin gene expression by glucose. Biochemical J. 285, 563-568

8. Nielsen, D. A., Welsh, M., Casadaban, M. J., and Steiner, D. F. 1985 Control of insulin gene expression in pancreatic beta-cells and in an insulin-producing cell line, RIN-5F cells. I. Effects of glucose and cyclic AMP on the transcription of insulin mRNA. J. Biol. Chem. 260, 13585-13589
9. Permutt, M. A., and Kipnis, D. M. 1972 Insulin biosynthesis. II. Effect of glucose on ribonucleic acid synthesis in isolated rat islets. J. Biol. Chem. 247, 1194-1199
10. Permutt, M. A. 1974 Effect of glucose on initiation and elongation rates in isolated rat pancreatic islets. J. Biol. Chem. 249, 2738-2742
11. Skelly, R. H., Schuppin, G. T., Ishihara, H., Oka, Y., and Rhodes, C. J. 1996 Glucose-regulated translational control of proinsulin biosynthesis with that of the proinsulin endopeptidases PC2 and PC3 in the insulin-producing MIN6 cell line. Diabetes 45, 37-43
12. Guest, P. C., Bailyes, E. M., Rutherford, N. G., and Hutton, J. C. 1991 Insulin secretory granule biogenesis. Co-ordinate regulation of the biosynthesis of the majority of constituent proteins. Biochem. J. 274, 73-78
13. Grimaldi, K. A., Siddle, K., and Hutton, J. C. 1987 Biosynthesis of insulin secretory granule membrane proteins. Control by glucose. Biochem. J. 245, 567-573
14. Guest, P. C., Rhodes, C. J., and Hutton, J. C. 1989 Regulation of the biosynthesis of insulin-secretory-granule proteins. Co-ordinate translational control is exerted on some, but not all, granule matrix constituents. Biochem. J. 257, 431-437
15. Pause, A., Belsham, G. J., Gingras, A.-C., Donze, O., Lin, T.-A., Lawrence, J. C., Jr. , and Sonenberg, N. 1994 Insulin-dependent stimulation of protein synthesis by phosphorylation of a regulator of 5'-cap function. Nature 371, 762-767
16. Lin, T.-A., Kong, X., Haystead, T. A. J., Pause, A., Belsham, G., Sonenberg, N., and Lawrence, J. C., Jr., 1994 PHAS-I as a link between mitogen-activated protein kinase and translation initiation. Science 266, 653-656
17. Grammer, T. C., Cheathem, L., Chou, M. M., and Blenis, J. 1996 The $p70^{s6k}$ signaling pathway: a novel signaling system involved in growth regulation. Cancer Surv 27, 271-291
18. Pullen, N., and Thomas, G. (1997) The modular phosphorylation and activation of $p70^{s6k}$. FEBS Lett 410, 78-82
19. Shah, O. J., Anthony, J. C., Kimball, S. R., and Jefferson, L. S. 2000 4E-BP1 and S6K1: translational integration sites for nutritional and hormonal information in muscle. Am. J. Physiol. Endocrinol. Metab. 279, E715-E729
20. Peterson, R. T., and Schreiber, S. L. 1998 Translation control: Connecting mitogens and the ribosome. Curr. Biol. 8, R248-R250
21. Stewart, M. J., and Thomas, G. 1994 Mitogenesis and protein synthesis: A role for ribosomal protein S6 phosphorylation? BioEssays 16, 1-7
22. Jefferies, H. B. J., Fumagalli, S., Dennis, P. B., Reinhard, C., Pearson, R. B., and Thomas, G. 1997 Rapamycin suppresses 5'TOP mRNA translation through inhibition of p70s6k. EMBO J. 12, 3693-3704
23. Dufner, A., and Thomas, G. 1999 Ribosomal S6 kinase signaling and the control of translation. Exp. Cell Res. 253, 100-109
24. Haystead, T. A., Haystead, C. M., Hu, C., Lin, T.-A., and Lawrence, J. C., Jr., 1994 Phosphorylation of PHAS-I by mitogen-activated protein (MAP) kinase. Identification of a site phosphorylated by MAP kinase in vitro and in response to insulin in rat adipocytes. J. Biol. Chem. 269, 23185-23191
25. Brunn, G. J., Williams, J., Sabers, C., Wiederrecht, G., Lawrence, J. C., Jr., and Abraham, R. T. 1996 Direct inhibition of the signaling functions of the mammalian target-of rapamycin by the phosphoinositide 3-kinase inhibitors, wortmannin and LY294002. EMBO J. 15, 5256-5267
26. Beretta, L., Gingras, A.-C., Svitkin, Y. V., Hall, M. N., and Sonenberg, N. 1996 Rapamycin blocks the phosphorylation of 4E-BP1 and inhibits cap-dependent initiation of translation. EMBO J. 15, 658-664

27. Brunn, G. J., Hudson, C. C., Sekulic, A., Williams, J. M., Hosoi, H., Houghton, P. J., Lawrence, J. C., Jr.,, and Abraham, R. T. 1997 Phosphorylation of the translational repressor PHAS-I by the mammalian target of rapamycin. Science 277, 99-101
28. Hara, K., Yonezawa, K., Kozlowski, M. T., Sugimoto, T., Andrabi, K., Weng, Q.-P., Kasuga, M., Nishimoto, I., and Avruch, J. 1997 Regulation of eIF-4E BP1 phosphorylation by mTOR. J. Biol. Chem. 272, 26457-26463
29. Schuppin, G. T., Pons, S., Hugl, S., Aiello, L. P., King, G. L., White, M., and Rhodes, C. J. 1998 A specific increased expression of insulin receptor substrate 2 in pancreatic beta-cell lines is involved in mediating serum-stimulated beta-cell growth. Diabetes 47, 1074-1085
30. Withers, D. J., Gutierrez, J. S., Towery, H., Burks, D. J., Ren, J. M., previs, S., Zhang, Y., Bernal, D., Pons, S., Shulman, G. I., Bonner-Weir, S., and White, M. F. 1998 Disruption of IRS-2 causes type 2 diabetes in mice. Nature 391, 900-904
31. Harbeck, M. C., Louie, D. C., Howland, J., Wolf, B. A., and Rothenberg, P. L. 1996) Expression of insulin receptor mRNA and insulin receptor substrate 1 in pancreatic islet beta-cells. Diabetes 45, 711-717
32. Xu, G., Marshall, C. A., Lin, T.-A., Kwon, G., Munivenkatappa, R. B., Hill, J. R., Lawrence, J. C., Jr.,, and McDaniel, M. L. 1998 Insulin mediates glucose-stimulated phosphorylation of PHAS-I by pancreatic beta cells. J. Biol. Chem. 273, 4485-4491
33. Xu, G., Kwon, G., Marshall, C. A., Lin, T.-A., Lawrence, J. C., Jr.,, and McDaniel, M. L. 1998 Branched-chain amino acids are essential in the regulation of PHAS-1 and p70 S6 kinase by pancreatic b-cells. J. Biol. Chem. 273, 28178-28184
34. Akiyama, T., Ishida, J., Nakagawa, S., Ogawara, H., Watanabe, S., Itoh, N., Shibuya, M., and Fukami, Y. 1987 Genistein, a specific inhibitor of tyrosine-specific protein kinases. J. Biol. Chem. 262, 5592-5595
35. Rao, G. N., Delafontaine, P., and Runge, M. S. 1995 Thrombin stimulates phosphorylation of insulin-like growth factor-1 receptor, insulin receptor substrate-1, and phospholipase C-gamma 1 in rat aortic smooth muscle cells. J. Biol. Chem. 270, 27871-27875
36. Corbett, J. A., Kwon, G., Misko, T. P., Rodi, C. P., and McDaniel, M. L. 1994 Tyrosine kinase involvement in IL-1 beta-induced expression of iNOS by beta-cells purified from islets of Langerhans. Am. J. Physiol. 267, C48-C54
37. Kwon, G., Corbett, J. A., Rodi, C. P., Sullivan, P., and McDaniel, M. L. 1995 Interleukin-1 beta-induced nitric oxide synthase expression by rat pancreatic beta-cells: evidence for the involvement of nuclear factor kappa B in the signaling mechanism. Endocrinology 136, 4790-4795
38. Patti, M. E., Barmbilla, E., Luzi, L., Landaker, E. J., and Kahn, C. R. 1998 Bidirectional modulation of insulin action by amino acids. J. Clin. Invest. 101, 1519-1592
39. May, M. E., and Buse, M. G. 1989 Effects of branched-chain amino acids on protein turnover. Diabetes Metab. Rev. 5, 227-245
40. Li, J. B., and Jefferson, L. S. 1978 Influence of amino acid availability on protein turnover in perfused skeletal muscle. Biochim. Biophys. Acta 544, 351-359
41. Shigemitsu, K., Tsujishita, Y., Hara, K., Nanahoshi, M., Avruch, J., and Yonezawa, K. 1999 Regulation of translational effectors by amino acid and mammalian target of rapamycin signaling pathways: possible involvement of autophagy in cultured hepatoma cells. J. Biol. Chem. 274, 1058-1065
42. Mortimore, G. E., Poso, A. R., Kadowake, M., and Wert, J., J. J., 1987 Multiphasic control of hepatic protein degradation by regulatory amino acids: general features and hormonal modulation. J. Biol. Chem. 262, 16322-16327
43. Swenne, I. 1992 Pancreatic beta-cell growth and diabetes mellitus. Diabetologia 35, 193-201
44. Xu, G., Kwon, G., Cruz, W., Marshall, C. A., and McDaniel, M. L. 2001 Metabolic regulation by leucine of translation initiation through the mTOR-signaling pathway by pancreatic b-cells. Diabetes 50, 353-360.

5

TRANSLATIONAL INITIATION OF ORNITHINE DECARBOXYLASE mRNA

Lo Persson and Koichi Takao*

*University of Lund, Lund, Sweden and Josai University, Saitama, Japan**

The polyamines (putrescine, spermidine and spermine) are a group of small intracellular aliphatic amines essential for a variety of growth-related processes in the cell. The importance of the polyamines in cell function is reflected in a strict regulatory control of their intracellular levels. High polyamine concentration can be toxic to the cell, whereas low concentrations may negatively affect anabolic events such as the synthesis of DNA, RNA and protein, eventually giving rise to cell growth arrest. Ornithine decarboxylase (ODC) catalyzes the first and what is often considered as the rate-limiting step in the biosynthesis of the polyamines. The enzyme is highly regulated at a multitude of levels, including the translational level. Some of the mechanisms involved in the regulation of ODC are unique and resembling those found in the control of various protooncogenes. Due to the extremely fast turnover of ODC (half-life of minutes to an hour), the cellular level of ODC protein and thus the enzyme activity is rapidly altered when the synthesis and/or degradation of the enzyme is changed.

TRANSLATIONAL CONTROL OF ODC

Cellular ODC activity usually reflects the growth state of the cell. Quiescent cells have very low ODC activity, whereas proliferating cells have high enzyme activity (1). ODC is induced by a variety of treatments and agents affecting cell growth or metabolism. Part of the induction can usually be explained by an increased transcription of the gene resulting in elevated levels of ODC mRNA (2). However, the increase in ODC activity is often several fold larger than that observed in the ODC mRNA level, indicating the involvement of other post-transcriptional mechanisms. During induction of ODC activity, a marked stabilization of the enzyme against degradation is usually also observed, which may contribute to the increase in ODC activity. Nevertheless, sometimes the induction of ODC cannot be explained only by an increase in ODC mRNA together with a stabilization of the enzyme protein. This indicates that a change in the translational efficiency of the ODC mRNA occurs.

ODC is subject to a strong feedback control by the polyamines (1,3,4). In the presence of an excess of polyamines, ODC is rapidly down-regulated, whereas when cellular polyamine levels are depleted there is a compensatory up-regulation of ODC activity. Part of this feedback control of ODC is explained by changes in the turnover of the enzyme. The mechanisms involved in the polyamine-mediated control of ODC degradation have recently been unraveled and shown to be unique to this enzyme (please see a more detailed description below). In addition to changes in ODC turnover the feedback regulation of ODC appears to involve changes in the ODC synthesis rate (1,3,4). However, these polyamine-mediated changes in enzyme synthesis are not accompanied by changes in the amount of ODC mRNA indicating that the effects of the polyamines are on the translational efficiency rather than on the transcription or stability of the ODC mRNA (1,3).

Also the osmolarity of the growth medium has been show to strongly affect the cellular expression of ODC (5,6,7). When cells are exposed to a hypertonic medium the ODC activity is rapidly reduced, whereas when cells are exposed to a hypotonic medium there is a dramatic increase in ODC activity within a very short time after the onset of the osmotic shock. The mechanisms involved are not fully understood. In some systems, the phenomenon is explained by a change in the ODC mRNA content (6). However, in other systems it appears to involve mainly, if not exclusively, translational and posttranslational mechanisms (5,7).

α-Difluoromethylornithine (DFMO) is a specific enzyme-activated irreversible inhibitor of ODC. The inhibitor has a strong antiproliferative effect when used *in vitro* (8). However, the therapeutic effect of DFMO against various forms of cancer has so far been rather disappointing. This is most likely due to compensatory effects neutralizing the inhibition of ODC (increased cellular uptake of exogenous polyamines, decreased polyamine degradation, increased expression and decreased turnover of ODC) (1). However, in spite of the discouraging results against tumors, DFMO has been shown to be highly effective against the West African form of sleeping sickness and is now an established drug ("eflornithine") against this disease (9).

STRUCTURE OF ODC mRNA

Like many mRNAs encoding various growth-related proteins, mammalian ODC mRNA belongs to the rare group of mRNAs that has a very long 5' untranslated region (UTR) (10,11). For ODC mRNA this region is about 300 nucleotides long. It has a very high ratio of GC to AU, especially in its first half. Thus, this region of the mRNA may form strong secondary structures that could negatively affect the translation of the message (Fig. 1). In addition to potentially strong secondary structures, the 5' UTR of ODC mRNA contains an upstream open reading frame which also may hamper the translation. Using various DNA constructs in transient and stable expression systems, it has been demonstrated that the 5' UTR of ODC mRNA strongly inhibits the translation of subsequent reporter genes (12,13,14). Most of this repressive effect on translation is mapped to the first part of the 5' UTR, which is particularly G/C-rich and is expected to form a stable stem loop. Interestingly, tissue extracts has been demonstrated to contain a 58 kDa protein that specifically binds to this part of the ODC mRNA (15).

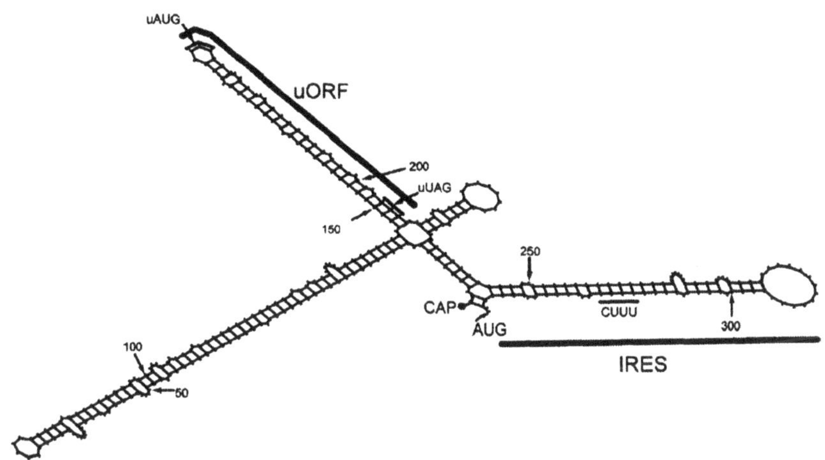

Figure 1. Predicted secondary structure of human ODC mRNA 5' UTR. *uORF, upstream open reading frame. IRES, internal ribosomal entry site (with pyrimidine-rich element).*

However, the function of that protein has not been clarified yet. That the 5' UTR of ODC mRNA is indeed inhibiting translation *in vivo* is indicated by the finding that most of the ODC mRNA is found associated with fractions containing ribosomal subunits and monosomes in polysome profiles (16,17).

The 3' UTR of ODC mRNA is also relatively long (> 300 nucleotides), but with a lower frequency of G and C. Thus, this region of the ODC mRNA has less potential to form stable secondary structures than the 5' UTR (18). Nevertheless, it has been shown that the 3' UTR may interact with the 5' UTR of ODC mRNA in such a way that the inhibitory effect of the 5' UTR on translation is relieved (12,14). The underlying mechanism is, however, not known.

EFFECT OF eIF4E ON ODC mRNA TRANSLATION

The translation initiation process is well understood for the majority of cellular mRNAs. The process is dependent on several protein factors, i.e. the eukaryotic initiation factors (eIFs), of which most have been characterized (19). The initiation factor 4E fulfils an important function in the early process of initiation by bringing the 5' methylated guanosine cap structure of the mRNA and the 40 S ribosomal subunit together. The eIF4E has a binding site for the cap structure of the mRNA as well as for another initiation factor, eIF4G, which in turn is linked to the 40S subunit by the initiation factor eIF3 (Fig. 2). Thus, the initiation factors eIF4E, eIF4G and eIF3 form a bridge between the cap structure and the 40S ribosomal subunit. eIF3 may also bind directly to the mRNA. The eIF4G is a large protein and binds the initiation factor eIF4A in addition to eIF4E and eIF3. The initiation factor eIF4A has a RNA helicase activity and is thus important for the melting of secondary structures of the 5' UTR. The large complex consisting of eIF4E, eIF4G and eIF4A is sometimes called eIF4F. When all components needed for the initiation are brought together the ribosomal subunit "scans" the mRNA for the correct initiation codon, starting from the cap end (20). This scanning procedure may be hampered by strong secondary structures (21). However, the helicase activity of eIF4A may facilitate a continued scanning (19).

The initiation factor eIF4E has been proposed as one of the limiting factors for the initiation of translation (22). The translation of mRNAs with long highly structured 5' UTRs is especially affected by low levels of eIF4E, since these mRNAs compete poorly with the mRNAs having short unstructured 5' UTRs for the initiation factors (23). The majority of mRNAs coding for protooncogenes or other growth-related proteins belong to group of mRNAs having long G/C rich 5' UTRs and thus are especially sensitive to the eIF4E level. The importance of eIF4E in the control of cell growth and proliferation is supported by the observation that increased expression of eIF4E stimulates DNA synthesis and cell cycle progression, but inhibits apoptosis (24,25). Forced overexpression of eIF4E induces transformation of a variety rodent cell lines and thus eIF4E has also been suggested to play a role in tumorigenesis (26,27).

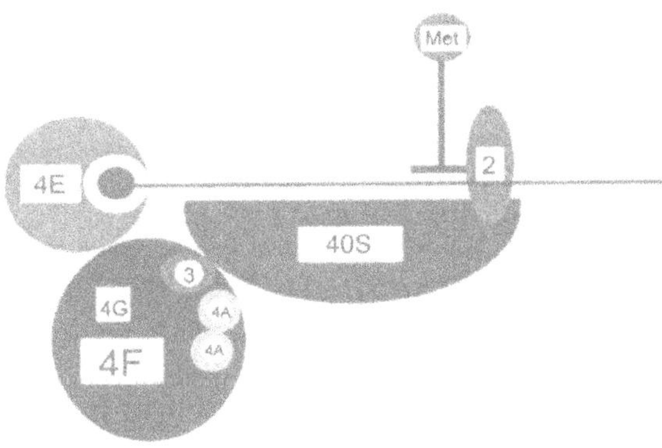

Figure 2. Schematic model of the cap-dependent initiation complex. *eIF2, eIF3, eIF4A, eIF4E and eIF4G are shown. eIF4F consists of eIF3, eIF4A and eIF4G.*

As mentioned earlier, the ODC mRNA belongs to the group of mRNAs with a long and highly structured 5' UTR. Overexpression of ODC may induce transformation, and thus this enzyme may be considered as a protooncogene (28). Cells overexpressing eIF4E have markedly increased ODC activity, which appear to be due to the relief of translational repression exerted by the secondary structure of ODC mRNA 5' UTR (29,30). The expression of antisense eIF4E RNA, on the other hand, results in a translational repression of ODC with a decrease in cellular ODC activity (31). Interestingly, eIF4E-induced transformation may be reversed by expression of an ODC dominant negative mutant or by treatment with a specific inhibitor of ODC, indicating the involvement of ODC activity in the transformation mechanism (32,33).

eIF4E, which is considered as the rate-limiting factor in cap-dependent translational initiation, is regulated at a multitude of levels (25). Many growth factors have been shown to increase the transcription of eIF4E mRNA and the eIF4E promoter contains binding sites for the transcription factor c-*myc*. The activity of eIF4E is partly regulated by phosphorylation. Increased phosphorylation of eIF4E, which is often observed after stimulation with various growth factors, results in an increased translation rate possibly by an enhanced affinity of eIF4E for the cap structure of the mRNA. The phosphorylation of eIF4E is believed to be catalyzed by the

MAPK-activated protein kinase MNK1, which is part of the extracellular signal-regulated kinase (ERK) and the p38 MAPK signaling cascades. In addition eIF4E is regulated by a family of inhibitory binding proteins; 4E-BP1, 4E-BP2 and 4E-BP3. By binding to the eIF4E the 4E-BPs inhibit the association of eIF4E to eIF4G and thus the assembly of the functional eIF4F. The 4E-BPs are also regulated by phosphorylation, which reduces their affinity for the eIF4E. The phosphorylation of the 4E-BPs is inhibited by the antibiotic rapamycin, which suppresses cap-dependent translation.

IRES-DEPENDENT TRANSLATION OF ODC mRNA

The majority of cellular mRNAs are translated in a cap-dependent manner. However, a small number of mRNAs exist that seem to be capable of "internal initiation" without preceding cap-binding and scanning (34,35). In this process, the ribosomal complex binds directly to an internal ribosomal entry site (IRES) on the mRNA close to the initiation codon (Fig. 3). IRES was first demonstrated for the picornavirus mRNAs, which are uncapped mRNAs with high degrees of secondary structure (as well as upstream AUGs) in their 5' UTRs (36,37). During a picornavirus infection, the host cap-dependent translation is inhibited due to a viral protease cleavage of eIF4G (37). However, the IRES-dependent initiation is not affected resulting in an cellular adjustment to picorna protein production. The process of internal initiation appears to be dependent on virtually the same initiation factors as the cap-dependent initiation, except for the eIF4E and the amino-terminal part of eIF4G (which is cleaved off by the picorna protease).

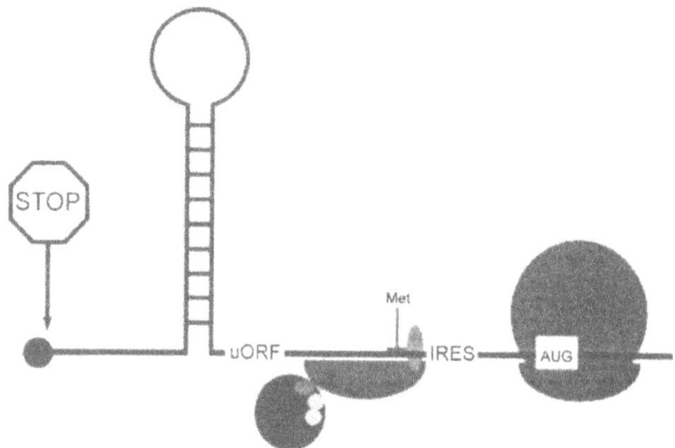

Figure 3. Schematic model of cap-independent initiation at an IRES.

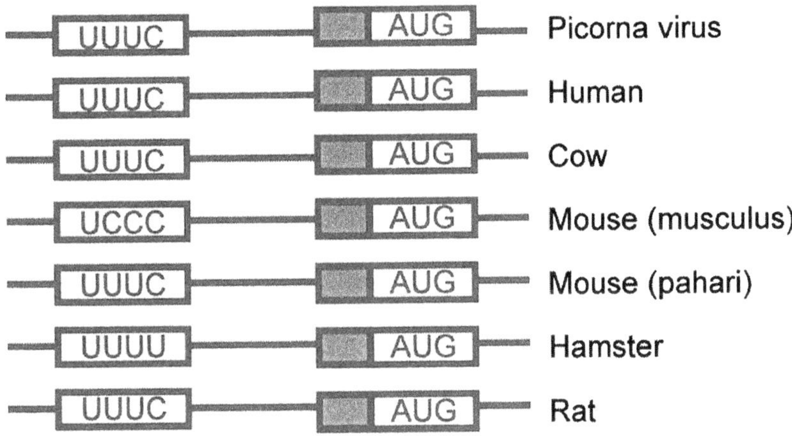

Figure 4. Comparison of pyrimidine-rich regions (left boxes) of ODC mRNA 5' UTRs from mammalian species with those of picorna virus mRNA 5' UTRs. *The pyrimidine-rich elements are located 25-30 nucleotides from the initiation codon.*

During the last decade an increasing number of cellular IRES-containing mRNAs have been identified. Many of these are coding for protooncogenes, growth factors or other proteins involved in cell growth or embryonic development (38). Characteristically, they also have long highly structured 5' UTRs often with one or several upstream AUGs. Recently, it was demonstrated that ODC mRNA also contains an IRES that functions exclusively at the G2/M phase of the cell cycle (39). ODC activity peaks two times during the cell cycle (40,41). Transient increases in ODC activity are observed during the G1/S boundary as well as during the G2/M transition. The first peak corresponds well with a general increase in protein synthesis and is most likely the result of normal cap-dependent translation. However, the second peak occurs at a time when the-cap dependent translation is greatly inhibited. In a series of experiments Pyronnet et al. (39) were able to demonstrate that the increase in ODC synthesis occurring during the G2/M phase of the cell cycle was the result of internal initiation at an IRES. The IRES was mapped close to the initiation codon and shown to contain a pyrimidine-rich sequence homologous to that of the picornavirus 5' UTR (Fig. 4). Polyamines produced during the G2/M phase of the cell cycle are believed to be important for the mitotic spindle formation as well as the chromatin condensation. Also c-*myc* mRNA, which contains an IRES, was shown to be translated during the G2/M phase of the cell cycle (39). Thus, IRES-dependent translation may be a general mechanism by which some essential proteins may be produced during phases of the cell cycle, e.g. mitosis, when cap-dependent translation is impaired.

ODC activity changes in relation to the phases of the cell cycle (40,41). The enzyme activity is increasing during the progression of the G1 phase with a peak in late G1/early S. A second peak of ODC activity is found during the G2 phase (due to cap-independent initiation at an IRES). Inhibition of ODC activity results in a marked reduction in the progression of cells through the S phase (42). The effect on DNA replication can be as early as within the first cell cycle after seeding the cells in the presence of the ODC inhibitor. A similar inhibition of DNA replication is observed when inhibitors of other enzymes in the polyamine biosynthetic pathway are used, indicating that the effect is related to the inhibitor-related decrease in cellular polyamine content (43). However, the mechanism(s) by which the polyamines affect DNA replication is still unknown.

FEEDBACK CONTROL OF ODC mRNA TRANSLATION

As mentioned earlier, the polyamines exert a strong feedback control of ODC (1,4). Part of this control appears to be at the translational level. The synthesis of the enzyme is down-regulated when cells are exposed to an excess of polyamines and up-regulated when cellular polyamine levels are depleted. The steady-state level of ODC mRNA is, however, not altered by changes in cellular polyamine levels, suggesting that the effect is on the efficiency by which the mRNA is translated.

The finding that ODC mRNA had a long G/C rich 5' UTR raised the question whether this part of the mRNA was involved in the feedback control of ODC synthesis. However, results from experiments using various constructs, in which the ODC mRNA 5' UTR was subcloned in front of mRNAs coding for reporter genes like β-galactosidase, CAT and luciferase, indicated that the polyamine-mediated effects on ODC synthesis occurred independent of the 5' UTR (12,17). Furthermore, no differences in polyamine-induced regulation of ODC were observed between stable transgenic CHO cells expressing the full-length ODC mRNA and those expressing the ODC mRNA devoid of most the 5' UTR, suggesting that the 5' UTR of the ODC mRNA is unimportant for the feedback control of ODC synthesis (44).

In fact, it has not been conclusively established that the polyamines, directly or indirectly, affect the translation of ODC mRNA. No effects have been seen on the ribosomal distribution of ODC mRNA by treatments that affect the cellular polyamine levels and thus activate the feedback regulation of ODC synthesis. However, since a large fraction of the ODC mRNA is in the region of the polysome profile containing ribosomal subunits and monosomes a minor shift in the distribution of the mRNA may represent a large change in the translation of the message. The conclusion that polyamines regulate ODC mRNA translation is based on results in which the synthesis of ODC is measured by pulse-labeling with ^{35}S-methionine. Since ODC has a very rapid turnover it is essential that the labeling time is short enough to avoid any degradation during the pulse-labeling. As mentioned earlier, the turnover of ODC protein is also greatly affected by the cellular polyamine levels (4,45). The polyamines induce the synthesis of a specific protein, named ODC antizyme, which binds strongly to the enzyme and stimulates its degradation by the 26S proteasome (4,45). In contrast to the degradation of most other proteins by the 26S proteasome the degradation of ODC by this proteolytic system is not triggered by ubiquitination. ODC is the first example of a non-ubiquitinated protein that is degraded by the 26S proteasome (46). Instead, the degradation of ODC is stimulated by the binding of antizyme (4,45). In the presence of a large excess of polyamines the half-life of ODC can be as short as a few minutes. It has been suggested that the observed polyamine-mediated changes in incorporation of labeled methionine into ODC are actually a result of rapid degradation (induced by antizyme) of newly synthesized ODC at or close to the ribosome (17).

Synthesis of antizyme is occurring through a unique mechanism involving ribosomal frameshifting. The antizyme mRNA contains two major reading frames, of which both are too small to encode the entire protein (47). Conventional translation of antizyme mRNA arrests at a premature termination signal. However, in the presence of polyamines a +1 frameshift is induced at this termination codon resulting in continued translation of the second reading frame, giving rise to the full-length active antizyme (48,49). In the absence or in the presence of low concentrations of polyamines this is a very rare event. Thus, it appears that polyamines induce the production of antizyme through a unique mechanism, ribosomal frameshifting. So far, this is the only example of mammalian ribosomal frameshifting.

OSMOTIC EFFECTS ON ODC mRNA TRANSLATION

As mentioned earlier, cellular ODC activity is strongly induced when cells are exposed to a hypotonic growth medium (5,6,7). The increase is fast and occurs in spite of a general inhibition of protein synthesis. Measurements of synthesis and turnover of ODC show that both these processes are affected. Usually there is no change in the amount of ODC mRNA suggesting that the change in synthesis rate is purely a translational effect. However, the exact mechanisms are not yet understood. In a series of experiments, we have used stable transfectants of CHO cells expressing ODC mRNAs with various truncations in the 5' UTR and 3' UTR to investigate the importance of these mRNA regions in the translational induction of ODC synthesis by hypotonic shock (unpublished results). It was demonstrated that cells expressing ODC mRNAs with a major truncation or a complete deletion in the 5' UTR still induced ODC when exposed to a hypotonic medium. Instead, the hypotonic induction of ODC was found to be highly dependent on the presence of the 3' UTR. Cells expressing ODC mRNAs lacking the 3' UTR (with and without 5' UTR) did not, or only slightly, induce ODC after the hypotonic shock. Thus, it appears that the 3' UTR of ODC mRNA in some way may affect the translation of the message. The 3' UTR of ODC mRNA has been suggested to be involved in the translational control of ODC by being able to partially neutralize the inhibition exerted by the 5' UTR (12,14). However, in the case of hypotonic induction of ODC, it appears that such an interaction is not a prerequisite since the phenomenon occurred also in cells expressing ODC mRNA devoid of the 5' UTR. Nevertheless, it is conceivable that the effect of the 3' UTR of ODC mRNA on the hypotonic induction of ODC synthesis involves an interaction with some other region of the mRNA.

That the 3' UTR of an mRNA may be involved in the translational control has attracted increased attention during recent years. A variety of examples of sequence elements as well as protein factors involved have been described (50-52). However, the exact mechanism by which the 3' UTR of ODC mRNA interferes with the translational induction of ODC by hypotonic shock remains to be established.

Summary of key concepts

- *ODC, which catalyses the first step in the synthesis of a group of intracellular growth factors called polyamines, has the characteristics of a protooncogene and is highly regulated at a multitude of levels (including translational control).*
- *Like many other mRNAs coding for protooncogenes and growth-related proteins, ODC mRNA has a long G/C rich 5' UTR that strongly inhibits cap-dependent translation.*
- *Overexpression of the initiation factor eIF4E results in a derepression of ODC mRNA translation.*
- *Inhibition of ODC activity may reverse eIF4E-induced transformation.*
- *ODC mRNA contains an internal ribosomal entry site (IRES) close to the initiation codon, which is used for cap-independent translation during the G2/M phase of the cell cycle when normally cap-dependent translation is inhibited.*
- *Hypotonic shock appears to induce ODC mRNA translation by a mechanism involving the 3' UTR of the message.*
- *ODC mRNA translation may be regulated by a feedback mechanism.*

Study Guide Questions

1) Why do mRNAs coding for protooncogenes often have long G/C rich 5' UTRs?
2) How can eIF4E stimulate the translation of mRNAs containing long 5' UTRs with strong secondary structures?
3) What is the physiological function of IRESs?
4) In what situations is cap-dependent, but not cap-independent, translation inhibited?
5) Speculate how the 3' UTR of an mRNA can affect translational initiation.

Acknowledgements

Work done in the authors' laboratory is supported by the Swedish National Cancer Society and the Medical Faculty at Lund University.

REFERENCES

1. Heby O., Persson L. 1990 Molecular genetics of polyamine synthesis in eukaryotic cells. Trends Biochem Sci 15: 153-8

2. Law G.L., Li R.S., Morris D.R. 1995 "Transcriptional control of the ODC gene." In *Polyamines: regulation and molecular interaction*, R. A. Casero Jr, ed. Austin: R.G. Landes Company, 5-26
3. Persson L., Svensson F., Lövkvist Wallström E. 1996 "Regulation of polyamine metabolism." In *Polyamines in cancer: basic mechanisms and clinical approaches*, K. Nishioka, ed. Austin: R.G. Landes Company, 19-43
4. Hayashi S., Murakami Y., Matsufuji S. 1996Ornithine decarboxylase antizyme: A novel type of regulatory protein. Trends Biochem Sci 21: 27-30
5. Poulin R., Pegg A.E. 1990 Regulation of ornithine decarboxylase expression by anisosmotic shock in difluoromethylornithine-resistant L1210 cells. J Biol Chem 265: 4025-32
6. Lundgren D.W. 1992 Effect of hypotonic stress on ornithine decarboxylase mRNA expression in cultured cells. J Biol Chem 267: 6841-7
7. Lövkvist-Wallström E., Stjernborg-Ulvsbäck L., Scheffler I.E., Persson L. 1995 Regulation of mammalian ornithine decarboxylase - Studies on the induction of the enzyme by hypotonic stress. Eur J Biochem 231: 40-4
8. McCann P.P., Pegg A.E. 1992 Ornithine decarboxylase as an enzyme target for therapy. Pharmacol Ther 54: 195-215
9. Barrett S.V., Barrett M.P. 2000 Anti-sleeping sickness drugs and cancer chemotherapy. Parasitol Today 16: 7-9
10. Kozak M. 1987 An analysis of 5'-noncoding sequences from 699 vertebrate messenger RNAs. Nucleic Acids Res 15: 8125-48
11. Brabant M., McConlogue L., van Daalen Wetters T., Coffino P. 1988 Mouse ornithine decarboxylase gene: cloning, structure, and expression. Proc Natl Acad Sci USA 85: 2200-4
12. Grens A., Scheffler I.E. 1990 The 5'- and 3'-untranslated regions of ornithine decarboxylase mRNA affect the translational efficiency. J Biol Chem 265: 11810-6
13. Manzella J.M., Blackshear P.J. 1990 Regulation of rat ornithine decarboxylase mRNA translation by its 5'-untranslated region. J Biol Chem 265: 11817-22
14. Lorenzini E.C., Scheffler I.E. 1997 Co-operation of the 5' and 3' untranslated regions of ornithine decarboxylase mRNA and inhibitory role of its 3' untranslated region in regulating the translational efficiency of hybrid RNA species via cellular factor(s). Biochem J 326 : 361-7
15. Manzella J.M., Blackshear P.J. 1992 Specific protein binding to a conserved region of the ornithine decarboxylase mRNA 5'-untranslated region. J Biol Chem 267: 7077-82
16. Holm I., Persson L., Stjernborg L., Thorsson L., Heby O. 1989 Feedback control of ornithine decarboxylase expression by polyamines. Analysis of ornithine decarboxylase mRNA distribution in polysome profiles and of translation of this mRNA in vitro. Biochem J 258: 343-50
17. van Daalen Wetters T., Macrae M., Brabant M., Sittler A., Coffino P. 1989 Polyamine-mediated regulation of mouse ornithine decarboxylase is post-translational. Mol Cell Biol 9: 5484-90
18. Kahana C., Nathans D. 1985 Nucleotide sequence of murine ornithine decarboxylase mRNA. Proc Natl Acad Sci USA 82: 1673-7
19. Pain V.M. 1996 Initiation of protein synthesis in eukaryotic cells. Eur J Biochem 236: 747-71
20. Kozak M. 1999 Initiation of translation in prokaryotes and eukaryotes. Gene 234: 187-208
21. Kozak M. 1989 Circumstances and mechanisms of inhibition of translation by secondary structure in eucaryotic mRNAs. Mol Cell Biol 9: 5134-42
22. McKendrick L., Pain V.M., Morley S.J. 1999 Translation initiation factor 4E. Int J Biochem Cell Biol 31: 31-5

23. Koromilas A.E., Lazaris-Karatzas A., Sonenberg N. 1992 mRNAs containing extensive secondary structure in their 5' non-coding region translate efficiently in cells overexpressing initiation factor eIF-4E. EMBO J 11: 4153-8
24. Polunovsky V.A., Rosenwald I.B., Tan A.T., White J., Chiang L., Sonenberg N., Bitterman P.B. 1996 Translational control of programmed cell death: eukaryotic translation initiation factor 4E blocks apoptosis in growth-factor-restricted fibroblasts with physiologically expressed or deregulated Myc. Mol Cell Biol 16: 6573-81
25. Raught B., Gingras A.C. 1999 eIF4E activity is regulated at multiple levels. Int J Biochem Cell Biol 31: 43-57
26. Lazaris-Karatzas A., Montine K.S., Sonenberg N. 1990 Malignant transformation by a eukaryotic initiation factor subunit that binds to mRNA 5' cap. Nature 345: 544-7
27. Zimmer S.G., DeBenedetti A., Graff J.R. 2000 Translational control of malignancy: the mRNA cap-binding protein, eIF-4E, as a central regulator of tumor formation, growth, invasion and metastasis. Anticancer Res 20: 1343-51
28. Auvinén M., Paasinen A., Andersson L.C., Hölttä E. 1992 Ornithine decarboxylase activity is critical for cell transformation. Nature 360: 355-8
29. Rousseau D., Kaspar R., Rosenwald I., Gehrke L., Sonenberg N. 1996 Translation initiation of ornithine decarboxylase and nucleocytoplasmic transport of cyclin D1 mRNA are increased in cells overexpressing eukaryotic initiation factor 4E. Proc Natl Acad Sci U S A 93: 1065-70
30. Shantz L.M., Hu R.H., Pegg A.E. 1996 Regulation of ornithine decarboxylase in a transformed cell line that overexpresses translation initiation factor eIF-4E. Cancer Res 56: 3265-9
31. Graff J.R., De Benedetti A., Olson J.W., Tamez P., Casero R.A., Jr., Zimmer S.G. 1997 Translation of ODC mRNA and polyamine transport are suppressed in *ras*-transformed CREF cells by depleting translation initiation factor 4E. Biochem Biophys Res Commun 240: 15-20
32. Shantz L.M., Pegg A.E. 1994 Overproduction of ornithine decarboxylase caused by relief of translational repression is associated with neoplastic transformation. Cancer Res 54: 2313-6
33. Shantz L.M., Coleman C.S., Pegg A.E. 1996 Expression of an ornithine decarboxylase dominant-negative mutant reverses eukaryotic initiation factor 4E-induced cell transformation. Cancer Res 56: 5136-40
34. Martinez-Salas E. 1999 Internal ribosome entry site biology and its use in expression vectors. Curr Opin Biotechnol 10: 458-64
35. Sachs A.B. 2000 Cell cycle-dependent translation initiation: IRES elements prevail. Cell 101: 243-5
36. Pelletier J., Sonenberg N. 1988 Internal initiation of translation of eukaryotic mRNA directed by a sequence derived from poliovirus RNA. Nature 334: 320-5
37. Sonenberg N. 1990 Measures and countermeasures in the modulation of initiation factor activities by viruses. New Biol 2: 402-9
38. van der Velden A.W., Thomas A.A. 1999 The role of the 5' untranslated region of an mRNA in translation regulation during development. Int J Biochem Cell Biol 31: 87-106
39. Pyronnet S., Pradayrol L., Sonenberg N. 2000 A cell cycle-dependent internal ribosome entry site. Mol Cell 5: 607-16
40. Fredlund J.O., Johansson M.C., Dahlberg E., Oredsson S.M. 1995 Ornithine decarboxylase and S-adenosylmethionine decarboxylase expression during the cell cycle of Chinese hamster ovary cells. Exp Cell Res 216: 86-92
41. Heby O., Gray J.W. , Lindl P.A., Marton L.J., Wilson C.B. 1976 Changes in L-ornithine decarboxylase activity during the cell cycle. Biochem Biophys Res Commun 71: 99-105
42. Fredlund J.O., Oredsson S.M. 1996 Impairment of DNA replication within one cell cycle after seeding of cells in the presence of a polyamine-biosynthesis inhibitor. Eur J Biochem 237: 539-44

43. Fredlund J.O., Oredsson S.M. 1997 Ordered cell cycle phase perturbations in Chinese hamster ovary cells treated with an S-adenosylmethionine decarboxylase inhibitor. Eur J Biochem 249: 232-8
44. Lövkvist Wallström E., Persson L. 1999 No role of the 5' untranslated region of ornithine decarboxylase mRNA in the feedback control of the enzyme. Mol Cell Biochem 197: 71-8
45. Hayashi S., Murakami Y. 1995 Rapid and regulated degradation of ornithine decarboxylase. Biochem J 306: 1-10
46. Murakami Y., Matsufuji S., Kameji T., Hayashi S., Igarashi K., Tamura T., Tanaka K., Ichihara A. 1992 Ornithine decarboxylase is degraded by the 26S proteasome without ubiquitination. Nature 360: 597-9
47. Miyazaki Y., Matsufuji S., Hayashi S. 1992 Cloning and characterization of a rat gene encoding ornithine decarboxylase antizyme. Gene 113: 191-7
48. Rom E., Kahana C. 1994 Polyamines regulate the expression of ornithine decarboxylase antizyme *in vitro* by inducing ribosomal frame-shifting. Proc Natl Acad Sci USA 91: 3959-63
49. Matsufuji S., Matsufuji T., Miyazaki Y., Murakami Y., Atkins J.F., Gesteland R.F., Hayashi S. 1995 Autoregulatory frameshifting in decoding mammalian ornithine decarboxylase antizyme. Cell 80: 51-60
50. Black B.L., Lu J., Olson E.N. 1997 The MEF2A 3' untranslated region functions as a cis-acting translational repressor. Mol Cell Biol 17: 2756-63
51. Piecyk M., Wax S., Beck A.R., Kedersha N., Gupta M., Maritim B., Chen S., Gueydan C., Kruys V., Streuli M., Anderson P. 2000 TIA-1 is a translational silencer that selectively regulates the expression of TNF-alpha. EMBO J 19: 4154-63
52. Mbella E.G., Bertrand S., Huez G., Octave J.N. 2000 A GG nucleotide sequence of the 3' untranslated region of amyloid precursor protein mRNA plays a key role in the regulation of translation and the binding of proteins. Mol Cell Biol 20: 4572-9

6

REGULATION OF mRNA STABILITY BY AUF1

Gerald M. Wilson and Gary Brewer
UMDNJ-Robert Wood Johnson Medical School, Piscataway, NJ

A+U-rich elements (AREs) are potent cis-acting determinants of rapid cytoplasmic mRNA turnover in mammalian cells. Regulation of mRNA decay rates by these sequences is mediated by interaction with cellular factors. Association of the protein AUF1 with an ARE-containing transcript targets the mRNA for decay, involving the assembly or recruitment of a multi-subunit trans-acting complex. In this chapter, recent evidence is described which indicates that recognition of ARE sequences by AUF1 induces dynamic protein oligomerization, which may serve as a scaffold for association of other cytoplasmic factors leading to catabolism of the RNA substrate. This mechanism of targeted trans-acting complex assembly may be regulated at several points, either involving differential expression of individual subunits or through the activity of selected signal transduction pathways. Finally, specific examples are described where alterations in the distribution of AUF1 isoforms lead to differential gene expression during development, and where accelerated mRNA turnover associated with enhanced AUF1 protein levels may contribute to the pathogenesis of congestive heart failure.

BACKGROUND

In eukaryotes, gene expression is a highly regulated process exhibiting control at many levels, all directed to ensure that gene products (protein or RNA) are maintained within levels appropriate for cellular growth, maintenance, function, and even programmed self-destruction. A critical determinant governing the synthetic rates of proteins are the concentrations of cytoplasmic mRNAs encoding them. As with any biological system, the steady-state level of a cytoplasmic mRNA is dependent on its rates of both synthesis and degradation. The production rate of a cytoplasmic mRNA is a cumulative function of transcription, pre-mRNA processing, and nucleocytoplasmic export, each of which may be subject to independent regulatory control.

Cytoplasmic mRNA turnover is also tightly regulated, with mammalian mRNAs displaying a spectrum of constitutive decay rates spanning up to two orders of magnitude. Differences in mRNA half-lives contribute to regulated gene expression by two principal means. First, unstable mRNAs approach new steady-state levels more quickly than stable transcripts following changes in the synthetic rate (1), thus decreasing the response time between a transcriptional stimulus and phenotypic output. Second, cells may vary the turnover rates of specific transcripts in response to diverse stimuli, allowing for increased or decreased rates of protein production independent of changes in transcriptional activity. Generally, determinants of both constitutive and inducible mRNA turnover rates are present as *cis*-acting sequences within individual mRNAs. In an increasing number of cases, multiple determinants of stability have been identified within individual transcripts that may operate constitutively, redundantly, or be modulated in response to specific stimuli, thus providing many options for fine-tuning gene expression.

Rapid mRNA Decay Associated with A+U-Rich Elements (AREs)

The most extensively characterized determinants of rapid constitutive mRNA turnover in mammalian systems are the adenosine + uridine-rich elements (AREs) localized to the 3'-untranslated regions (3'-UTRs) of many labile transcripts. In mammals, a major pathway for mRNA turnover is initiated by shortening of the poly(A) tail, followed by rapid digestion of the mRNA body (2). The deadenylation phase of mRNA turnover appears to be rate-limiting in most cases, since deadenylated mRNA decay intermediates are difficult to detect *in vivo.* In general, the presence of an ARE increases the deadenylation rate of an mRNA, thus accelerating its turnover (3).

The first functional demonstration of the mRNA-destabilizing activity of an ARE was reported by Shaw and Kamen (4), who observed that insertion of a highly conserved 51-nucleotide A+U-rich sequence from the 3'-UTR of granulocyte-macrophage colony-stimulating factor (GM-CSF) mRNA into the 3'-UTR of the stable β-globin transcript dramatically shortened its cytoplasmic half-life. Subsequently, AREs were identified in other labile mRNAs encoding cytokines, oncoproteins, and G protein-coupled receptors (reviewed in Refs. 2, 3), which contribute to the rapid decay of these mRNAs *in vivo.* Clinical significance of ARE-directed mRNA turnover was observed with the mRNA encoding the oncoprotein *c-fos*. Mutations in which the c-*fos* ARE is deleted significantly stabilize the transcript (5) and greatly increase its oncogenic potential (6).

The sequences comprising AREs from individual mRNAs are

frequently conserved between species, yet AREs from different transcripts often show remarkable diversity in primary structure. In general, an ARE consists of a 40- to 150-nucleotide U-rich sequence, often containing one or more repeats of the sequence AUUUA. These pentameric motifs may be overlapping or dispersed. While some studies have demonstrated that sequences of the form UUAUUUA(U/A)(U/A) are sufficient to accelerate mRNA turnover in *cis* (7, 8), other AREs have been characterized as potent destabilizing sequences in the absence of AUUUA pentamers (9). This sequence heterogeneity observed among functional AREs likely contributes to differential regulation of mRNA turnover, including modulation of decay rates for selected transcripts in response to specific stimuli. Several cases have been characterized in which the stability of mRNAs containing AREs may be regulated in response to extrinsic factors. For example, the rapid ARE-directed turnover of interleukin-3 mRNA in mast cells is inhibited by Ca^{2+} influx (10). Similarly, mRNAs containing AREs are stabilized during heat shock (11). By contrast, the decay of interferon-β mRNA is accelerated in response to glucocorticoid treatment in an ARE-dependent manner (12). Specific intracellular signaling pathways have also been identified which contribute to the regulation of ARE-directed mRNA turnover. In particular, components of the p38 mitogen-activated protein (MAP) kinase and c-*jun* N-terminal kinase (JNK) pathways have been implicated in stabilization of ARE-containing mRNAs associated with the inflammatory response (13-15) and tumor cell metastasis (16). By contrast, ARE-dependent stabilization of cyclooxygenase-2 mRNA by $G\alpha_q$-coupled receptor signaling in smooth muscle is mediated by the p42/p44 MAP kinases and is independent of p38 MAP kinase activity (17). Both tyrosine and p38 MAP kinase activities are required for the stabilization of interleukin-1β and GROα mRNAs induced by monocyte adhesion (18). These examples illustrate that regulatory control of ARE-directed mRNA turnover is mediated by a complex array of intracellular signaling systems, resulting in both mRNA- and cell type-specific modulation of mRNA decay rates.

Identification of AUF1 as an Activity That Accelerates c-*myc* mRNA Decay *In Vitro*

Concomitant with studies focusing on the characterization of AREs and the signaling pathways regulating their mRNA-destabilizing activities, additional research has been directed at the identification and characterization of the *trans*-acting factors which carry out these functions. A common approach used to discover potential *trans*-acting factors has been to identify proteins capable of directly associating with RNA substrates containing AREs (reviewed in Ref. 19). In the following sections, however, we describe how a functional screen for cytoplasmic mRNA degradation activities led to the identification and cloning of the ARE-dependent mRNA-destabilizing protein, AUF1.

The mRNA encoded by the proto-oncogene c-*myc* is another example of a labile transcript containing an ARE in its 3'-UTR. Mechanisms contributing to c-*myc* mRNA turnover were characterized using cell-free mRNA decay assays (reviewed in Ref. 20). In this system, rapid decay of polysome-associated c-*myc* mRNA required the addition of a cytosolic post-polysomal S130 fraction containing a labile mRNA-destabilizing component (21). Consistent with the decay of ARE-containing transcripts *in vivo*, this activity promoted rapid deadenylation, followed by 3'→5' decay of the mRNA body (22). Subsequently, additional purification steps were employed to identify the *trans*-acting factor(s) present in the S130 extract contributing to the rapid turnover of c-*myc* mRNA. Using sucrose density gradients, an activity with sedimentation of 7S was purified which consisted principally of two polypeptides with molecular weights of 37 and 40 kDa, collectively termed AUF1 (23). When added to cell-free mRNA decay reactions, this fraction significantly accelerated the rate of polysomal c-*myc* mRNA decay, but did not affect the turnover of the polysomal mRNA encoding γ-globin. Furthermore, this activity directly associated with the ARE located in the 3'-UTR of c-*myc* mRNA, but did not bind the γ-globin mRNA 3'-UTR. Taken together, the correlation of specific ARE-binding activity and destabilization of an ARE-containing mRNA by this fraction indicated that AUF1 constituted a *trans*-acting factor promoting ARE-directed mRNA turnover (reviewed in Ref. 19).

Cloning and Domain Structure of AUF1

A cDNA encoding the p37 isoform of AUF1 was cloned from an expression library using antibodies raised against the purified p37/p40 proteins (24). Interestingly, this antiserum also recognized a 45 kDa nuclear protein, indicating that additional isoforms of AUF1 may exist. Subsequent cDNA cloning experiments identified a total of four isoforms of AUF1 from both human (25) and murine sources (26), differing by sequence insertions near the N- and C-termini (Fig. 1). Based on their apparent molecular weights, they have been designated $p37^{AUF1}$, $p40^{AUF1}$, $p42^{AUF1}$, and $p45^{AUF1}$. Recently, the gene encoding human AUF1 was cloned and characterized as a single copy gene on chromosome 4 consisting of 10 exons (25, 27). Alternative splicing of exons 2 and 7 from a common pre-mRNA generates the four protein isoforms (Fig. 1).

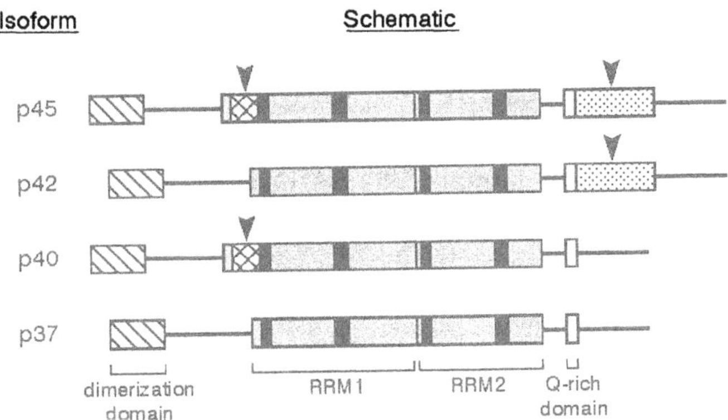

Figure 1. ***Isoforms of AUF1 generated by alternative pre-mRNA splicing.*** *Each isoform of AUF1 contains two tandem non-identical RRMs (shaded boxes) with characteristic RNP-2 and RNP-1 motifs (black boxes). The N-terminal dimerization (hatched) and C-terminal Q-rich (open) domains are also common to each isoform. Arrowheads indicate peptide sequences encoded by alternatively spliced exons 2 (cross-hatched) and 7 (stippled). Structural and functional consequences of these inserted sequences are discussed in the text.*

Sequence comparisons between AUF1 and other RNA-binding proteins demonstrated the presence of two tandem non-identical RNA-recognition motifs (RRMs) within all AUF1 isoforms. These binding motifs assume a characteristic βαβ-βαβ folding pattern that presents a four-stranded anti-parallel β-sheet to the RNA substrate (28). Adjacently positioned near the center of this β-sheet are a highly conserved octameric sequence (RNP-1) and slightly less conserved hexamer (RNP-2). Co-crystals of RRMs contained within the spliceosomal protein U1A (29) or the splicing regulator *sex-lethal* (30) with specific RNA targets revealed that the RNP-1 and –2 sequences interact extensively with RNA substrates. Stacking interactions between planar amino acid side chains within these motifs and adjacent bases of the RNA contribute considerable binding energy to this association. However, the diversity of RNA sequences recognized by different proteins containing RRMs indicates that structural determinants involved in the discrimination of RNA substrates are likely encoded by intradomain loops and sequences flanking the RRMs. In the AUF1 locus, exon 2 encodes a 19-amino acid sequence located within RRM1 of $p40^{AUF1}$ and $p45^{AUF1}$, positioned immediately N-terminal of the RNP-2 element. This sequence may exert significant influence over the binding affinity and/or specificity of these AUF1 isoforms for nucleic acid substrates. While initial evidence suggests that this may be the case (25), more investigations will be necessary to assess the role of this inserted element in the regulation of AUF1 function.

Analyses of AUF1 deletion mutants have been instrumental in assessing the functions of other AUF1 protein domains (31). In solution, $p37^{AUF1}$

exists as a dimer, dependent on the presence of a 29-amino acid alanine-rich region at the N-terminus common to all isoforms (Fig. 1). Sequences downstream of the RRMs are dispensable for dimerization activity. Also common to all AUF1 isoforms is a short domain rich in glutamine residues (Q-rich domain) located C-terminal of RRM2. While dispensable for RNA-binding activity, this domain is important for sequence-specific oligomerization of AUF1 on RNA substrates (discussed below). The additional amino acid residues encoded by exon 7, present in $p42^{AUF1}$ and $p45^{AUF1}$, are located immediately downstream of the Q-rich sequence. Unlike the p37 and p40 isoforms, which are localized to both the nucleus and cytoplasm, $p42^{AUF1}$ and $p45^{AUF1}$ appear to be exclusively nuclear (24, 32). Recent studies have identified a binding motif consisting of the Q-rich domain and the adjacent protein sequence encoded by exon 7 within $p42^{AUF1}$ and $p45^{AUF1}$ which functions as a nuclear localization signal, involving association with scaffold attachment factor-B, a protein associated with the nuclear matrix (32).

In addition to its involvement in the regulation of cytoplasmic mRNA turnover, some groups have also implicated nuclear roles for selected isoforms of AUF1. The transcription factor E2BP, which regulates the EII enhancer element of hepatitis B virus, is identical to $p40^{AUF1}$ (33). Based on cross-hybridization to E2BP cDNA fragments and reactivity with anti-hnRNP D antibodies, clones encoding $p40^{AUF1}$, $p42^{AUF1}$, and $p45^{AUF1}$ were independently isolated and denoted hnRNP D0 (34). These studies also indicated that AUF1 proteins associated with DNA and RNA sequences corresponding to telomeric repeats. More recently, nuclear AUF1 proteins have been identified as transcription factors regulating human genes. A transcriptional activator specific to B cells, termed LR1, is comprised of a heterodimer of nucleolin and $p40^{AUF1}$ (35). Nuclear AUF1 proteins have also been implicated in the transcriptional regulation of the complement receptor 2 gene in B cells (36). This combination of nuclear and cytoplasmic roles for AUF1 isoforms involving association with both DNA and RNA substrates highlights the significance of AUF1 as a multi-functional regulator of gene expression in mammals.

Role of AUF1 in Promoting Rapid mRNA Decay: AUF1 Association with AREs Induces Protein Oligomerization

An early indication that AUF1 was directly involved in the destabilization of c-*myc* mRNA *in vitro* was the observation that AUF1 specifically associated with the c-*myc* ARE, but not with RNA sequences from stable transcripts (23). Assays of RNA substrate specificity indicated that AUF1 could associate with AREs from many different mRNAs, showing preference for sequences rich in uridylate residues (24). Studies using recombinant $p37^{AUF1}$ demonstrated that the affinity of AUF1 for an ARE closely correlated with the ability of the ARE to destabilize an mRNA in *cis*. AUF1-binding affinity to potent mRNA-destabilizing sequences such as those contained within the c-*fos* and c-*myc* 3'-UTRs was very strong, while binding affinity to mutant AREs defective in inducing rapid mRNA decay was proportionally weaker (37). Furthermore, a number of cases have been described in which the rapid decay of an ARE-containing transcript *in vivo* is dependent on the concentration and/or binding activity of AUF1.

The application of fluorescence-based methods for monitoring the association of macromolecules has allowed the investigation of protein-RNA binding equilibria to be performed in much greater detail. Using small, fluorescent RNA substrates, the association of AUF1 proteins with an ARE was observed to proceed *via* a multi-step pathway, with AUF1 dimers binding sequentially to generate an oligomeric AUF1:RNA complex (38). On 38-base RNA substrates encoding either the ARE from tumor necrosis factor α (TNFα) mRNA or a poly(U) sequence, two AUF1 dimers associate to generate a tetrameric AUF1 complex on the RNA, described by serial equilibrium constants K_1 and K_2 (Fig. 2). Off-rate analyses demonstrated that both phases of the binding mechanism are highly dynamic in solution (38), which may contribute to the efficiency of ARE recognition in complex RNA populations.

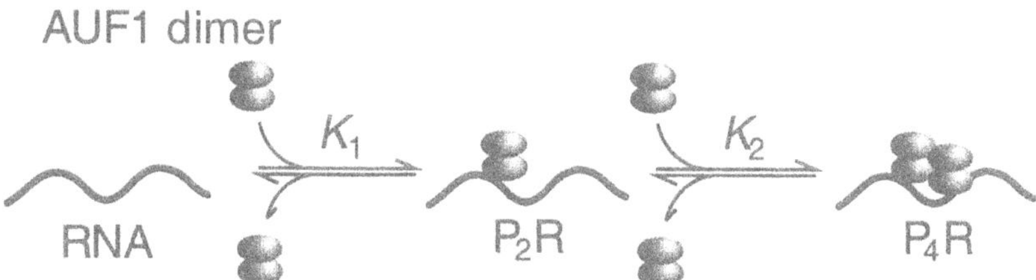

Figure 2. RNA-dependent formation of AUF1 oligomers by sequential association of protein dimers. *Interaction of an AUF1 dimer with an ARE generates the P_2R complex with affinity described by the equilibrium constant K_1. The P_2R complex may then serve as a substrate for association of an additional AUF1 dimer, described by K_2. Total complex size is limited to tetramers on short RNA targets (<40 bases), but additional AUF1 dimers may associate with larger RNA substrates (described in text).*

Additional experiments using deletion mutants of AUF1 indicated that C-terminal protein sequences contributed to the regulation of tetramer formation on an ARE, since K_2 was inhibited by loss of the Q-rich region of $p37^{AUF1}$. Using the larger c-*fos* ARE, measuring 75 bases in length, hydrodynamic studies demonstrated that AUF1 complexes as large as hexamers could form (31). Concomitant with protein oligomerization, these complexes exhibited increasing deviation from a spherical shape, consistent with a tendency towards maximization of surface area. The present model for induction of ARE-directed mRNA turnover by AUF1 is by the targeted assembly of a *trans*-acting complex that marks the mRNA for decay (Fig. 3).

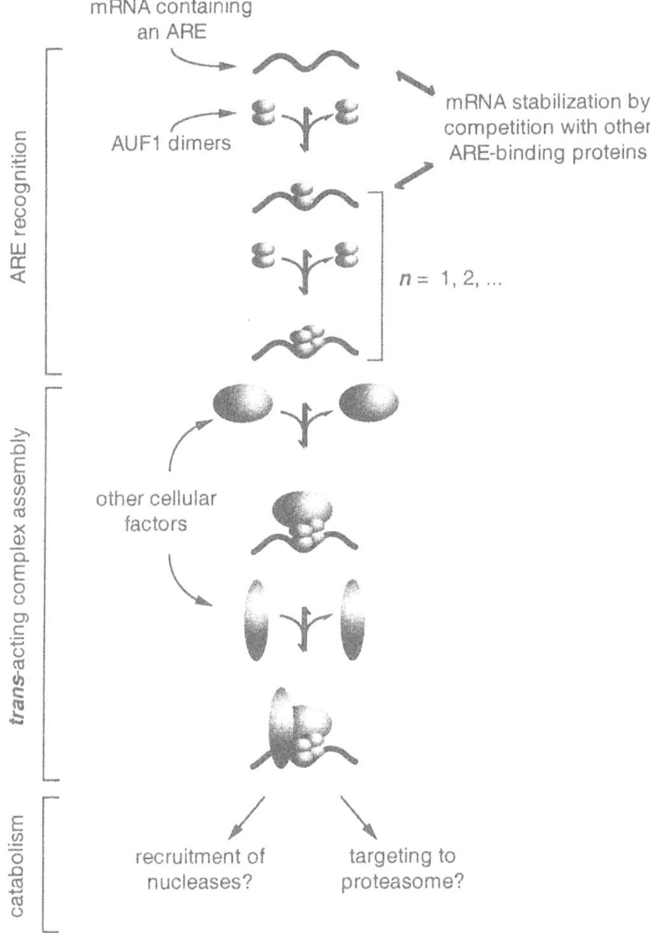

Figure 3. AUF1 as a targeting system for assembly or recruitment of a trans-acting mRNA destabilizing complex. *By this model, recognition of an ARE-containing mRNA by AUF1 dimers initiates the dynamic formation of AUF1 oligomers on the mRNA. Other ARE-binding proteins with stabilizing functions may operate by competing with AUF1 for the RNA substrate. The number of AUF1 dimers binding to the ARE is likely dependent on the size of the ARE (described in text). Following construction of the AUF1 multimer, additional cytoplasmic factors may be recruited based on the presentation of binding surfaces on the AUF1 oligomer and/or structural changes in the mRNP induced by AUF1 binding. Minimization of off-rates at one or more stages of assembly would function to stabilize the complex. Ultimately, 3'→5' decay of the mRNA poly(A) tail is induced, possibly by the recruitment or activation of specific exoribonucleases, or by ubiquitination of protein components of the mRNP and targeting to the proteasome.*

The first phase of this process is the recognition of ARE sequences in the bulk mRNA population. This process must be very efficient, since ARE-containing mRNAs may exhibit cytoplasmic half-lives of 10 minutes or less. Dynamic AUF1 dimer binding and oligomerization (described above) may be sufficient for this phase of assembly. Other ARE-binding proteins, including the Hu family of RNA-binding proteins, are reported to function as mRNA stabilizers (described in the accompanying chapter by Keene). Such factors may operate by competing for AUF1 binding to the ARE at this stage, thus preventing AUF1 oligomerization and subsequent factor recruitment.

In this section, we describe evidence that recruitment of ancillary factors by AUF1 is essential for promotion of ARE-directed mRNA decay. First, the assembly of AUF1 multimers on an ARE is not sufficient to induce mRNA turnover. Neither recombinant AUF1 nor cellular AUF1 purified to homogeneity exhibit nucleolytic activity, nor are they capable of accelerating the decay of polysomal mRNAs *in vitro* (G. Brewer, *unpublished observations*). However, the presence of AUF1 in a 7S mRNA destabilizing complex indicates that other factors present in the 7S complex and/or associated with polysomes are essential for the decay process. Furthermore, these observations demonstrated that the ARE-binding and mRNA-degrading activities of the 7S complex are separable, and led to the conclusion that AUF1 functions as a component of a multi-subunit *trans*-acting complex *in vivo.*

The association of additional cytoplasmic proteins with AUF1 was confirmed by co-immunoprecipitation experiments (24). Pulse labeling with ^{32}P also revealed that $p37^{AUF1}$ and $p40^{AUF1}$, along with several associated factors, were phosphorylated *in vivo*. A subset of the proteins associated with AUF1 in the cytoplasm have been identified immunologically as the translation initiation factor eIF4G, poly(A)-binding protein, the heat shock protein Hsp70, and the cognate heat shock protein Hsc70 (11). While the contributions of these factors to ARE-directed mRNA turnover are currently unknown, it is likely that the association of at least some of them with AUF1 stabilizes the *trans*-acting complex. This possibility is evidenced by the observations that the AUF1:ARE equilibrium is highly dynamic in solution (38), yet multi-subunit complexes containing AUF1 are sufficiently stable to be detected by co-immunoprecipitation (11, 24) or following fractionation through sucrose density gradients (23).

The mechanisms linking AUF1 oligomerization and complex assembly to decay of the mRNA substrate are largely unknown, but some studies have proffered a few hints. Early observations indicated 3'→5' shortening of the poly(A) tail as the initial catabolic step in ARE-directed mRNA turnover. Intuitively, recruitment or activation of one or more cytoplasmic 3'→5' exoribonucleolytic activities on the mRNA would seem a probable mechanism for acceleration of mRNA decay following *trans*-acting complex assembly. Indeed, a human cytoplasmic poly(A)-specific exoribonuclease, PARN, has been identified and cloned (39, 40). Some studies using *in vitro* mRNA decay systems have implicated this enzyme in 5'-cap-dependent deadenylation processes (41, 42). However, observations that AUF1 may be ubiquitinated *in vivo*, and that inhibition of proteasome activity results in stabilization of ARE-containing transcripts, have prompted an alternate hypothesis in which ubiquitinated, RNA-associated AUF1 is degraded by the proteasome, together with the bound RNA substrate. Schneider presents further discussion of a role for the proteasome in ARE-directed mRNA turnover in an accompanying chapter.

In addition to its role in promoting mRNA decay, AUF1 has also been identified as a component of the α-globin mRNA stability complex, which associates with a pyrimidine-rich element in the 3'-UTR of the stable α-globin mRNA (see accompanying chapter by Liebhaber). Contained within this complex are two poly(C)-binding proteins, termed αCP1 and αCP2, which associate with AUF1 *in vitro* and *in vivo* (43). The role of AUF1 in this mRNA-stabilizing complex remains unknown, but may be indicative of an additional cytoplasmic function independent of an ARE.

Signaling Systems Contributing to Regulation of ARE-Directed mRNA Decay Involving AUF1

The model of AUF1-targeted assembly of a *trans*-acting complex presents a plethora of potential mechanisms for regulatory control of ARE-directed mRNA turnover. One signaling pathway involved in regulating the decay rates of some ARE-containing transcripts includes the p38 MAP kinase (13, 14). In monocytes, adhesion at sites of inflammation or tissue injury is accompanied by the induction of mRNAs encoding cytokines and inflammatory mediators (44, 45). For transcripts encoding interleukin-1β and GROα, this induction includes dramatic mRNA stabilization concomitant with the loss of an ARE-binding activity containing AUF1 (18). Inhibition of tyrosine kinases or the p38 MAP kinase pathway abrogated stabilization of the transcripts and prevented the release of this ARE-binding complex. In this example, a regulated decrease in AUF1-binding activity is associated with stabilization of ARE-containing mRNAs. To date, specific substrates of the p38 MAP kinase pathway have not been identified within protein complexes containing AUF1. However, post-translational modification of some factor(s) leading to decreased RNA-binding affinity or dissociation of complex subunits remains an appealing model for regulation of ARE-directed mRNA decay by this signaling pathway.

Another mechanism regulating ARE-directed mRNA turnover involves changes in the intracellular concentration or isoform distribution of AUF1. Specific examples have been described where modulation of AUF1 levels contributes to alterations in the decay of ARE-containing mRNAs in congestive heart failure and during development (*see highlighted text*). Similarly, relationships between AUF1 expression and mRNA turnover rates have been reported in studies using cultured cell model systems. For example, GM-CSF mRNA, containing an ARE, is relatively stable in the bladder carcinoma cell line 5637 (46), coincident with the absence of detectable $p37^{AUF1}$ and $p40^{AUF1}$ (47). In K562 erythroleukemia cells, induction of erythroid differentiation by hemin is accompanied by inhibition of ARE-directed mRNA turnover concomitant with sequestration of endogenous AUF1 into a hemin-induced protein complex (48). Rapid decay of ARE-containing transcripts was restored, however, by ectopic expression of either $p37^{AUF1}$ or $p42^{AUF1}$. Based on the model of *trans*-acting complex assembly presented earlier, these examples may be interpreted in terms of the dynamic equilibrium between dimeric AUF1 and ARE-containing mRNA substrates. If AUF1 multimerization on an ARE is requisite for initiation of the mRNA decay cascade, the binding equilibrium leading to formation of AUF1 oligomers would be largely dependent on the concentration of free AUF1. In cases where the availability or accessibility of AUF1 is limiting, the net rate of AUF1 multimer formation may thus represent the rate-limiting step of mRNA decay. As such, increasing the concentration of free AUF1 proteins by ectopic expression would be expected to enhance the decay rate by increasing the rate of RNA-dependent AUF1 oligomerization.

While this model of ARE-targeting and *trans*-acting complex assembly remains speculative, a growing body of evidence supports such a role for AUF1 in the initiation of ARE-directed mRNA decay, as well as the obligate involvement of AUF1-associated factors in mediating the link between ARE-binding and mRNA-degradation activities. The identification of additional factors participating in these processes and the characterization of the molecular architecture of the *trans*-acting complexes will be critical steps towards furthering our understanding of the control of mRNA turnover mediated by AREs.

Desensitization of the adrenergic receptor-signaling pathway in congestive heart failure associated with overexpression of AUF1

During congestive heart failure, increases in the circulating levels of the hormone norepinepherine result in desensitization of the β-adrenergic receptor (β-AR)/ G protein/ adenylyl cyclase signal transduction pathway (49, 50). Desensitization of this pathway is characterized by a decrease in the abundance of β_1-ARs, the major myocardial adrenergic receptor subtype, owing to a decrease in the levels of β_1-AR mRNA (50, 51). In a cultured cell model, exposure to the β-AR agonist (-)isoproteronol similarly decreased hamster β_2-AR transcript levels, involving acceleration of β_2-AR mRNA turnover (52). In this system, the increased decay rate of β_2-AR mRNA is accompanied by a 1.5- to 2.5-fold increase in the levels of AUF1 proteins (53). Parallel increases in AUF1 levels were observed comparing ventricular myocardium from failing *versus* healthy human hearts. Since recombinant $p37^{AUF1}$ specifically associates with an ARE located in the 3'-UTR of human β_1-AR mRNA (53), a modest increase in the cellular concentration of AUF1 may improve the efficiency of *trans*-acting complex assembly, leading to accelerated decay of the β_1-AR transcript. These studies illustrate an example where modest changes in AUF1 levels may enhance mRNA turnover rates *in vivo*, with serious pathophysiological consequences.

Distribution of AUF1 isoforms contributes to differences in cytokine expression in adult versus neonatal mononuclear cells

In activated mononuclear cells (MNCs) taken from neonates, expression of GM-CSF is approximately seven-fold lower than in comparable cells from adult sources (54), which is largely due to a three-fold increase in the turnover rate of this transcript in neonatal MNCs (55). In binding reactions containing cytoplasmic extracts from neonatal cells, the GM-CSF ARE associates with a protein complex containing AUF1 (47). A similar binding activity is observed in cytoplasmic extracts from adult MNCs, but at approximately 35-fold lower abundance. Interestingly, both adult and neonatal MNCs contain AUF1 proteins, but the distribution of AUF1 isoforms expressed is remarkably different. In neonatal cells, the principal AUF1 isoforms are $p37^{AUF1}$ and $p40^{AUF1}$ (47). Previously, these isoforms were characterized as the primary cytoplasmic AUF1 isoforms (24), and were identified in the 7S complex necessary for acceleration of ARE-directed decay of polysomal mRNAs in cell-free decay assays (23). By contrast, adult MNCs express primarily $p45^{AUF1}$ (47), an isoform generally localized to the nucleus (24, 32). In cell-free mRNA assays containing cytoplasmic proteins from neonatal MNCs, mRNA substrates containing the ARE from GM-CSF mRNA decayed more rapidly than in decay reactions containing adult MNC proteins (56). However, immunodepletion of AUF1 from the neonatal extract abrogated rapid decay of this mRNA, indicating an essential role for the p37 and/or p40 isoforms of AUF1 in regulating the expression of GM-CSF in MNCs.

Summary of key concepts

- AREs are 3'-untranslated mRNA sequences that accelerate the cytoplasmic turnover of mRNAs but may be subject to regulation.
- Rapid decay of ARE-containing transcripts in cell-free mRNA decay systems and *in vivo* is associated with AUF1-binding activity.
- The AUF1 gene encodes four protein isoforms of varying functions that are generated by alternative splicing of a common pre-mRNA.
- ARE-containing RNA substrates associate directly with $p37^{AUF1}$ with high affinity and specificity; this interaction induces oligomerization of the protein.
- Association of AUF1 with an ARE may contribute to destabilization of the mRNA by nucleating the assembly or recruitment of a multi-subunit *trans*-acting complex on the mRNA.
- Modulation of the cellular concentration or binding activity of AUF1 proteins alters the decay rates of ARE-containing mRNAs.

Study Guide Questions

1) Signal transduction pathways involving phosphorylation events appear to regulate the ability of AUF1 to target ARE-containing mRNAs for decay. Since AUF1 is phosphorylated *in vivo*, what molecular events resulting from changes in the phosphorylation status of AUF1 might contribute to alterations in mRNA decay rates?
2) Overexpression of putative *trans*-acting factors in cultured cells is a common tactic for examining the functional significance of regulatory proteins. Given the model of AUF1 function presented in Fig. 3, how might overexpression of AUF1 lead to stabilization, rather than destabilization, of ARE-containing mRNAs?
3) In congestive heart failure, a two-fold increase in cardiac AUF1 levels accompanies a two-fold decrease in β_1-adrenergic receptor levels due to enhanced turnover of the β_1-AR transcript, effectively decreasing cardiac output by half. What clinical consequences might be expected to result from a tissue-specific decrease in AUF1 expression?
4) Pharmacological agents capable of modulating the stability of ARE-containing mRNAs may provide promising treatments for chronic inflammatory syndromes and some cancers. Based on the model for *trans*-acting factor assembly presented in Figure 3, identify candidate molecular targets for these drugs, and propose modes of action leading to enhancement or inhibition of mRNA decay. Also, consider how the effects of these therapies might be made specific for a subset of ARE-containing mRNAs.

Acknowledgements
Work in the authors' laboratory is supported by grant CA52443 from the National Institutes of Health and a grant from the Arthritis Foundation.

REFERENCES

1. Hargrove, J.L. and F.H. Schmidt. 1989. The role of mRNA and protein stability in gene expression. *FASEB J.* 3: 2360-2370.
2. Ross, J. 1995. mRNA stability in mammalian cells. *Microbiol. Rev.* 59: 423-450.
3. Chen, C.-Y.A. and A.-B. Shyu. 1995. AU-rich elements: characterization and importance in mRNA degradation. *Trends Biochem. Sci.* 20: 465-470.
4. Shaw, G. and R. Kamen. 1986. A conserved AU sequence from the 3' untranslated region of GM-CSF mRNA mediates selective mRNA degradation. *Cell* 46: 659-667.
5. Wilson, T. and R. Treisman. 1988. Removal of poly(A) and consequent degradation of c-*fos* mRNA facilitated by 3' AU-rich sequences. *Nature* 336: 396-399.
6. Meijlink, F., T. Curran, A.D. Miller, and I.M. Verma. 1985. Removal of a 67-base-pair sequence in the noncoding region of protooncogene fos converts it to a transforming gene. *Proc. Natl. Acad. Sci. USA* 82: 4987-4991.
7. Zubiaga, A.M., J.G. Belasco, and M.E. Greenberg. 1995. The nonamer UUAUUUAUU is the key AU-rich sequence motif that mediates mRNA degradation. *Mol. Cell. Biol.* 15: 2219-2230.
8. Lagnado, C.A., C.Y. Brown, and G.J. Goodall. 1994. AUUUA is not sufficient to promote poly(A) shortening and degradation of an mRNA: the functional sequence within AU-rich elements may be UUAUUUA(U/A)(U/A). *Mol. Cell. Biol.* 14: 7984-7995.
9. Peng, S.S.Y., C.-Y.A. Chen, and A.-B. Shyu. 1996. Functional characterization of a non-AUUUA AU-rich element from the c-*jun* proto-oncogene mRNA: evidence for a novel class of AU-rich elements. *Mol. Cell. Biol.* 16: 1490-1499.
10. Stoecklin, G., S. Hahn, and C. Moroni. 1994. Functional hierarchy of AUUUA motifs in mediating rapid interleukin-3 mRNA decay. *J. Biol. Chem.* 269: 28591-28597.
11. Laroia, G., R. Cuesta, G. Brewer, and R.J. Schneider. 1999. Control of mRNA decay by heat shock-ubiquitin-proteosome pathway. *Science* 284: 499-502.
12. Peppel, K., J.M. Vinci, and C. Baglioni. 1991. The AU-rich sequences in the 3' untranslated region mediate the increased turnover of interferon mRNA induced by glucocorticoids. *J. Exp. Med.* 173: 349-355.
13. Winzen, R., M. Kracht, B. Ritter, A. Wilhem, C.-Y.A. Chen, A.-B. Shyu, M. Muller, M. Gaestel, K. Resch, and H. Holtmann. 1999. The p38 MAP kinase pathway signals for cytokine-induced mRNA stabilization via MAP kinase-activated protein kinase 2 and an AU-rich region-targeted mechanism. *EMBO J.* 18: 4969-4980.
14. Lasa, M., K.R. Mahtani, A. Finch, G. Brewer, J. Saklatvala, and A.R. Clark. 2000. Regulation of cyclooxygenase 2 mRNA stability by the mitogen-activated protein kinase p38 signaling cascade. *Mol. Cell. Biol.* 20: 4265-4274.
15. Ming, X.-F., M. Kaiser, and C. Moroni. 1998. *c-jun* N-terminal kinase is involved in AUUUA-mediated interleukin-3 mRNA turnover in mast cells. *EMBO J.* 17: 6039-6048.
16. Montero, L. and Y. Nagamine. 1999. Regulation by p38 mitogen-activated protein kinase of adenylate- and uridylate-rich element-mediated urokinase-type plasminogen activator (uPA) messenger RNA stability and uPA-dependent in vitro cell invasion. *Cancer Res.* 59: 5286-5293.
17. Xu, K., A.M. Robida, and T.J. Murphy. 2000. Immediate-early MEK-1-dependent stabilization of rat smooth muscle cell cyclooxygenase-2 mRNA by $G\alpha_q$-coupled receptor signaling. *J. Biol. Chem.* 275: 23012-23019.
18. Sirenko, O.I., A.K. Lofquist, C.T. DeMaria, J.S. Morris, G. Brewer, and J.S. Haskill. 1997. Adhesion-dependent regulation of an A+U-rich element-binding activity associated with AUF1. *Mol. Cell. Biol.* 17: 3898-3906.

19. Wilson, G.M. and G. Brewer. 1999. The search for trans-acting factors controlling messenger RNA decay. *Prog. Nucleic Acids Res. Mol. Biol.* 62: 257-291.
20. Brewer, G. and J. Ross. 1990. Messenger RNA turnover in cell-free extracts. *Methods Enzymol.* 181: 202-209.
21. Brewer, G. and J. Ross. 1989. Regulation of c-*myc* mRNA stability in vitro by a labile destabilizer with an essential nucleic acid component. *Mol. Cell. Biol.* 9: 1996-2006.
22. Brewer, G. and J. Ross. 1988. Poly(A) shortening and degradation of the 3' AU-rich sequences of human c-*myc* mRNA in a cell-free system. *Mol. Cell. Biol.* 8: 1697-1708.
23. Brewer, G. 1991. An A+U-rich element RNA-binding factor regulates c-*myc* mRNA stability in vitro. *Mol. Cell. Biol.* 11: 2460-2466.
24. Zhang, W., B.J. Wagner, K. Ehrenman, A.W. Schaefer, C.T. DeMaria, D. Crater, K. DeHaven, L. Long, and G. Brewer. 1993. Purification, characterization, and cDNA cloning of an AU-rich element RNA-binding protein, AUF1. *Mol. Cell. Biol.* 13: 7652-7665.
25. Wagner, B.J., C.T. DeMaria, Y. Sun, G.M. Wilson, and G. Brewer. 1998. Structure and genomic organization of the human AUF1 gene: alternative pre-RNA splicing generates four protein isoforms. *Genomics* 48: 195-202.
26. Ehrenman, K., L. Long, B.J. Wagner, and G. Brewer. 1994. Characterization of cDNAs encoding the murine A+U-rich RNA-binding protein AUF1. *Gene* 149: 315-319.
27. Wagner, B.J., L. Long, P.N. Rao, M.J. Pettenati, and G. Brewer. 1996. Localization and physical mapping of genes encoding the A+U-rich element RNA-binding protein AUF1 to human chromosomes 4 and X. *Genomics* 34: 219-222.
28. Nagai, K., C. Oubridge, N. Ito, J. Avis, and P. Evans. 1995. The RNP domain: a sequence-specific RNA-binding domain involved in processing and transport of RNA. *Trends Biochem.Sci.* 20: 235-240.
29. Oubridge, C., N. Ito, P.R. Evans, C.-H. Teo, and K. Nagai. 1994. Crystal structure at 1.92A resolution of the RNA-binding domain of the U1A spiceosomal protein complexed with an RNA hairpin. *Nature* 372: 432-438.
30. Handa, N., O. Nureki, K. Kurimoto, I. Kim, H. Sakamoto, Y. Shimura, Y. Muto, and S. Yokoyama. 1999. Structural basis for recognition of the *tra* mRNA precursor by the Sex-lethal protein. *Nature* 398: 579-585.
31. DeMaria, C.T., Y. Sun, L. Long, B.J. Wagner, and G. Brewer. 1997. Structural determinants in AUF1 required for high affinity binding to A+U-rich elements. *J. Biol. Chem.* 272: 27635-27643.
32. Arao, Y., R. Kuriyama, F. Kayama, and S. Kato. 2000. A nuclear matrix-associated factor, SAF-B, interacts with specific isoforms of AUF1/hnRNP D. *Arch. Biochem. Biophys.* 380: 228-236.
33. Tay, N., S.H. Chan, and E.C. Ren. 1992. Identification and cloning of a novel heterogeneous nuclear ribonucleoprotein C-like protein that functions as a transcriptional activator of the hepatitis B virus enhancer II. *J. Virol.* 66: 6841-6848.
34. Kajita, Y., J. Nakayama, M. Aizawa, and F. Ishikawa. 1995. The UUAG-specific RNA-binding protein, heterogeneous nuclear ribonucleoprotein D0. *J. Biol. Chem.* 270: 22167-22175.
35. Dempsey, L.A., L.A. Hanakahi, and N. Maizels. 1998. A specific isoform of hnRNP D interacts with DNA in the LR1 heterodimer: canonical RNA binding motifs in a sequence-specific duplex DNA binding protein. *J. Biol. Chem.* 273: 29224-29229.
36. Tolnay, M., J.D. Lambris, and G.C. Tsokos. 1997. Transcriptional regulation of the complement receptor 2 gene: role of a heterogeneous nuclear ribonucleoprotein. *J. Immunol.* 159: 5492-5501.
37. DeMaria, C.T. and G. Brewer. 1996. AUF1 binding affinity to A+U-rich elements correlates with rapid mRNA degradation. *J. Biol. Chem.* 271: 12179-12184.
38. Wilson, G.M., Y. Sun, H. Lu, and G. Brewer. 1999. Assembly of AUF1 oligomers on U-rich RNA targets by sequential dimer association. *J. Biol. Chem.* 274: 33374-33381.

39. Körner, C.G. and E. Wahle. 1997. Poly(A) tail shortening by a mammalian poly(A)-specific 3'-exoribonuclease. *J. Biol. Chem.* 272: 10448-10456.
40. Körner, C.G., M. Wormington, M. Muckenthaler, S. Schneider, E. Dehlin, and E. Wahle. 1998. The deadenylating nuclease (DAN) is involved in poly(A) tail removal during the meiotic maturation of *Xenopus* oocytes. *EMBO J.* 17: 5427-5437.
41. Dehlin, E., M. Wormington, C.G. Körner, and E. Wahle. 2000. Cap-dependent deadenylation of mRNA. *EMBO J.* 19: 1079-1086.
42. Gao, M., D.T. Fritz, L.P. Ford, and J. Wilusz. 2000. Interaction between a poly(A)-specific ribonuclease and the 5' cap influences mRNA deadenylation rates in vitro. *Mol. Cell* 5: 479-488.
43. Kiledjian, M., C.T. DeMaria, G. Brewer, and K. Novick. 1997. Identification of AUF1 (heterogeneous nuclear ribonucleoprotein D) as a component of the α-globin mRNA stability complex. *Mol. Cell. Biol.* 17: 4870-4876.
44. Haskill, S., C. Johnson, D. Eierman, S. Becker, and K. Warren. 1988. Adherence induces selective mRNA expression of monocyte mediators and proto-oncogenes. *J. Immunol.* 140: 1690-1694.
45. Sporn, S.A., D.F. Eierman, C.E. Johnson, J. Morris, G. Martin, M. Ladner, and S. Haskill. 1990. Monocyte adherence results in selective induction of novel genes sharing homology with mediators of inflammation and tissue repair. *J. Immunol.* 144: 4434-4441.
46. Ross, H., N. Sato, Y. Ueyama, and H. Koeffler. 1991. Cytokine messenger RNA stability is enhanced in tumor cells. *Blood* 77: 1787-1795.
47. Buzby, J.S., S. Lee, P. van Winkle, C.T. DeMaria, G. Brewer, and M.S. Cairo. 1996. Increased granulocyte-macrophage colony-stimulating factor mRNA instability in cord versus adult mononuclear cells is translation-dependent and associated with increased levels of A+U-rich element binding factor. *Blood* 88: 2889-2897.
48. Loflin, P., C.-Y.A. Chen, and A.-B. Shyu. 1999. Unraveling a cytoplasmic role for hnRNP D in the *in vivo* mRNA destablization directed by the AU-rich element. *Genes Dev.* 13: 1884-1897.
49. Cohn, J.N., T.B. Levine, M.T. Olivari, V. Garberg, D. Lura, G.S. Francis, A.B. Simon, and T. Rector. 1984. Plasma norepinepherine as a guide to prognosis in patients with chronic congestive heart failure. *N. Engl. J. Med.* 311: 819-823.
50. Bristow, M.R., W. Minobe, R. Rasmussen, P. Larrabee, L. Skerl, J.W. Klein, F.L. Anderson, J. Murray, L. Mestroni, S.W. Karwande, M. Fowler, and R. Ginsburg. 1992. β-adrenergic neuroeffector abnormalities in the failing human heart are produced by local rather than systemic mechanisms. *J. Clin. Invest.* 89: 803-815.
51. Bristow, M.R., W.A. Minobe, M.V. Raynolds, J.D. Port, R. Rasmussen, P.E. Ray, and M. Feldman. 1993. Reduced beta 1 receptor messenger RNA abundance in the failing human heart. *J. Clin. Invest.* 92: 2737-2745.
52. Hadcock, J.R., H. Wang, and C.C. Malbon. 1989. Agonist-induced destabilization of beta-adrenergic receptor mRNA. Attenuation of glucocorticoid-induced up-regulation of beta-adrenergic receptors. *J. Biol. Chem.* 264: 19928-19933.
53. Pende, A., K.D. Tremmel, C.T. DeMaria, B.C. Blaxall, W.A. Minobe, J.A. Sherman, J.D. Bisognano, M.R. Bristow, G. Brewer, and J.D. Port. 1996. Regulation of the mRNA-binding protein AUF1 by activation of the β-adrenergic receptor signal transduction pathway. *J. Biol. Chem.* 271: 8493-8501.
54. English, B.K., W.P. Hammond, D.B. Lewis, C.B. Brown, and C.B. Wilson. 1992. Decreased granulocyte-macrophage colony-stimulating factor production by human neonatal blood mononuclear cells and T cells. *Pediatr. Res.* 31: 211-216.
55. Lee, S.M., E. Knoppel, C. van de Ven, and M.S. Cairo. 1993. Transcriptional rates of granulocyte-macrophage colony-stimulating factor, granulocyte colony-stimulating factor, interleukin-3, and macrophage colony-stimulating factor genes in activated cord *versus* adult mononuclear cells: alteration in cytokine expression may be secondary to posttranscriptional instability. *Pediatr. Res.* 34: 560-564.

56. Buzby, J.S., G. Brewer, and D.J. Nugent. 1999. Developmental regulation of RNA transcript destabilization by A+U-rich elements is AUF1-dependent. *J. Biol. Chem.* 274: 33973-33978.

7

RNA BINDING BY MEMBERS OF THE 70-kDa FAMILY OF MOLECULAR CHAPERONES

Christine Zimmer, Eszter Nagy, John Subjeck*, Tamás Henics

*INTERCELL, Vienna, Austria, and * Roswell Park Cancer Inst., Buffalo, NY*

Research on heat shock and other stress proteins (hsps) has revealed a number of intriguing aspects about this remarkable class of molecules. First, hsps are highly abundant proteins in cells even under non-stress conditions. This abundance holds for virtually all organisms or cell types examined. Second, hsps are the most phylogenetically conserved proteins known to biology with an overall primary amino acid sequence homology of some 50 % between Escherichia coli and man. Third, hsps have been implicated in a myriad of cellular processes throughout the years. Although the majority of these biological functions delineate hsps as molecular chaperones, recent evidence suggests that certain heat shock proteins possess a likely ancient, evolutionarily conserved role pointing beyond their classical chaperoning function. This chapter deals with a recently described novel feature of the mammalian 70-kDa super-family of molecular chaperones (hsp70, hsc70, hsp110 and grp170), their inherent RNA binding properties. The monograph highlights the RNA sequence preference as well as the RNA-binding domain of these proteins. The influence of ATP and a peptide substrate on RNA-binding –all key components in chaperoning function- will also be detailed. Although the data were obtained using both hsp/hsc70 and hsp110 as the RNA binding partner, since most of the supporting evidence deals with hsp70/hsc70, discussions of possible in vivo functions will mainly be extended to these widely characterized stress proteins.

BACKGROUND

The 70-kDa Family of Molecular Chaperones

Hyperthermia exposures, oxidative stress and several other conditions, which can be proteotoxic, have been long recognized to increase the expression of a small group of proteins, known historically as heat shock proteins (hsps). The most strongly inducible hsps in mammals are observed at approximately 28, 60, 70, 90 and 110 kDa. The 70-kDa member is easily seen as two closely migrating species on gels, which are frequently called hsc70 (constitutive and inducible) and hsp70 (strongly inducible, but not usually constitutive). While both of these hsp70s are located in the cytoplasm, several additional hsp70 family members are known to occupy every major cellular organelle of eukaryotic cells. All of these different, but highly homologous proteins, are called the hsp70 family. These hsp70 family members usually share more than 60% identity in amino acid sequence.

Hsp70 family members have been shown to play essential roles in a variety of cellular activities, for example: folding of nascent polypeptides, protein translocation across intracellular membranes, and regulation of the activities of steroid hormone receptors and kinases (1-3). The unifying mechanism for their action is based on their chaperoning activity, i.e., their ability to recognize and bind to peptide segments which are not usually exposed to the aqueous environment because they are normally buried in the protein's interior or are hidden by interactions with other proteins (4-6).

Based on quantity and inducibility, hsp110 has also been recognized as a major heat shock protein for nearly as long as hsp70, specifically so in mammals. Despite this, hsp110 has only been recently cloned from a variety of organisms, as diverse as yeast and man (7, reviewed in 8). Surprisingly, the cloning of hsp110 family members has indicated that this family does not represent a genetically unique stress protein group, as previously seen with other heat shock protein families - such as hsp90, 60 or 28, but that they are clearly related to the hsp70 family (7). However, in addition to their significantly increased mass compared to the hsp70s, hsp110s exhibit significant sequence divergence from the archetypical hsp70s (7), being only 25 to 30% identical in amino acid sequence. Statistical analysis of the hsp70 and hsp110 stress protein groups demonstrates that they comprise two independent stress protein families with a common ancestral relationship. The appearance of the highly diverged hsp110 family, which exists in parallel with the hsp70s in the cytoplasm and nucleus of highly diverse organisms from yeast to human argues for differential function(s) for hsp110 compared to hsp70 that are essential in eukaryotic cells.

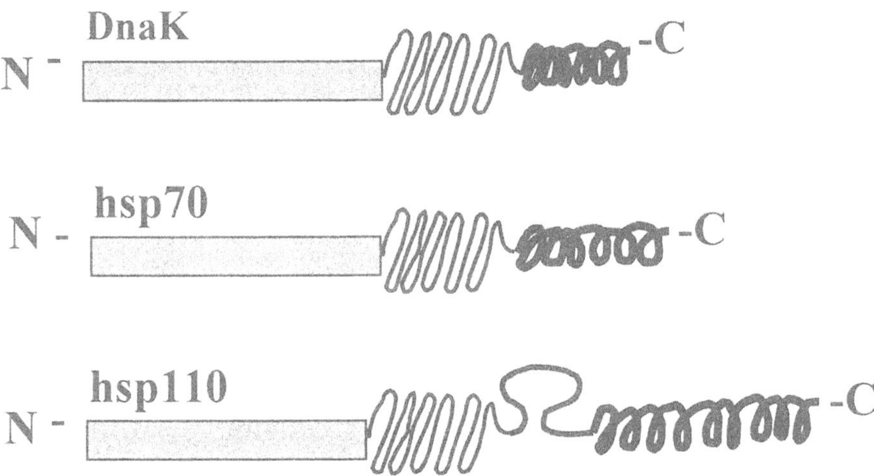

Figure 1. Schematic comparison of the domain composition of DnaK (the Escherichia coli hsp70 homologue), mammalian hsp70 and hsp110. *The N-terminal ATP-binding domain is shown in gray and the composing amino acids are 1-389 for DnaK, 1-383 for hsp70 and 1-375 for hsp110. The β-sheeted peptide or substrate binding middle domain has a respective size of 134, 129 and 133 amino acids. The C-terminal α-helical lid domain of DnaK and hsp70 are also comparable in size (117 and 130 aa, respectively), whereas the identical domain of hsp110 is considerably longer with 250 amino acids. Moreover, in hsp110, the peptide binding domain and the C-terminal-most lid domain are separated by a 100 aa long loop structure of unknown function.*

Secondary structure analysis of hsp110 demonstrates that it exhibits significant similarity to the secondary structure of hsp70 and DnaK (the hsp70 homologue of *E. Coli*), whose structures are well studied by several methods including crystallography. Hsp70/DnaK is composed of 1) an amino terminal ATP binding domain, followed by 2) an approximately 130 amino acid β-sheet region which has been identified as the peptide binding domain, and 3) a carboxyl terminal α-helical region which is involved in regulation of hsp70/DnaK function and appears as a lid covering the peptide binding domain (9). Analysis of the sequence, secondary structure, and peptide binding properties of hsp110 demonstrates that it has the same basic 1) ATP binding domain, 2) β-sheet (peptide binding) domain, and 3) carboxyl end α-helical region (Fig. 1). However, the ATP binding domain of hsp110 binds ATP poorly relative to hsc70 (10). The β-sheet configuration (i.e. the peptide binding cleft) of hsp110 has significant sequence homology to corresponding regions in the structure of DnaK and appears to be of similar overall size and organization. The regulatory, α-helical "lid" region of hsp110 is similar in secondary structure to that of DnaK, but exhibits little sequence similarity. In the case of DnaK, this region is seen as a lid covering the β-sheet peptide binding cleft. In the case of hsp110, this domain is essential for peptide binding (10). The predominant structural difference between hsp110 and DnaK/hsp70 is that the α-helical "lid" region of hsp110 is connected by a 100 amino acid "loop" to the β-sheet peptide binding domain, a structure which is virtually absent in hsp70 (8-9) and which is responsible for much of the increased size of hsp110 (Fig. 1). Thus, at first sight, hsp110 appears to be a very large version of hsp70. In addition, however, the major structural differences in the loop and enlarged "lid" domains and the variance in ATP binding affinity properties between hsp110 and hsp70 are predictive of important functional differences. Indeed, hsp110 has been observed to be a much stronger chaperone in binding peptide chains when compared to hsp70 (11).

Finally, in addition to hsp110 and its family members, there is yet another large relative of the hsp70-hsp110 "superfamily" which resides in the endoplasmic reticulum (ER), known as grp170 (12, reviewed in 8). Because of its cellular location in the ER, it will not be discussed in detail here. However, grp170 does exhibit similar structural and functional properties to hsp110, although being almost as diverged from hsp110 as it is from hsp70. Statistical analysis demonstrates that hsp70, hsp110 and grp170 represent three segregated families of stress proteins interrelated by a common ancestral protein.

Earlier Implications of Stress Proteins in RNA Metabolism

Several lines of evidence are now available suggesting that in cells, members of the various stress protein families also recognize and bind to non-protein macromolecules (13-15). Relative to this current chapter, heat shock proteins have recently been associated with various aspects of RNA metabolism. For instance, GroEL, the *E. coli* hsp60 homologue, is part of a protein complex that protects bacterial transcripts from RNase E mediated degradation (16-17). Intriguingly, it was later demonstrated that DnaK, and, in some instances, GroEL co-purify with the bacterial degradosome, a multi-component prokaryotic complex of regulated RNA decay (18). However, in these studies, no functional implications have been proposed for these stress-proteins in mRNA metabolism (18). The first direct evidence for functional involvement of a stress protein in a specific RNA metabolic process came with the identification of the 60-kDa chaperonin of the thermophilic archaeon, *Sulfolobus solfataricus* as an RNA-binding protein specifically interacting with the 16S rRNA (13). Additionally, *in vitro* reconstituted endonuclease activity was associated with this chaperonin, which cleaved pre-rRNA at a specific 5' site (13). Earlier studies on heat response regulation in *Drosophila* demonstrated that expression of the major heat-induced protein, hsp70, is self-regulated at both the transcriptional and post-transcriptional levels (19-21). Binding of hsp70 to its own mRNA has been proposed to explain the rapid alterations in mRNA stability observed in the cytoplasm (22). Additionally, a previously described 102-kDa RNA-binding protein, conserved throughout higher plants, has recently been identified as hsp101 (15). Through direct binding and complex formation with the 5` leader sequence of tobacco mosaic viral RNA, Hsp101 acts as a translational enhancer, a function that could be recapitulated in yeast and which is independent of its role in conferring thermoresistance (15). Zuotin, a yeast hsp40 (DnaJ) homologue has clearly been shown as a nuclear tRNA-binding protein (23). As an example for indirect involvement of hsp70 in the regulation of the cytoplasmic fate of a given mRNA, hsp70 is a potent modulator of complex formation between the 3`UTR of erythropoietin (Epo) mRNA and its specific binding protein, ERBP (24). Assembly of this complex is necessary for hypoxia-induced stabilization of Epo mRNA. Sequestration of ERBP by hsp70 in normoxic conditions is proposed to be a potent mechanism to prevent complex formation between ERBP and Epo mRNA 3'UTR leading to rapid decay of the message (24). A similar sequestration phenomenon by hsp70 has also been reported with regards to an RNA-binding protein, AUF1, a necessary factor for AU-rich 3'UTR instability sequence-mediated mRNA degradation. AUF1 complexes with hsp/hsc70, translation initiation factor eIF4G and poly(A) binding protein. Elimination of AUF1 by ubiquitination-coupled degradosomal activity is required for AU-rich RNA decay (25). Intriguingly, prevention of these processes as well as heat shock leads to nuclear-perinuclear sequestration of AUF1 by hsp70 and therefore, mRNA and AUF1 degradation is impaired (25). Taken together, these studies independently implicate various heat shock proteins in regulatory mechanisms of RNA metabolism through their direct or indirect association with RNA. They also highlight the potential biological relevance of such macromolecular associations in various organisms.

RNA BINDING BY MAMMALIAN hsp110 PROTEINS

Previous label transfer (UV-crosslinking) experiments performed in a number of laboratories indicated the formation of RNA-protein complexes with approximate MWs of 40-, 60-, 70- or 100-kDa (26-27). Indeed, using a variety of cell extracts as the protein source and AU-rich 3'UTR RNAs of various origins as RNA probes, we demonstrated such complexes in a number of systems (27-28). Our label transfer studies with cytoplasmic lysates of human lymphocytes demonstrated that a nearly 70-kDa complex was invariably formed with AU-rich 3'UTR RNA sequences of various lymphokine and proto-oncogene mRNAs (27-28). In addition, during our attempts of expression and purification of an AU-rich RNA-binding fusion protein in *E. coli*, an intense 70-kDa activity was co-purified with this protein (Nagy and Henics, unpublished observation). Based on these results, we tested the AU-rich RNA binding capacity of mammalian hsp/hsc70 proteins and of the more distantly related family member, hsp110. Results unequivocally showed direct RNA binding by the mammalian hsp/hsc70 and hsp110 molecules. We demonstrated that these proteins preferentially bound AU-rich 3'UTR sequences of various cytokine and proto-oncogene mRNAs (29). Interestingly, similar RNA-binding properties have been recently demonstrated with BiP, the endoplasmic reticular hsp70 homologue (30a). Importantly, the observation that hsp70 proteins were immunoprecipitated from protein-RNA complexes formed between cytoplasmic proteins of human lymphocytes and AU-rich RNA probes (29) provided the first indirect evidence that these hsp-RNA interactions might also be relevant *in vivo*.

RNA Sequence Preference

The question remained whether the binding of hsps is restricted to the instability determinant AU-rich elements found in the 3'UTR of intrinsically labile mRNAs (30) or if other sequences may bind as well. Studies were initiated to test *in vitro* RNA binding of human hsp70 as well as hsc70 proteins using different RNA probes, including AU-rich 3'UTR sequences of various cytokine and proto-oncogene mRNAs as well as an ORF sequence and tRNAs. These experiments conferred AU-rich RNA sequence preference for these chaperone molecules (29). Recently, we generated RNA probes with AUUUA motifs in 5x or 15x tandem repeats as well as a probe of ACCCA in 5x repeats. *In vitro* RNA binding assays using these probes further confirmed that hsp70 proteins preferentially recognize and bind AU-rich RNA stretches and that similarly positioned nucleotide sequences lacking U residues fail to serve as a ligand for these proteins (30a). Additionally, we characterized *in vitro* hsp-RNA interactions in competition experiments where various homoribopolymers were used as cold competitors in label transfer assays. As expected, no competition of AU-rich RNA binding was observed with poly(A) or poly(C), but poly(U) competed very well (29). From these results it was concluded that the AUUUA motif-containing 3'UTR destabilizing elements as well as poly(U) stretches also found within the 3'UTR of a number of mRNAs might indeed serve as binding sites for hsps.

THE RNA BINDING DOMAIN

After the initial observation that hsp70 and hsp110 are proteins capable of RNA binding, it was tempting to determine and identify the region of these molecules responsible for this feature. The construction of various deletion mutants of hsp110 was used to initially resolve this important question (29). These mutants were expressed and purified as His-tagged proteins in *E. coli* (10). The domain structure of hsp110 and hsp70 was described earlier (see also Fig. 1). Briefly, deletion of the N-terminal 375 aa region which covers the whole ATP-binding domain of hsp110 resulted in the complete loss of RNA-binding activity, regardless of whether the adjacent substrate- or peptide-binding region (aa 375-508) was retained or not. However, deletion mutants in which the loop domain (aa 508-608) or the C-terminal helical lid domain (aa 608-858) were separately, but not simultaneously deleted, exhibited considerably higher RNA-binding activity than that of the wild type hsp110 (29). These results demonstrated two points. One is that the RNA-binding segment of the molecule resides at the N-terminus and is identical with the nucleotide (ATP)-binding domain. This observation was further supported by direct competition of both hsp70 and hsp110 RNA-binding by ATP at nearly physiological concentrations (0.5-1 mM) (29). Second, that the C-terminal structures of the chaperone are not involved directly in RNA-binding but are likely to transmit sterical (structural) and/or functional information for the N-terminal ATP-binding domain, which then leads to the modulation of RNA-binding by the molecule (Fig. 2).

More recent work in our laboratory has extended these studies to hsp70. We demonstrated that the N-terminal 44-kDa ATPase domain of hsp70 is itself sufficient to bind RNA with the intensity comparable to that of the full-length hsp70. A somewhat unexpected but not entirely unknown feature has also emerged from this study. When RNA binding by the ATPase domain was compared to the full-length protein in Northwestern analysis (i.e. under conditions when the protein partners were immobilized on a membrane prior to RNA binding), it was seen that the ATPase domain bound to AC-rich RNA just as avidly to the AU-rich RNA. This loss in RNA sequence specificity was assumed to be attributable- at least in part- to perturbed inter-domain interactions between the N- and C-terminal domains of the chaperone as a result of immobilization on the membrane (30a).

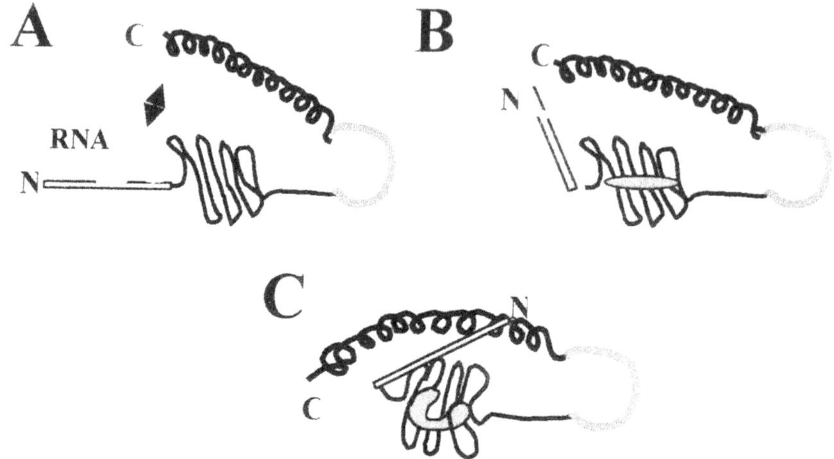

Figure 2. Schematic modeling of RNA-binding by hsp110. RNA (shown in yellow) is bound by the N-terminal ATPase domain (shown in light blue) of hsp70/110 *(A). The exact residues involved in RNA recognition and binding are not known. The organizational pattern of hsp70/110 ATPase domain (five-stranded antiparallel β-sheet enclosed by α-helices on both surfaces to form an ATP-binding cleft) is also present in actin (38-39), a molecule that does not bind RNA (40). This together with the previous demonstration of the hsp70/110 ATP-binding domain as the RNA-binding motif and the recent data that the ATP-binding domain of hsp70 alone loses RNA-binding sequence specificity, highlights the importance of the C-terminal domains of the molecule in potentiating RNA-binding. The functional role of the 100 amino acid-long loop structure unique to hsp110 is not known. A mechanistically possible function is that this unit may serve as a flexible 'connector' between the large N-terminal region harboring both the ATP- and peptide-binding domains and the C-terminal-most helical lid structure. This putative function would allow flexible communication of the C-terminus with the domains that mediate ATP- and peptide-binding (peptide substrate shown in magenta) as well as serve to transfer conformational or functional information between the protein units (B). Data presented in the text of this chapter appear to be consistent with the view that intra-molecular interactions between the ATP-binding and peptide-binding domains trigger conformational rearrangements within the chaperone molecule, which influence RNA-binding. Under certain conformational conditions, the ATPase domain is not available for RNA-binding (C).*

A similar gain of promiscuity in sequence recognition was demonstrated earlier with another RNA-binding protein, glyceraldehyde-3-phosphate dehydrogenase, when its N-terminal 43 amino acid single domain - corresponding to the first mononucleotide-binding fold, and which was shown to confer RNA-binding- was analyzed alone for RNA-binding (31). Together, these results seem to further support the idea that the C-terminal portions of the molecule may not only function to modulate RNA binding but are also required to potentiate this function by the N-terminal domain. The existence of intra-molecular transmission of functional information has been demonstrated in earlier studies, where the classical chaperoning mechanisms of DnaK, Ssa/Ssb (yeast hsp70 homologues) and hsp70 were probed. This led to the discovery of dramatic structural rearrangements of chaperone structure upon binding of a peptide substrate (32-37). These intra-molecular conformational changes might share influential effects with those observed upon deletion mutation of various C-terminal domains as presence of a peptide substrate in the RNA-binding reaction was shown to abrogate RNA-binding by both hsp70 and hsp110 (29). Indeed, as it will further be discussed later on in this chapter, this observation may model an important regulatory intersection of peptide chaperoning and RNA-binding functions of hsp70/hsp110 *in vivo* (see also Fig. 2).

POSSIBLE BIOLOGICAL SIGNIFICANCE

There is little doubt that the primary and most fundamental biological role of molecular chaperones is their overall and constant survey and maintenance of proper protein conformation required for physiological function. This includes assistance in initial folding of newly translated proteins, protection of energetically favored conformations as well as restoration of aberrant protein structures imposed by various harmful stimuli (4-6). The newly discovered RNA-binding properties of some members of the 70-kDa hsp family together with a number of recently reported findings outlined below all seem to point to the possibility that these molecular chaperones might also be regulatory components in specific processes of RNA metabolism. As nascent chain-binding chaperones, hsp70 homologues have been located in polysomal compartments, that is, the sites of translation in both yeast and mammalian cells (41-43). Moreover, re-evaluation of the subcellular location of zuotin, a novel yeast DnaJ co-chaperone revealed that this molecule also binds to ribosomes (44). Recently, in connection with its modulatory activity on the α subunit of eukaryotic initiation factor 2 (HRI), eukaryotic hsc70 has also been indirectly linked to the translation complex (45). Moreover, recent findings show that Sis1, the essential hsp40 homologue of *Saccharomyces cerevisiae*, together with Pab1 (poly(A)-binding protein) interact with Ssa, one of the yeast hsp70 homogues. Importantly, this interaction preferentially occurs on translating ribosomes, and is required for translation (46). We propose that hsp/hsc70 and hsp110 molecules are not only involved in co- and post-translational folding of newly synthesized polypeptides, but also may potentiate translational activity by directly chaperoning mRNA molecules and/or assisting in the assembly of RNA-protein complexes to modulate localization or stability of the transcripts (see also Fig. 3). The observations on the involvement of hsp/hsc70 in the regulation of Epo mRNA-ERBP or AU-rich RNA-AUF1 complexes, and thus, in regulation of mRNA stability, seem to be consistent with this hypothesis (24-25). Recently, native co-compartmentalization of hsc70 and hsp110 proteins has been demonstrated in mammalian cells, a function also observed *in vitro* between hsp110 and hsp70 (47).

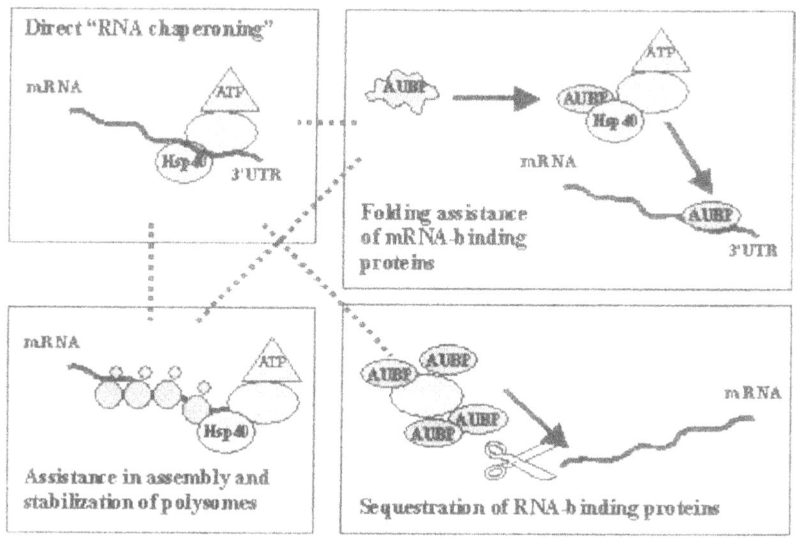

Figure 3. Illustration of possible in vivo functions of hsp/hsc70 molecules in eukaryotic mRNA metabolism. *Direct RNA-binding by members of the 70-kDa family of molecular chaperones together with a number of indirect observations suggest that such function may be relevant in vivo. According to this model, direct "RNA chaperoning" function of hsp/hsc70 or hsp110 would help in the formation and maintenance or stabilization of RNA sterical (secondary) structures facilitating the engagement of RNA in subsequent metabolic pathways. An indirect possibility of hsp involvement in mRNA metabolism can be that the chaperones would assist in the folding of RNA-binding proteins, hence facilitating the formation of RNA-protein complexes required for proper metabolic activity. Additionally, as supported by experimental data, transient sequestration of RNA-binding proteins (such as AUF1 or ERBP, indicated as AUBP "AU-rich sequence-binding protein" in the figure) may lead to alteration of mRNA decay and/or stability. Finally, assistance in the assembly and stabilization of various macromolecular complexes, such as polysomes (ribosomes shown in green) may also contribute to the regulation of translation and other metabolic processes. Dotted lines indicate potential regulatory interplay between the various putative physiological functions.*

Two additional relevant observations came from studies in the yeast, *Sacharomyces cerevisiae.* Analysis of a newly identified regulator of the yeast decapping enzyme, Dcp1p, revealed that mutation of this protein (Vps16p) results in the reduction of decapping activity as well as in increased mRNA stability. Interestingly, enhanced interaction of the abundant yeast hsp70 homologue, Ssa/2p was demonstrated with Dcp1p in cell extracts of mutant strains in decapping activity, including *vps16p* (48). Together, these results raise the possibility that this yeast hsp70 homologue may be involved in the regulation of mRNA decapping and consequently, in mRNA stability. Another study showed that mRNAs encoding for an insulinase-like endoprotease, Axl1p, are less abundant as well as less stable in mutant strains for Ydj1p, a yeast homologue of the hsp40 co-chaperone (49). Since Ydj1p acts to regulate Ssa protein function, these authors suggest that modulation of mRNA stability by the yeast hsp70/hsp40 system may occur by regulating the expression of certain protein factors. In both of these instances the possibility is clearly open that the yeast hsp70 homologue mediates the described phenomena through direct association with RNA.

In light of the data presented in this monograph, we propose that hsp/hsc70 proteins might share functional capacity between their major chaperoning activities (nascent protein folding, prevention of aggregation, protein re-folding) and RNA conformational assistance. The primary loci of these functions might be at or in close association with various polysomal compartments. These functional activities are not exclusive, but rather complementary depending on the primary environment of the hsp/hsc70 protein pools (Fig. 3). Such a diversity in function appears to be economically amenable as cells would not need to adopt and express distinct chaperone systems for the protection and folding assistance of each macromolecular subclass but rather re-direct or re-compartmentalize pre-existing pools of the same chaperone family when required.

FUTURE PERSPECTIVES

Research in the past few years on molecular chaperones in general, and on members of the hsp70 family in particular, has considerably changed our view on this remarkable class of proteins since the original introduction of the term by Laskey *et al.* in 1978. We are confident that uncovering the physiological relevance and exact mechanisms of RNA recognition and binding by molecular chaperones will create an exciting new field of research. In the light of available data, a number of significant and potentially interesting lines of investigations may arise. One will be to identify those RNA species that associate with hsps *in vivo*. This would help us deduce cellular mechanisms of RNA metabolism with key involvement of regulatory steps that are directly or indirectly linked to chaperones. Detailed characterization of identified hsp-RNA complexes may also contribute to better understanding of yet unknown features of RNA-binding proteins. An equally pertinent issue is to investigate in detail the role of co-chaperones in the regulation of RNA-binding by the principal hsp molecule. Another yet unanswered question is whether similarly to other mRNAs, hsps are involved in the regulation of stability, localization and/or translation of their own mRNAs. This may provide an explanation for the long-observed post-transcriptional component of heat-shock and other stress responses at the molecular level (50-53). These are all questions that are currently being addressed in our laboratory.

An exciting new perspective will be to understand how the general stress sensor and protection mechanisms are co-ordinately regulated in both the signal receiving and responsive phases. To this end, important studies have recently identified highly conserved regions (HCRs) within 3'UTR sequences of certain messages to be sensitive to oxidative and mitogenic stress (54). As elevated expression of hsps is also associated with such conditions, it will be important to test if certain sets of these stress proteins could also serve as sensors in direct RNA-binding to HCRs and, thus, modulate the expression of relevant gene products at the mRNA level. Finally, we believe that further investigation of the novel RNA-binding property of stress proteins of the 70-kDa family as well as other chaperones (44, 55-57) will also contribute to the emergence of a number of yet unsuspected mechanisms of how these ancient and conserved proteins integrate into basic physiological processes of cellular metabolism.

Summary of key concepts

- ❖ *Direct and indirect association of various heat shock proteins with RNA has implicated these proteins in the regulation of RNA metabolism.*
- ❖ *Hsp/hsc70 and hsp110 directly and preferentially bind AU-rich RNA sequences within the 3'UTR of various cytokine and proto-oncogene mRNAs. The RNA-binding motif locates to the N-terminal ATP-binding domain.*
- ❖ *The C-terminal structures of the chaperone do not directly interact with RNA but can influence RNA-binding affinity and sequence specificity of the N-terminal ATP-binding domain.*
- ❖ *Hsp70 immunoprecipitates from RNA-protein complexes formed between cytoplasmic proteins of human lymphocytes and AU-rich RNA probes. This, together with other evidence suggests that members of the 70-kDa family of molecular chaperones might also be regulatory components in vivo in specific processes of RNA metabolism.*
- ❖ *A proposed model suggests that hsp/hsc70 and hsp110 molecules not only assist in co-and post-translational folding of newly synthesized polypeptides, but may also chaperone mRNA in the polysomal compartment to potentiate translation. Also, by assisting in the assembly of various RNA-protein complexes they may modulate localization or stability of the transcripts. Hsp molecules may dynamically traffic in between chaperone pools engaged in these various functions.*

Study Guide Questions

1) What is the general definition of molecular chaperones? What are the main subclasses of inducible heat shock proteins?
2) What are the major structural (domain) determinants of the 70-kDa class of molecular chaperones? What are the major structural differences between hsp110 and the DnaK/hsp70 family?
3) List a few examples of hsp involvement in association with RNA metabolism.
4) What is the sequence preference of RNA-binding by members of the mammalian hsp70 family? What is the RNA binding domain?
5) Describe the potential biological implications of RNA-binding by hsps in the cell.

Acknowledgements

This work was supported by Public Health Service grant GM45994, the Roswell Park Canter Institute Core Grant CA16065, and sources from INTERCELL, Inc and Cistem Biotechnologies.

REFERENCES

1. Glick, B. S. (1995) Can Hsp70 proteins act as force-generating motors? *Cell*, 80: 11-14.
2. Craig, E. A., Gambill, B. D., and Nelson, R. J. (1993) Heat shock proteins: molecular chaperones of protein biogenesis. *Microbiol. Rev.*, 57: 402-414.
3. Rutherford, S. L., and Zuker, C. S. Protein folding and the regulation of signaling pathways. (1994) *Cell*, 79: 1129-1132.
4. Bukau, B., and Horwich, A. L. (1998) The Hsp70 and Hsp60 chaperone machines. *Cell*, 92: 351-366.
5. Hartl, F. U. (1996) Molecular chaperones in cellular protein folding. *Nature*, 381: 571-579.
6. Johnson, J. L., and Craig, E. A. (1997) Protein folding in vivo: unraveling complex pathways. *Cell*, 90: 201-204.
7. Lee-Yoon, D., Easton, D., Murawski, M., Burd, R. and Subjeck, J. R. (1995) Identification of a major subfamily of large hsp70-like proteins through the cloning of the mammalian 110-kDa heat shock protein. *J. Biol. Chem.*, 270: 15725-15733.
8. Easton, D.P., Kaneko, Y. and Subjeck, J.R. The hsp110 and grp170 stress proteins: Newly recognized relatives of the hsp70s. *Cell Stress and Chaperones*, in press.
9. Zhu, X., Zhao, X., Burkholder, W. F., Gragerov, A., Ogata, C. M., Gottesman, M. E. and Hendrickson, W. A. (1996) Structural analysis of substrate binding by the molecular chaperone DnaK. *Science*, 272: 1606-1614.
10. Oh, H.J., Easton, D.P., Murawski, M. Kaneko, Y. and Subjeck, J.R. The chaperoning activity of hsp110: Identification of functional domains by use of targeted deletions. *J. Biol. Chem.*, 274: 15712-15718.
11. Oh, H. J., Chen, X. and Subjeck, J. R. (1997) hsp110 protects heat-denatured proteins and confers cellular thermoresistance. *J. Biol. Chem.*, 272: 31636-31640.

12. Chen, X., Easton, D., Oh, H. J., Lee-Yoon, D., Liu, X. and Subjeck, J. (1996) The 170 kDa glucose regulated stress protein is a large HSP70-, HSP110-like protein of the endoplasmic reticulum. *FEBS Lett.*, 380: 68-72.
13. Ruggero, D., Ciammaruconi, A. and Londei, P. (1998) The chaperonin of the archaeon, *Sulfolobus solfataricus* is an RNA-binding protein that participates in ribosomal RNA processing. *EMBO J.*, 17: 3471-3477.
14. Török, Zs., Horváth, I., Goloubinoff, P., Kovács, E., Glatz. A., Balogh, G. and Vígh, L. (1997) Evidence for a lipochaperonin: association of active protein-folding GroESL oligomers with lipids can stabilize membranes under heat shock conditions. *Proc. Natl. Acad. Sci. USA*, 94: 2192-2197.
15. Wells, D. R., Tanguay, R. L., Le, H. and Gallie, D. R. (1998) Hsp101 functions as a specific translational regulatory protein whose activity is regulated by nutrient status. *Genes Develop.*, 12: 3236-3251.
16. Sohlberg, B., Lundberg, U., Hartl, F-U. and von Gabain, A. (1993) Functional interaction of heat shock protein GroEL with an RNase E-like activity in *Escherichia coli. Proc. Natl. Acad. Sci. USA*, 90: 277-281.
17. Georgellis, D., Sohlberg, B., Hartl, F-U. and von Gabain, A. (1995) Identification of GroEL as a constituent of an mRNA-protection complex in *Escherichia coli. Molec. Microbiol.*, 16: 1259-1268.
18. Miczak, A., Kaberdin, V. R., Wei, C-L. and Lin-Chao, S. (1996) Proteins associated with Rnase E in a multicomponent ribonucleolytic complex. *Proc. Natl. Acad. Sci. USA*, 93: 3865-3869.
19. DiDomenico, B. J., Bugaisky, G. E. and Lindquist, S. (1982) The heat shock response is self-regulated at both the transcriptional and posttranscriptional levels. *Cell*, 31: 593-603.
20. Simcox, A. A., Cheney, C. M., Hoffman, E. P. and Shearn, A. (1985) A deletion of the 3' end of the *Drosophila melanogaster* hsp70 gene increases stability of mutant mRNA during recovery from heat shock. *Mol. Cell. Biol.*, 5: 3397-3402.
21. Theodorakis, N. G. and Morimoto, R. (1987) Posttranscriptional regulation of *hsp70* expression in human cells: Effects of heat shock, inhibition of protein synthesis, and adenovirus infection on translation and mRNA stability. *Mol. Cell. Biol.*, 7: 4357-4368.
22. Yost, H. J., Petersen, R. B. and Lindquist, S. (1990) Posttranscriptional regulation of heat shock protein synthesis in *Drosophila.* In: *Stress proteins in biology and medicine.* Ed by D. Pauli and A. Tissières. pp 379-409.
23. Wilhelm, M. L., Reinholt, J., Gangloff, J., Dirheimer, G. and Wilhelm, F. X. (1994) Transfer RNA binding protein in the nucleus of *Saccharomyces cerevisiae. FEBS Lett.*, 349: 260-264.
24. Scandurro, A. B., Rondon, I. J., Wilson, R. B., Tenenbaum, S. A., Garry, R. and Beckman, B. (1997) Interaction of erythropoetin RNA binding protein with erythropoetin RNA requires an association with heat shock protein 70. *Kidney Internatl.*, 51: 579-584.
25. Laroia, G., Cuesta, R., Brewer, G. and Schneider, R. J. (1999) Control of mRNA decay by heat shock-ubiquitin-degradosome pathway. *Science*, 284: 499-502.
26. Malter, J. S. (1989) Identification of an AUUUA-specific messenger RNA binding protein. *Science*, 246: 664-666.
27. Henics, T. , Sanfridson, A., Hamilton, B. J., Nagy, E. and Rigby, W. F. C. (1994) Enhanced stability of interleukin-2 in MLA 144 cells: Possible role of cytoplasmic AU-rich sequence binding proteins. *J Biol. Chem.*, 269: 5377-5383.
28. Soós, H., Bujáky, Cs., Kiss, Á., Kovács, É., Somoskeöy, Sz. and Henics, T. (1998) Distribution profile and *in vivo* RNA association of cytoplasmic AU-rich sequence binding proteins in various mammalian cells: effect of the organizational state of cellular architecture. *Physiol. Chem. Phys. Med. NMR.*, 30: 163-174.

29. Henics, T., Nagy, E., Oh, H. J., Csermely, P., von Gabain, A. and Subjeck, J. R. (1999) Mammalian Hsp70 and Hsp110 proteins bind to RNA motifs involved in mRNA stability. *J Biol. Chem.*, 274: 17318-17324.
30. Chen, C.-Y. A. and Shyu, A.-B. (1995) AU-rich elements: characterization and importance in mRNA degradation. *Trends Biochem. Sci.*, 20: 465-470.
30a. Zimmer, C., von Gabain, A. and Henics, T. (2001) Analysis of sequence-specific binding of RNA to Hsp70 and its various homologues indicates the involvement of N- and C-terminal interactions. RNA, in press.
31. Nagy, E., Henics, T., Eckert, M., Miseta, A., Lightowlers, R. N. and Kellermayer, M. (2000) Identification of the NAD^+-binding fold of glyceraldehyde-3-phosphate dehydrogenase as a novel RNA-binding domain. *Biochem Biophys Res Commun.*, 275: 253-260.
32. Liberek, K., Skowyra, D., Zylicz, M., Johnson, C. and Georgopoulos, C. (1991) The *Escherichia coli* DnaK chaperone, the 70-kDa heat shock protein eukaryotic equivalent, changes conformation upon ATP hydrolysis, thus triggering its dissociation from a bound target protein. *J Biol. Chem.*, 266: 14491-14496.
33. Buchberger, A., Theyssen, H., Schröder, H., McCarty, J. S., Virgallita, J., Milkereit, P., Reinstein, J. and Backau, B. (1995) Nucleotide-induced conformational changes in the ATPase and substrate binding domains of the DnaK chaperone provide evidence for interdomain communication. *J Biol. Chem.*, 270: 16903-16910.
34. Fung, K. L., Hilgenberg, L., Wang, N. M. and Chirico, W. J. (1996) Conformations of the nucleotide and polypeptide binding domains of a cytosolic Hsp70 molecular chaperone are coupled. *J. Biol. Chem.*, 271: 21559-21565.
35. James, P., Pfund, C. and Craig, E. A. (1997) Functional specificity among Hsp70 molecular chaperones. *Science*, 275: 387-389.
36. Lopez-Buesa, P., Pfund, C. and Craig, E. A. (1998) The biochemical properties of the ATPase activity of a 70-kDa heat shock protein (Hsp70) are governed by the C-terminal domains. *Proc. Natl. Acad. Sci. USA*, 95: 15253-15258.
37. Davis, J. E., Voisine, C. and Craig, E. A. (1999) Intragenic suppressors of Hsp70 mutants: Interplay between the ATPase- and peptide-binding domains. *Proc. Natl. Acad. Sci. USA*, 96: 9269-9276.
38. Morshauser, R. C., Wang, H., Flynn, G. C. and Zuiderweg, E. R. P. (1995) The peptide-binding domain of the chaperone protein Hsc70 has an unusual secondary structure topology. *Biochemistry*, 34: 6261-6266.
39. Sheterline, P., Clayton, J. and Sparrow, J. C. (1996) Actin structure *in* Actins, 3rd Edition, Academic Press, Ltd. pp 15-51.
40. Henics, T., Nagy, E. and Szekeres-Barthó, J. (1997) Interaction of AU-rich sequence binding proteins with actin: Possible involvement of the actin-cytoskeleton in lymphokine mRNA turnover. *J Cell. Physiol.*, 173: 19-27.
41. Nelson, R. J., Ziegelhoffer, T., Nicolet, C., Werner-Washburne, M. and Craig, E. A. (1992) The translation machinery and 70 kd heat shock protein cooperate in protein synthesis. *Cell*, 71: 97-105.
42. Pfund, C., Lopez-Hoyo, N., Ziegelhoffer, T., Schilke, B. A., Lopez-Buesa, P., Walter, W. A., Wiedmann, M. and Craig, E. A. (1998) The molecular chaperone Ssb from *Saccharomyces cerevisiae* is a component of the ribosome-nescent chain complex. *EMBO J.*, 17: 3981-3989.
43. Beck, S. C. and De Maio, A. (1994) Stabilization of protein synthesis in thermotolerant cells during heat shock. *J. Biol. Chem.*, 269: 21803-21811.
44. Yan, W., Schilke, B., Pfund, C., Walter, W., Kim, S. and Craig, E. A. (1998) Zuotin, a ribosome-associated DnaJ molecular chaperone. *EMBO J.*, 17: 4809-4817.
45. Uma, S., Thulasiraman, V. and Matts, R. L. (1999) Dual role for Hsc70 in the biogenesis and regulation of the heme-regulated kinase of the α subunit of eukaryotic translation initiation factor 2. *Mol. Cell. Biol.*, 19: 5861-5871.

46. Horton, L.E., James, P., Craig, E.A., and Hensold, J.O. (2001) The yeast hsp70 homologue Ssa is required for translation and interacts with Sis1 and Pab1 on translating ribosomes. J Biol Chem 276:14426-14433.
47. Wang, X-Y., Chen, X., Oh, H. J., Repsky, E., Kazim, L. and Subjeck, J. (2000) Characterization of native interaction of hsp110 with hsp25 and hsc70. *FEBS Lett.*, 465: 98-102.
48. Zhang, S., Williams, C. J., Hagan, K. and Peltz, S. W. (1999) Mutations in *VPS16* and *MRT1* stabilize mRNAs by activating an inhibitor of the decapping enzyme. *Mol. Cell. Biol.*, 19: 7568-7576.
49. Meacham, G. C., Browne, B. L., Zhang, W., Kellermayer, R., Bedwell, D. M. and Cyr, D. M. (1999) Mutations in the yeast Hsp40 chaperone protein Ydj1 cause defects in Axl1 biogenesis and pro-a-factor processing. *J Biol. Chem.*, 274: 34396-34402.
50. Moseley, P. L., Wallen, E. S., McCafferty, J D., Flanagan, S and Kern, J. A. (1993) Heat stress regulates the human 70-kDa heat-shock gene through the 3'-untranslated region. *Am. J. Physiol.*, 264: L533-537.
51. Brennecke, T., Gellner, K. and Bosch, T. C. G. (1998) The lack of a stress response in *Hydra oligactis* is due to reduced *hsp70* mRNA stability. *Eur. J. Biochem.*, 255: 703-709.
52. Lee, M. G-S. (1998) The 3' untranslated region of the hsp 70 genes maintains the level of steady state mRNA in *Trypanosoma brucei* upon heat shock. *Nucl. Acids Res.*, 26: 4025-4033.
53. Kaarniranta, K., Elo, M., Sironen, R., Lammi, M. J., Goldring, M. B., Erikson, J. E., Sistonen, L. and Helminen, H. J. (1998) Hsp70 accumulation in chondrocytic cells exposed to high continuous hydrostatic pressure coincides with mRNA stabilization rather than transcriptional activation. *Proc. Natl. Acad. Sci.*, 95: 2319-2324.
54. Spicher, A., Guicherit, O. M., Duret, L., Aslanian, A., Sanjines, E. M., Denko, N. C., Giaccia, A. J. and Blau, H. M. (1998) Highly conserved RNA sequences that are sensors of environmental stress. *Mol. Cell. Biol.*, 18: 7371-7382.
55. Korber, P., Zander, T., Herschlag, D. and Bardwell, J. C. A. (1999) A new heat shock protein that binds nucleic acids. *J Biol. Chem.*, 274: 249-256.
56. Korber, P., Stahl, J. M., Nierhaus, K. H. and Bardwell, J. C. A. (2000) Hsp15: a ribosome-associated heat shock protein. *EMBO J.*, 19: 741-748.
57. Staker, B. L., Korber, P., Bardwell, J. C. A. and Saper, M. (2000) Structure of Hsp15 reveals a novel RNA-binding motif. *EMBO J.*, 19: 749-757.

8

POST-TRANSCRIPTIONAL CONTROL OF TYPE-1 PLASMINOGEN ACTIVATOR INHIBITOR mRNA

Joanne H. Heaton and Thomas D. Gelehrter
University of Michigan Medical School, Ann Arbor, MI

In this chapter we will describe the identification of a sequence in the 3'UTR of the type-1 plasminogen activator inhibitor (PAI-1) mRNA that mediates the cyclic nucleotide regulation of mRNA stability. This will be followed by a discussion of the isolation and characterization of a novel RNA-binding protein that interacts with the PAI-1 cyclic nucleotide responsive sequence.

BACKGROUND

Plasminogen activators are serine proteases that catalyze the conversion of the zymogen, plasminogen, to the broad-spectrum endopeptidase, plasmin. Plasmin, the major enzyme responsible for intravascular fibrinolysis, also plays a role in other biological processes involving tissue remodeling, including angiogenesis and wound healing as well as tumor cell invasion and metastasis (1, 2). As the major biological regulator of plasminogen activation, PAI-1 is an important component in the regulation of physiological and pathological processes. PAI-1 is synthesized in a variety of cell types and its expression is regulated by growth factors, cytokines and hormones, including agents that elevate cAMP levels (3-5).

The role of cAMP as a second messenger in the hormonal regulation of gene transcription has been extensively studied (6). However, agents that elevate cAMP levels have also been reported to regulate gene expression by regulating mRNA turnover, stabilizing some RNAs and destabilizing others (7-14).

Several mRNA sequences that appear to mediate cyclic nucleotide regulation of mRNA degradation have been identified and in some cases binding experiments have demonstrated that cellular proteins interact specifically with these regulatory sequences. For example, cyclic nucleotide stabilization of phosphoenolpyruvate carboxykinase (PEPCK) mRNA is associated with decreased binding of a 100 kDa protein to a predicted stem-loop structure in the PEPCK 3'UTR (8). The stabilization of lactate dehydrogenase A subunit is associated with binding of cytoplasmic proteins to a 30 nt U-rich sequence (7, 15). Incubation of LLC-PK_1 cells with IBMX, which elevates cellular cAMP by inhibiting phosphodiesterase activity, stabilizes the Na^+/Glucose cotransporter (SGLT1) mRNA. A U-rich region in the 3'UTR of SGLT1 is able to confer this regulation onto the β-globin construct and to bind a 38 kDa IBMX-inducible protein (16). The LH/hCG receptor is down regulated by LH concomitant with an increase in binding of a 50 kDA protein to a C-rich sequence in the coding region of the LH receptor mRNA (13, 17). Isoproterenol elevates cellular cAMP levels, destabilizes β-adrenergic receptor mRNA and increases the binding of both AUF-1 and the 35kDa βARB protein to an AU-rich sequence in the 3'UTR of the β_2-adrenergic receptor mRNA (10, 11). Mass spectrometric analysis indicates that the βARB complex includes two other known RNA binding proteins, the ELAV-like protein, HuR, and hnRNP A1 (18). Interestingly, binding of HuR is thought to stabilize transcripts (19, 20). hnRNP A1, one of many complexes associated with pre-RNA and thought to be involved in RNA splicing, has not been previously implicated in regulation of mRNA decay although it has been shown to shuttle between the nucleus and cytoplasm (21). Proteins that bind the other RNA sequences implicated in cAMP-regulated message decay have been observed in binding experiments but have not been identified.

Cyclic Nucleotide Regulation of PAI-1 mRNA

Cyclic nucleotide regulation of PAI-1 mRNA accumulation, which was described by us in HTC cells (5), has also been reported in a number of cell systems, including rat testicular peritubular cells (22), astrocytes (23), and osteoblasts (24), mink lung epithelial cells (25), human fibrosarcoma cells (26) umbilical vein endothelial cell (27) and synovial cells (28). In most cases the mechanisms by which cAMP decreases PAI-1 mRNA accumulation have not been analyzed.

Our studies have shown that while incubation with cyclic nucleotides results in a modest decrease in PAI-1 gene transcription, the greater effect is the 3-fold decrease in PAI-1 mRNA stability. Using the transcriptional inhibitor, actinomycin D, we have determined the half-life of HTC cell PAI-1 mRNA to be 4.0-4.5 h. As illustrated in Fig. 1, a similar half-life is found by measuring the return to steady state after washing out the transcriptional inducer, dexamethasone. If, at the time the dexamethasone is removed, the culture is incubated with 1mM 8-bromo-cAMP plus 1mM

IBMX (referred to as cA), PAI-1 mRNA is destabilized, displaying a half-life of about 1.5 h (29).

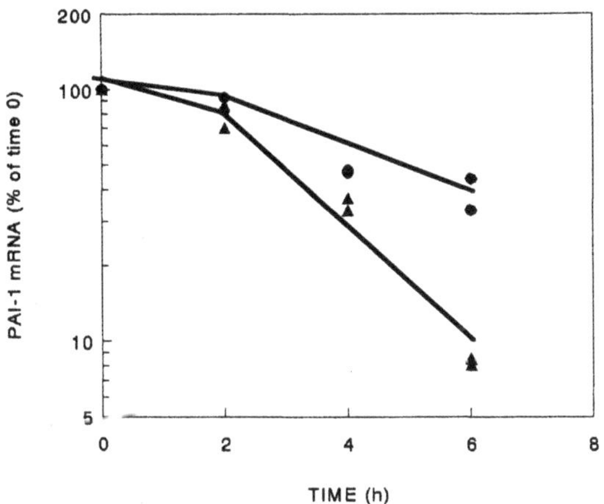

Figure 1. Effect of 8-bromo-cAMP on HTC cell PAI-1 mRNA decay. *Monolayer cultures of HTC cells were incubated with 0.1μM dexamethasone for 9h to increase endogenous PAI-1 transcription. Cultures were washed to remove dexamethasone, and incubated with or without 1mM IBMX and 1mM 8-bromo-cAMP (cA). Cells were harvested at the times indicated, total RNA was isolated and slot blot analysis carried out using 5μg RNA per sample. The signal in PAI-1 mRNA was normalized to that in glyceraldehyde-3-phosphate dehydrogenase (GAPDH) mRNA and the data are presented as a percentage of the amount at time zero. Duplicate data points represent duplicate cultures. (●), Washout control; (▲), cA*

IDENTIFICATION OF THE CYCLIC NUCLEOTIDE RESPONSIVE SEQUENCE

To locate the sequence within the PAI-1 mRNA that is necessary for cyclic nucleotide regulation of PAI-1 mRNA stability, we stably transfected HTC cells with chimeric gene constructs and examined both basal and cyclic nucleotide-regulated message decay. The vector pSVL was chosen for these experiments for two important reasons. First, unlike some other viral promoters, the SV40 promoter does not appear to be regulated by cAMP, allowing examination of the effect on mRNA decay rates independent of any transcriptional effect. Second, expression from the SV40 late promoter is reversibly induced by cycloheximide (30). When cycloheximide is withdrawn, expression is dramatically decreased and degradation of the transcript can be followed without use of actinomycin D. Although actinomycin D does not affect basal decay of endogenous PAI-1 mRNA, it does block the cyclic nucleotide-induced increase in PAI-1 mRNA decay.

To confirm that cyclic nucleotide regulation can be faithfully reproduced in this transfection system, we first constructed a PAI-1 mini-gene and subcloned it downstream from the SV40 promoter in pSVL. The mini-gene transcript can be distinguished from the endogenous transcript in an RNase protection assay (RPA) using an appropriate riboprobe. We demonstrated that the transcript from the transfected mini-gene construct decays with a half-life similar to that of the endogenous PAI-1 message and responds to cyclic nucleotides with an increased decay rate.

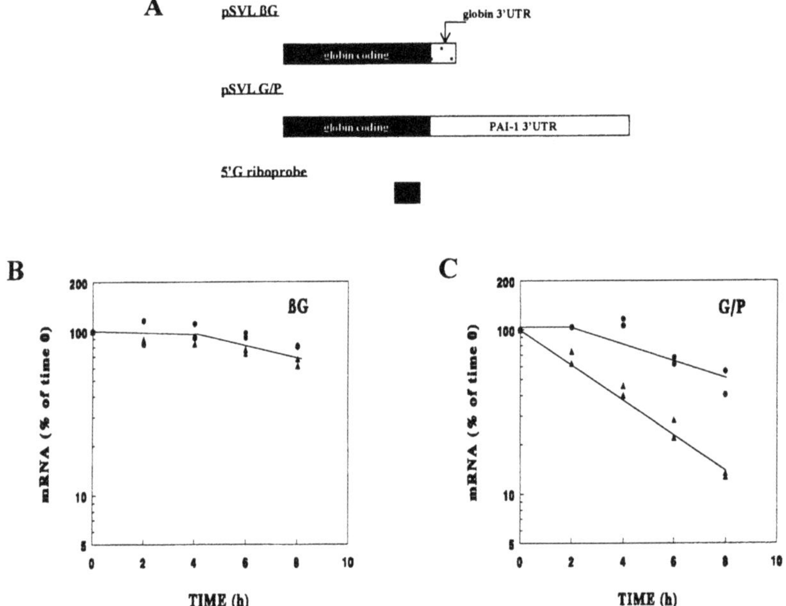

Figure 2. Effect of 8-bromo-cAMP on transcripts from chimeric constructs. *HTC cells were stably transfected with either the mouse β-globin gene (βG) or the chimera of the β-globin coding region and the rat PAI-1 3'UTR (G/P) subcloned behind the SV40 late promoter in the vector pSVL (Amersham-Pharmacia). Following a 10 h incubation in serum-free medium containing 0.1mM cycloheximide, which elevates expression from the SV40 late promoter, cultures were washed in PBS and incubated in serum-free medium with or without cA. At the times indicated the cells were harvested, total RNA isolated and ribonuclease protection assays were carried out using ^{32}P-labeled riboprobes for the globin coding sequence (5' G riboprobe) and for GAPDH. The signal in mRNA from the transfected construct was normalized to that in GAPDH mRNA and the data are presented as a percentage of the amount at time zero. (●), Control; (▲), cA*

We next prepared chimeric constructs of the rat PAI-1 cDNA and the mouse β-globin gene. The β-globin mRNA is normally very stable and its degradation is not regulated by cyclic nucleotides. However, the chimera of the globin coding region and the PAI-1 3'UTR (pSVL G/P) encodes a transcript that decays more rapidly and whose degradation is accelerated by incubating the cells with cA (Fig. 2). These results clearly demonstrate that sequences in the PAI-1 3'UTR are able to confer cA regulation of mRNA stability onto a heterologous (β-globin) gene.

In an effort to locate the region within the 1730 nt PAI-1 3'UTR responsible for the cA-induced destabilization, we used convenient restriction sites to delete selected portions of the PAI-1 cDNA in the context of the pSVLG/P construct. Surprisingly, none of these deletions resulted in significant loss of cAMP-induced destabilization, which led us to conclude that more than one region of the 3'UTR can confer cyclic nucleotide regulation. Significantly, the construct in which all but the 3'-most 134 nt of PAI-1 sequence was deleted, retains the cAMP regulation of message decay, suggesting that this sequence is sufficient to mediate this response (31).

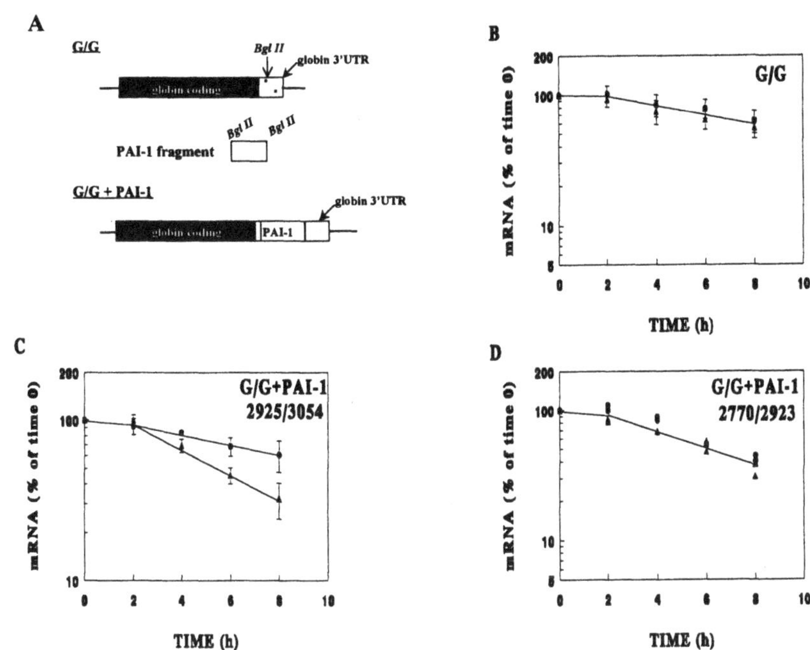

Figure 3. Effect of 8-bromo cAMP on decay rates of globin transcripts containing insertions of PAI-1 3'UTR sequence. A. Schematic representation of constructs. *The Bgl II site was engineered into the 3' UTR of the β-globin gene by PCR and the construct was cloned into the pSVL vector. PCR generated fragments of the PAI-1 3'UTR were ligated into globin at the BglII site. The constructs were designated by the region of the PAI-1 sequence inserted.* ***B, C and D.*** *HTC cells were stably transfected with each of the constructs and examined for cA regulation of mRNA stability as described in the legend to Figure 2. (●), Control; (▲), cA.*

This conclusion from the deletion experiments was confirmed by studies using β-globin constructs with insertions of portions of the PAI-1 sequence. A *Bgl*II recognition site was engineered into the 3'UTR of the β-globin gene and PCR generated portions of the PAI-1 sequence were inserted at this site. As illustrated in Fig. 3, insertion of the 3'-most 130 nt sequence (bp 2925-3054) resulted in a globin transcript whose degradation is accelerated by incubation with cA, whereas insertion of other portions of the PAI-1 sequence (e.g. bp 2770-2923) did not confer cA regulation (31).

The 134 nt PAI-1 cyclic nucleotide responsive sequence (CRS) is shown in Fig. 4. It includes a 75 nt U-rich region present at its 5' end and a 24-nt A-rich region at its 3' end. The final 6 As shown are the beginning of the poly A tail.

• 2926
CCAAAUGUGGGGUCACAUGGUCUUGAAUUUUGUGGGG

• 2966
UACAUAUGCCUUUGUUUGUUUGUUUUCACUUUUGAUAU

• 3010 • 3024
AUAAACAGGUAAAUGUGUUUUU*AAAAAAUAAUAAAAAU*

• 3040 • 3060
***AGAGAAUA*UGCAGACAAAAAAA**

Figure 4. Nucleotide sequence of the PAI-1 134 nt CRS. *The nucleotide sequence of the PAI mRNA from positions 2926-3060 is shown. The U-rich region is underlined and A-rich region is in italics.*

IDENTIFICATION OF *TRANS*-ACTING FACTORS

To define the mechanism of cyclic nucleotide regulation of message stability it was necessary to identify cytosolic factors that may play a role through interaction with the CRS. Figure 5 shows the results of binding studies using HTC cell cytoplasmic fractions (S10) as a protein source and ^{32}P-labeled PAI-1 CRS as RNA probe. In RNA electrophoretic mobility shift assays (RNA-EMSA) (Fig. 5C), in which complexes are separated on native gels, two slowly migrating complexes are seen. UV-crosslinking analyses demonstrated binding proteins ranging in size from 38kDa to 76kDa, with a major complex migrating at approximately 50-55kDa (Fig. 5A). Specificity of binding was demonstrated by competition studies. Unlabeled identical RNA sequence competes effectively, whereas unlabeled irrelevant RNA (a portion of the bacterial CAT sequence) does not compete. In addition, a portion of the PAI-1 sequence that fails to confer cyclic nucleotide responsiveness when inserted into the β-globin gene (nt 2125-2296) (31) also fails to compete for binding. That the complexes represent binding of protein to the labeled RNA is shown in Fig. 5B and D;

heat denaturing the S10 fractions or treating with proteinase K abolishes complex formation. Heat denaturing the RNA probe did not prevent binding. This result suggests, but does not prove, that RNA secondary structure is not critical for binding of these proteins.

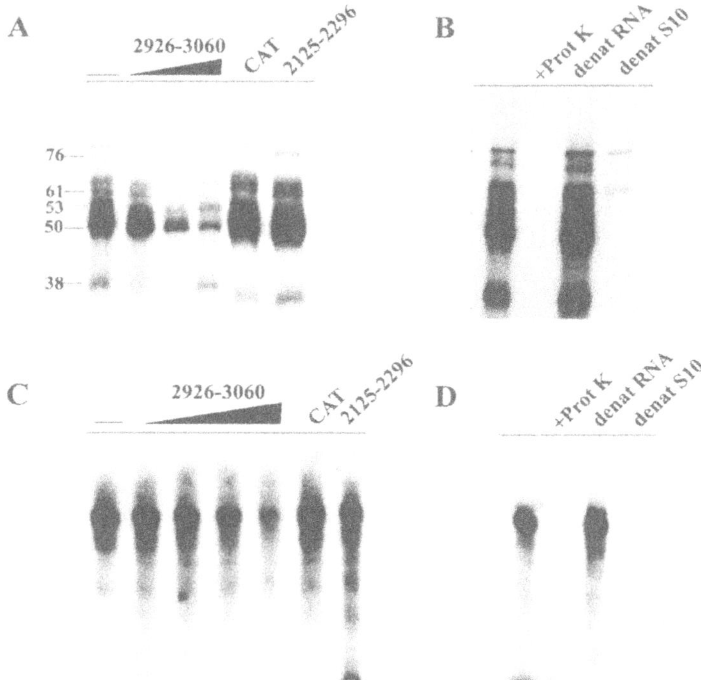

Figure 5. UV cross-linking and RNA-EMSA analyses of HTC cell cytoplasmic proteins and PAI-1 CRS. Panels A and C: *Specificity of binding. Unlabeled PAI-1 CRS (nt 2926-3060) (0,10-, 50-, 100- and 250-fold [panel C only] molar excess) or 250-fold molar excess of CAT nt 2528-2829 or PAI-1 nt 2125-2296 (a region that does not mediate cyclic nucleotide regulation) was pre-incubated with HTC cell cytoplasmic extracts prior to incubation with radiolabeled PAI-1 CRS. RNA-protein interactions were examined by UV cross-linking (panel A) or RNA-EMSA (panel C).* ***Panels B and D****: Characterization of complexes. HTC cell cytoplasmic extracts were treated with proteinase K (Prot K) or were heated to 85°C for 15 min prior to incubation with radiolabeled probe (denat S10). Where indicated labeled RNA (denat RNA) was heated to 85°C for 15 min and rapidly cooled prior to incubation with cytoplasmic extracts. RNA-protein interactions were examined by UV cross-linking analysis (panel B) or RNA-EMSA (panel D).*

To delineate the nucleotide sequence within the CRS required for binding, studies were carried out using portions of the PAI-1 CRS as labeled probe. The data shown in Fig. 6 demonstrate that, while the 38 kDa protein interacts with the U-rich region (nt 2926-3010), the major band and the higher molecular weight proteins all require the A-rich region for binding.

In fact, the sequence 3010-3040, which consists primarily of the A-rich portion, is able to form the major 50-55 kDa protein-RNA complex.

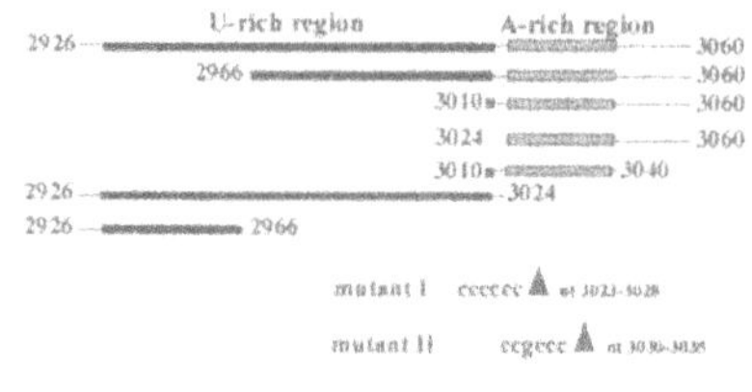

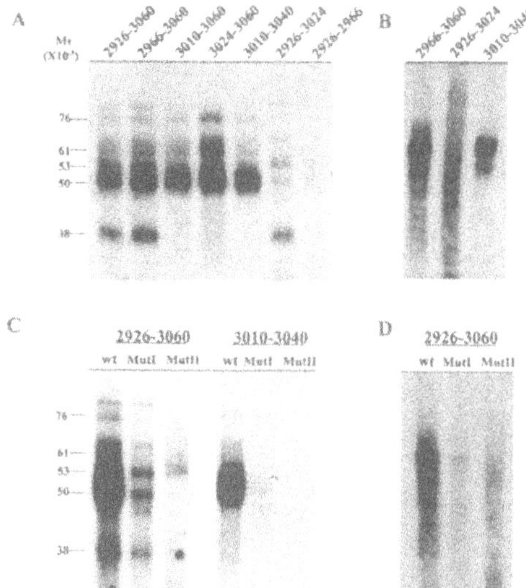

Figure 6. Identification and mutational analysis of protein binding sites. Top: *Diagrammatic representation of the PAI-1 CRS indicating the U-rich portion as a solid bar and the A-rich portion as a shaded bar. The mutations in the A-rich region were created in both the full length element (nt 2926-3060) and in the minimal binding region (nt 3010-3040).* ***Panels A and B****: Identification of protein binding sites: HTC cell cytoplasmic extracts were incubated with the radiolabeled RNAs indicated above each lane and the RNA-protein interactions were analyzed by UV-crosslinking (panel A) or RNA-EMSA (panel B).* ***Panels C and D****: Mutational analysis of the A-rich region: HTC cell cytoplasmic extracts were incubated with the wild type or the mutant radiolabeled RNAs indicated above the lanes and UV-crosslinking (panel C) or RNA-EMSA (panel D) analyses were carried out.*

More importantly, this 30nt sequence is also able to form the high molecular weight complex identified by RNA-EMSA. Mutation of the A-rich region (nts 3023-3028 and 3030-3035) in which As and Us were changed to Cs and Gs, respectively, abolished RNA-protein complex formation both in UV-crosslinking (Fig. 6C) and in RNA-EMSA (Fig. 6D), confirming that most of the proteins interact with the A-rich sequence.

To demonstrate the functional role of the binding proteins in the cyclic nucleotide regulation of stability, we prepared a PAI-1 CRS with both mutations shown in Fig. 6. This sequence was subcloned into the *Bgl*II site in the globin 3'UTR and HTC cells were stably transfected with the mutant construct. As shown in Fig. 7, the mutant construct failed to confer cyclic nucleotide regulation of message degradation. Thus, the same mutations that abolish complex formation also eliminate cyclic nucleotide regulation of message stability in transfected HTC cells, indicating that this sequence and one or more of its binding proteins play a critical role in regulation of message stability (32).

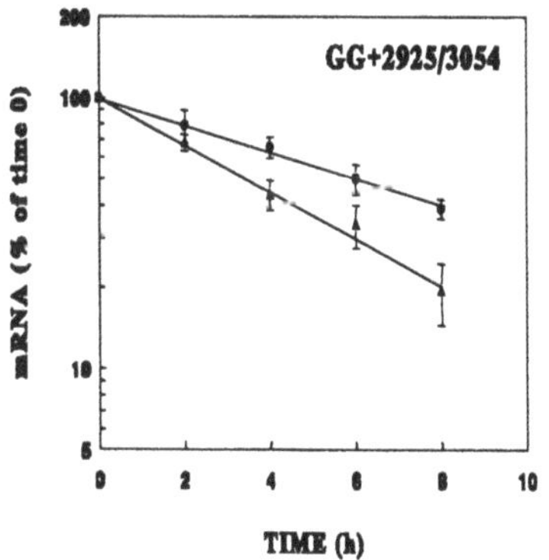

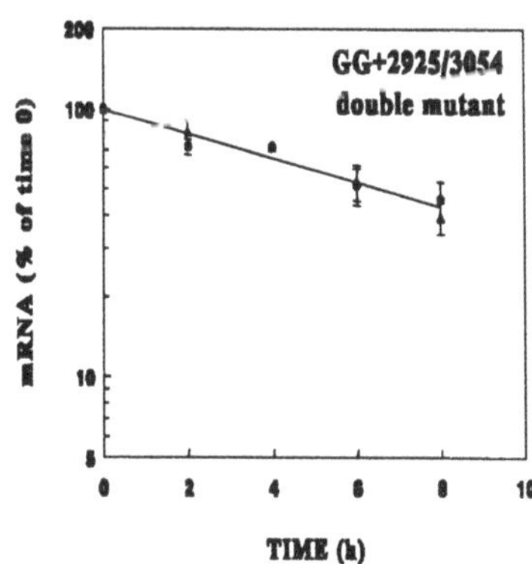

Figure 7. Mutation of the A-rich region abolishes cyclic nucleotide responsiveness. *The PAI-1 sequence from nt 2925/3054 was amplified by PCR using a 3' primer with mutations of the A-rich region (both mutations shown in Fig. 6). The product was ligated into the BglII site of G/G as diagrammed in Fig. 3 and designated G/G +2925/3054 double mutant. HTC cells stably transfected with this construct or with G/G +2925/3054 were analyzed for cA regulation of mRNA degradation as described in the legend to Fig. 2. (●), Control; (▲), cA.*

While the PAI-1 CRS has both U- and A-rich portions, it is not like the AU-rich element (ARE) instability determinants described in the 3'UTRs of many labile transcripts (33); in the PAI-1 CRS the U-rich and A-rich portions are separate and there are no AUUUA motifs. The PAI-1 CRS is also unlike sequences identified as responsible for cAMP regulation of stability in other transcripts. The 20 nt and 160 nt sequences identified in hamster and human β-AR transcripts, respectively, are primarily U-rich (11, 14). U-rich sequences are also implicated in cAMP regulation of the LDH and the Na+/Glucose cotransporter (SGLTI) mRNAs (7, 16). In contrast, cAMP-induced regulation of LH/hCG receptor and of PEPCK transcript stability appears to be mediated by a C-rich sequence or by alternating purine:pyrimidine bases, respectively (8, 34). It is evident from the mutation studies that the A-rich portion of the PAI-1 CRS is required both for cA regulation of message stability and for binding to cytoplasmic proteins.

ISOLATION AND cDNA CLONING OF A NOVEL PAI-1 RNA-BINDING PROTEIN

To define the role of PAI- 1 mRNA-binding proteins in cyclic nucleotide regulation of mRNA stability, it is necessary to purify and characterize the specific binding proteins. Fig. 8 shows diagrammatically the protocol we have used successfully to isolate one such binding protein (35).

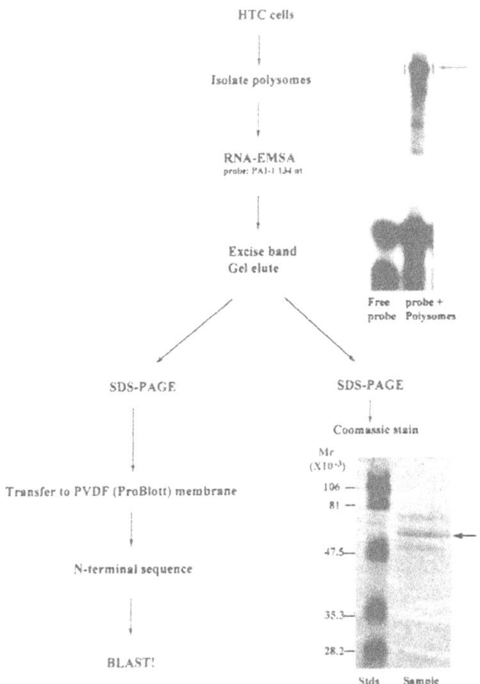

Figure 8. Isolation of PAI-1 mRNA binding protein. *Polysomal proteins were incubated with ^{32}P-labeled (8 reactions) or unlabeled (24 reactions) PAI-1 CRS and separated on native PAGE gels (RNA-EMSA). Gel slices were taken from the location corresponding to the more slowly migrating complex (as indicated by the bracket) and proteins were eluted in Tris/NaCl/SDS buffer (36). Eluted proteins were pooled, concentrated by acetone precipitation and separated on SDS-PAGE. Gels were then either stained with Coomassie Blue or transferred to a PVDF membrane. N-terminal sequencing was performed on the major band (as indicated by the arrow) at The University of Michigan Biomedical Research Core Facilities. The 19 amino acids N-terminal sequence was submitted to a BLAST search of protein databases.*

HTC cell polysomes were incubated with PAI-1 CRS and separated on non-denaturing polyacrylamide gels. To locate the high molecular weight complex in the gel, ^{32}P-labeled (10^6 dpm/reaction) probe was used in every 4th lane. The other reactions used unlabeled RNA at the same concentration. The location of the complex was determined by exposure of the gel to X-ray film for 1 hour at room temperature. High molecular weight complexes from 32 reactions (2mg total polysomal protein) were excised and the proteins were eluted from the gel by an overnight incubation with 50mM Tris, pH 7.9, 0.1 mM EDTA, 5mM DTT, 150mM NaCl and 0.1% SDS (36). The eluates were then pooled, concentrated by acetone precipitation and analyzed by 12% SDS-PAGE. The Coomassie Blue stained gel (Fig. 8) shows three bands of protein with an apparent molecular weight between 50 and 60kDa. In an identical experiment the proteins were separated by SDS-PAGE and electrophoretically transferred to a PVDF membrane. The membrane was stained with Coomassie Blue to locate the proteins and was submitted to the Protein and Carbohydrate Facility of the University of Michigan Biomedical Research Core Facilities, where N-terminal amino acid sequencing was carried out. We obtained 19 amino acids of N-terminal sequence from the major band indicated by an arrow in Fig. 8.

The N-terminal sequence was submitted to a BLAST search of the non-redundant protein database and two sequences with 18 identical N-terminal amino acids were found. One of the sequences, CGI-55 (GenBank™ accession number AF151813), is a composite of several ESTs. The other, "hypothetical protein" (GenBank™ accession number AL080119) is a cDNA cloned by the German Cancer Research Center (DKFZ) and was kindly provided to us by the Resource Center and Primary Database of the German Genome Project (RZPD, Berlin). The cDNA sequence indicates a putative protein of 42,400 daltons, but the protein had not been isolated nor expressed and its function was unknown. We were encouraged that the clone might encode an RNA binding protein because the amino acid sequence includes an RGG box as well as R-rich and RG-rich regions (see Fig. 12), motifs frequently found in RNA binding proteins.

Expression of the RNA-Binding Protein in *E.Coli*

To confirm that the isolated protein is a PAI-1 RNA binding protein, it was necessary to isolate sufficient amounts of protein to do binding studies. The most rapid and convenient method to accomplish this is by expression in bacteria. The coding region of the "hypothetical protein" was amplified by PCR and subcloned into the *Nde*I/*Bam*HI site in the pET-15b vector, downstream from the sequence coding for a histidine tag and a thrombin cleavage site. The construct was confirmed by PCR and the open reading frame was verified by DNA sequencing. The construct was transformed into *E.coli* BL21(DE3)pLysS, expression was induced with IPTG (isopropyl β-D-thiogalactopyanoside), and the recombinant protein was partially purified using a Ni^{++} chelation column. Fig. 9 shows a Coomassie Blue stained SDS-PAGE gel that shows the major protein as a triplet. That all three bands represent the "hypothetical protein" was confirmed by thrombin digestion, which cleaves the His-tag and approximately 2000 daltons of sequence (lane 4). The identity of the band was also confirmed by Western analysis using anti-His antibody (Fig. 9B), and by Northwestern analysis using ^{32}P-labeled PAI-1 CRS (Fig. 9C).

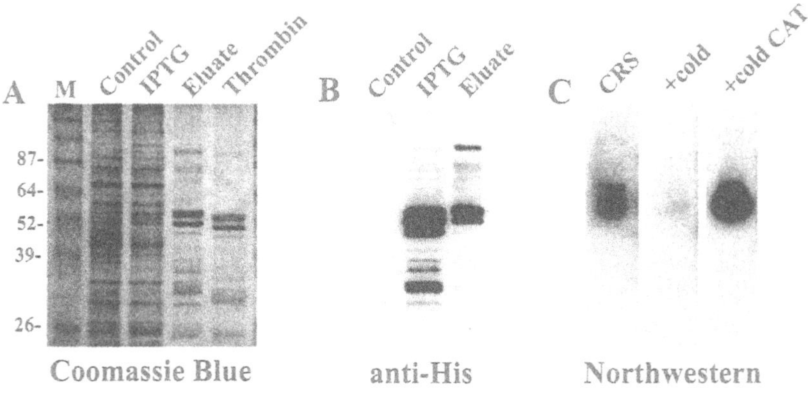

Figure 9. Analysis of partially purified PAI-1 mRNA binding protein. *The cDNA for hypothetical protein (DKFZp564M2425), subcloned in the vector pET-15b, was introduced into E.coli BL21(DE3)pLysS. Expression from the vector was induced with IPTG and the recombinant protein was purified by Ni^{++} chelation chromatography. A. Proteins were separated by 0.1% SDS-10% PAGE and the gel stained with Coomassie Blue. Lane M, prestained protein markers; Control, untreated bacterial extract (10 μg); IPTG, IPTG-induced bacterial extract (10 μg); Eluate, Ni^{++} column eluate (4 μg); Thrombin, column eluate (4 μg) was incubated for 3h at 22°C with thrombin (0.01 unit/μg protein, 0.04 units total). B. Following separation by SDS-PAGE, proteins were transferred to a nitrocellulose membrane and immunoblotting was carried out using monoclonal anti-His antibody. C. Ni^{++} column eluate (5μg/lane) was subjected to SDS-PAGE, transferred to a nitrocellulose membrane and Northwestern analysis carried out using ^{32}P-labeled PAI-1 CRS (CRS), labeled CRS plus unlabeled CRS (+cold), labeled CRS plus unlabeled CAT RNA (+cold CAT).*

RNA Binding Activity of the Recombinant Protein

The partially purified protein was incubated with ^{32}P-labeled PAI-1 CRS and binding experiments were carried out. Both in RNA-EMSA and in UV-crosslinking experiments (Fig. 10) the recombinant protein binds in a concentration dependent fashion (lanes 1-5). Binding is competed by increasing amounts of unlabeled identical RNA (CRS, lanes 6-8) but is not competed by a 100-fold molar excess of bacterial CAT RNA (lane 11). The unlabeled U-rich portion of the CRS (nt 2926/3024), which by itself does not form the major complex with HTC cell proteins, competes somewhat but much less than the full length CRS (lane 9). More significantly, a portion of the PAI-1 mRNA that we have shown by functional studies does not confer cA regulation of message decay (31), also does not compete for binding to the recombinant protein (nt 2125-2296, lane 10). The single RNA-protein complex seen in UV-crosslinking experiments migrates at a position consistent with a molecular weight of about 50kDa, similar in size to the major complex seen when HTC cell extracts are used as a source of protein. Results of these binding experiments clearly demonstrate that the protein we have isolated from the RNA-EMSA complex is, in fact, a PAI-1 RNA binding protein; we now refer to this protein as PAI-RBP1.

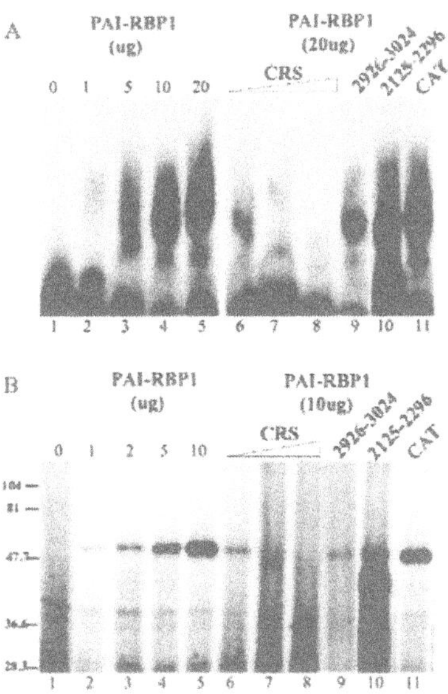

Figure 10. RNA-EMSA and UV cross-linking analyses of PAI-RBP1. *Recombinant human PAI-RBP1, expressed in E. coli and partially purified by* Ni^{++} *chelation column chromatography, was incubated at the concentrations indicated (lanes 1-5) with* ^{32}P*-labeled PAI-1 CRS and RNA-EMSA (panel A) or UV cross-linking (panel B) were carried out. To examine specificity of binding, PAI-RBP1 was incubated with unlabeled competitor RNA for 10 minutes prior to addition of the labeled probe. Unlabeled PAI-1 CRS was used at 10-, 50- or 100-fold molar excess (lanes 6-8). The U-rich portion of the PAI-1 CRS (nt 2926-3034, lane 9), a portion of the PAI-1 3'UTR that does not confer cAMP responsiveness (nt 2125-2296, lane 10) and bacterial CAT RNA (lane 11) were each added at 100-fold molar excess.*

To further compare the binding specificity of the PAI-RBP1 with that of the major HTC cell binding protein, we have done experiments using portions of the PAI-1 CRS as probe. As demonstrated by the experiment shown in Fig. 11, the U-rich region of the CRS does not bind the isolated protein either in RNA-EMSA (A, lanes 3 and 4) or UV-crosslinking (B, lanes 3 and 4). Mutation of the A-rich portion in the context of the full length CRS results in a transcript that binds weakly (mutation I, lanes 5 and 6) or not at all (mutation II, lanes 7 and 8). In UV-crosslinking either mutation abolishes binding (B, lanes 5-8). Neither labeled CAT RNA nor the labeled PAI-1 nt 2125-2296 forms complex with the PAI-RBP1. Thus, the nucleotide sequence requirement for binding to the newly isolated PAI-RBP1 is like that of the major PAI-1 RNA binding protein in HTC cells. PAI-RBP1 may, in fact, be the human homologue of that rat protein.

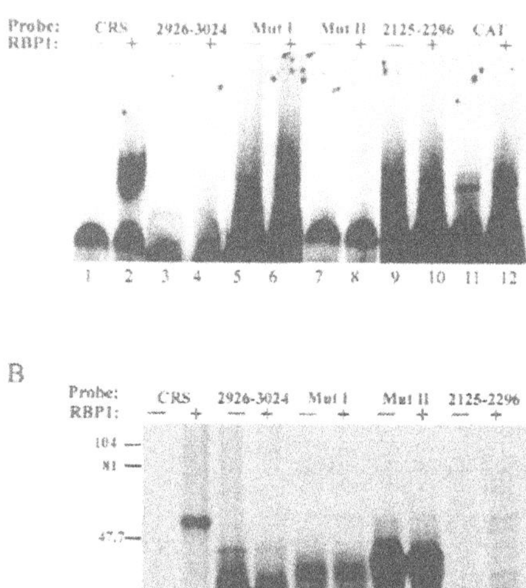

Figure 11. Analysis of RNA sequence requirement for PAI-RBP1 binding. *The ^{32}P-labeled RNA sequence indicated above each lane and shown schematically in Fig. 6 was incubated with buffer only (-) or with 7 μg of recombinant PAI-RBP1 (+) and RNA-protein interactions were examined by RNA-EMSA (panel A) or UV cross-linking (panel B).*

As noted above, partially purified RBP1 migrates as three bands in SDS-PAGE gel. This is probably the result of differences in codon usage between bacteria and higher organisms causing premature termination and thus C-terminal truncation. One of the arginines and 4 of the glycines in the C-terminal RGG box are coded by rare codons in bacteria and could cause premature termination (37).

Characterization of PAI-1 RNA Binding Protein (PAI-RBP1)

Figure 12 shows the 387 amino acid sequence of the human PAI-RBP1. It has an RGG box near its C-terminal (amino acid position #343-360) and R-rich (amino acid position #126-137) and RG-rich (amino acid position

#163-180) stretches in the N-terminal half of the protein. These are sequences commonly found in RNA binding proteins. It does not, however, have other recognizable RNA binding motifs such as KH-domains or RRMs. Submitting the sequence to a prosite scan (www.expasy.ch/tools/) reveals several potential phosphorylation sites including one protein kinase A site at serine 74. Thus, it is reasonable to suggest that PAI-RBP1 activity might be posttranslationally regulated by a cAMP-mediated event.

PAI-1 RNA-Binding Protein

1 MPGHLQEGFG CVVTNRFDQL FDDESDPFEV LKAAENKKKE AGGGGVGGPG

*

51 AKSAAQAAAQ TNSNAAGKQL RKESQKDRKN PLPPSVGVVD KKEETQPPVA

101 LKKEGIRRVG RRPDQQLQGE GKIID**RRPER RPPRERR**FEK PLEEKGEGGE

151 FSVDRPIID**R PIRGRGGLGR GRGGRGRGMG RG**DGFDSRGK REFDRHSGSD

201 RSGLKHEDKR GGSGSHNWGT VKDELTDLDQ SNVTEETPEG EEHHPVADTE

^ 6aa ^ 15aa

251 NKENEVEEVK EEGPKEMTLD EWKAIQNKDR AKVEFNIRKP NEGADGQWKK

301 GFVLHKSKSE EAHAEDSVMD HHFRKPANDI TSQLEINFGD L**GRPGRGGRG**

351 **GRGGRGRGGR** PNRGSRTDKS SASAPDVDDP EAFPALA

Figure 12. Amino acid sequence of PAI-RBP1. *The amino acid sequence of "hypothetical protein" was taken from GenBank™ locus CAB45718. The RG- and R-rich regions are shown in bold and the RGG box is shown as bold and underlined. The asterisk (*) indicates the potential protein kinase A phosphorylation site. The positions of the 6 and 15 amino acid inserts found in variants are indicated.*

The nucleotide sequence of CGI-55, which is a composite of ESTs, indicates a potential protein having a 6 amino acid insertion at position #203. In addition, an incomplete sequence from a rat homologue cloned from intestinal epithelium (clone 426 acc# U21718) indicates a potential protein with a 15 amino acid insertion at position #226 of the PAI-RBP1. We have used primers complementary to the human PAI-RBP1 to PCR amplify the rat homologue from an HTC cell cDNA library and have obtained clones with the 6 amino acid insertion (i.e. identical to CGI-55) and clones with both the 6 and 15 amino acid insertions. Our preliminary studies indicate that both of these variants are able to bind PAI-1 CRS; we have as yet no data on binding affinity. Thus, there are potentially 4 variants of PAI-RBP1. It will be interesting to learn whether these variants display different binding activities and/or different functions.

PAI-RBP1 Defines a Body of Proteins

The initial database search to identify the PAI-1 RNA binding protein was done with the 19 amino acids of N-terminal sequence. Having obtained the full amino acid and nucleotide sequence of a PAI-RBP1, we were able to carry out a more extensive search of protein and nucleic acid databases. This search revealed several proteins with a high degree of sequence similarity in the C-terminal half of the proteins. A computer generated alignment of these C-terminal regions shows that these proteins share several blocks of conserved sequence separated by regions that are less conserved (35). Representatives of this new family of proteins are found in a variety of multicellular organisms, including plants (spinach and Arabidopsis), insects (Drosophilia), birds (chicken) and mammals (rats, humans). A search of the human EST database indicates that the gene is expressed in many tissues including liver, pancreas, muscle, and brain. Thus, the PAI-RBP1 has defined a family of proteins with wide species and tissue distribution.

The RGG box in PAI-RBP1 is located in a gap between two conserved blocks. Most of the other proteins in this family have a similar RGG box or RG-rich stretch in the same region, but because these represent low complexity sequence, they do not appear in a computer-generated alignment. Because "hypothetical protein"/PAI-RBP1 is an RNA binding protein and because two other proteins in the family are annotated as RNA binding proteins, it is possible that PAI-RBP1 may define a family of proteins with a novel RNA binding motif. The precise RNA-binding motif, however, remains to be defined.

FUTURE DIRECTIONS

Our current studies are aimed at determining the possible role of this novel RNA binding protein in regulation of mRNA stability. The protein was isolated based on its ability to bind to the critical cyclic nucleotide responsive sequence in the 3'UTR of the PAI-1 mRNA. As determined by its amino acid sequence and by EMSA and UV-crosslinking studies, the recombinant protein is clearly a specific RNA binding protein.

There are several possible ways that RBP1 may alter PAI-1 mRNA degradation. Studies with HuR and AUF1 have demonstrated that the binding of the protein to a transcript may result in either stabilization or destabilization (19, 20, 38). If binding of RBP1 to the PAI-1 3'UTR were to stabilize the mRNA, we might expect that incubation with cyclic nucleotides would decrease its cellular concentration. Conversely, if binding resulted in destabilization, we might expect that cyclic nucleotides would increase the concentration. Alternatively, cyclic nucleotides could change the affinity of the protein for the RNA, perhaps by inducing a posttranslational modification, such as phosphorylation, resulting in allosteric modifications. It is also possible that it is not the binding of RBP1, per se, that affects the degradation rate, but rather the activity of the bound protein. In this case, the role of cAMP may be to affect the phosphorylation state of the bound protein and thereby alter its function. By analogy, phosphorylation of the cAMP response element-binding protein (CREB) by protein kinase A enhances transactivation of cAMP-responsive promoters, but does not affect binding of CREB to the cAMP response element (CRE) (39). As noted previously, PAI-RBP1 does have a potential protein kinase A phosphorylation site at serine 74. Finally, PAI-RBP1 may not directly alter message stability, but may act as a docking protein for other regulatory factors. These other factors, which may not bind directly to the PAI-1 CRS and would not be recognized by UV-crosslinking, might be induced or post-translationally modified by cyclic nucleotides.

We will begin studies on the role of PAI-RBP1 by examining regulation of its expression by cyclic nucleotides. Complexes formed between HTC cell cytoplasmic proteins and the PAI-1 CRS do not appear to be regulated by cyclic nucleotides. Incubation with 8-bromo-cAMP for different times and at different concentrations failed to change either the mobility of complexes or the intensity of signal in either UV-crosslinking or RNA-EMSA experiments (32). However, the number of binding proteins found in crude cytoplasm could mask subtle changes in any one of them.

The possible role of RBP1 in posttranscriptional regulation can be addressed either by analyzing the effect of overexpression of the protein in HTC cells or by addition/depletion of the protein in an *in vitro* RNA decay system. Either approach has potential benefits as well as pitfalls. Probably the best method to analyze the effect of overexpression is the tetracycline-regulated gene expression system (38, 40). If RBP1 has a role in message degradation, simply overexpressing the protein may change endogenous PAI-1 mRNA degradation, making determination of the effect of cA difficult. This result would, nevertheless, help establish the biological role for the protein. While it may not be possible to detect the cA effect on message stability in cells expressing a very high level of RBP1, it may be possible at intermediate levels. Because the tetracycline-regulated gene expression system allows the level of expression to be regulated by the concentration of tetracycline used, it will be possible to compare the cyclic nucleotide effect at different cellular concentrations of RBP1.

A polysome-free *in vitro* RNA decay system may be a more facile system for these studies because it allows both manipulation of protein components by addition or immunoprecipitation and examination of decay rates of synthetic transcripts (41-43). We are currently using this system to compare decay rates of a synthetic ARE-containing sequence, which decays very rapidly, with that of the non-ARE containing globin sequence, which decays more slowly, and with that of the PAI-1 CRS. We plan to examine the effect of added recombinant PAI-RBP1 on degradation of each of these synthetic transcripts. Failure to detect an effect of the bacterially expressed protein could be due to the absence of posttranslational modifications. Therefore, if we do not detect an effect, it will be necessary to express RBP1 under conditions where posttranslational modification occurs, such as the baculovirus system in insect cells.

Analogous with transcriptional regulation, regulation of message stability may require multiple proteins, some that bind directly to the RNA and some that interact indirectly by binding to the RNA-binding protein (44). UV-crosslinking of the PAI-1 CRS with HTC proteins suggests that multiple proteins interact with this sequence. The protein we have isolated may be only the major binding protein or perhaps the protein with the highest affinity. Therefore, it is important to continue experiments aimed at isolating other PAI-1 RNA binding proteins as well as proteins associated with PAI-1 RNA binding proteins. Such associated proteins could be identified either biochemically by co-immunoprecipitation studies or genetically by the yeast two-hybrid system (45, 46).

To identify the RNA binding domain of PAI-RBP1, we have prepared several constructs that encode truncated versions of the protein and are analyzing their RNA binding properties. If these studies reveal a novel RNA binding motif, that information may assist in identification of other RNA binding proteins. Further studies will be done to delineate domains of the protein necessary for its putative role in posttranscriptional regulation.

In conclusion, we have isolated and cloned a novel RNA binding protein that displays specific binding to a cyclic nucleotide responsive sequence in the 3'UTR of PAI-1 mRNA. Because sequences that share a high degree of similarity in the C-terminal half are found in many different species, (including ones not known to express PAI-1), and because RBP1 is expressed in a wide range of tissues, we expect that the protein may have a broad biological role. That role may involve regulation of message stability or some other process that requires interaction with RNA. It is also possible that the C-terminal sequence is required for the broader role of the protein and that the N-terminal portion dictates a more specific function.

Summary of key concepts

- *Regulation of mRNA degradation is an important mechanism of regulation of gene expression. cAMP, whose role in gene transcription is well characterized, also regulates gene expression by altering mRNA degradation.*
- *Specific cis-acting sequences can mediate regulation of mRNA stability. cAMP destabilization of PAI-1 mRNA is mediated in part by a specific cyclic nucleotide response sequence (PAI-CRS) in the 3' UTR of the transcript. The PAI-CRS is unlike the sequence of other known RNA stabilizing or destabilizing elements.*
- *RNA binding proteins are important components in regulation of mRNA degradation and may act to stabilize or destabilize mRNA. PAI-1 RNA binding protein (PAI-RBP1) interacts with the PAI-CRS, but its role in regulation of stability is not yet known.*
- *The amino acid sequence of PAI-RBP1 includes an RGG box and two R-rich stretches, but no other recognizable RNA binding motifs.*
- *PAI-RBP1 shares blocks of sequence similarity with a number of proteins and may define a family of proteins with a novel RNA binding motif.*

Study Guide Questions

1) How would you establish that specific RNA sequences are necessary and sufficient to mediate regulated degradation of the mRNA?
2) What are some of the techniques by which you could identify and isolate mRNA binding proteins?
3) What are potential mechanisms by which proteins might bring about or regulate mRNA degradation?
4) How would you establish that a particular mRNA binding protein is involved in the regulation of degradation of its cognate mRNA?
5) How would you identify other RNA targets of an RNA binding protein?

Acknowledgements

Figures reproduced with the permission of the Journal of Biological Chemistry. The authors would like to acknowledge and thank Wendy Dlakic, Mensur Dlakic, Nancy Leff, Maribeth Tillmann-Bogush and Owen Wittekindt for their contributions to the work described here. These studies were supported by Public Health Service Grant CA22729 from the National Cancer Institute (to T.D.G.). We also acknowledge National Institutes of Health Grants 5 P60 DK-20572 to the University of Michigan Diabetes Research and Training Center and 2 P30 CA46592 to the University of Michigan Comprehensive Cancer Center for support of core services.

REFERENCES

1. Vassalli, J.D., A.P. Sappino, and D. Belin. 1991. The plasminogen activator/plasmin system. *J Clin Invest.* 88:1067-72.
2. Vassalli, J.D., and M.S. Pepper. 1994. Tumour biology. Membrane proteases in focus. *Nature.* 370:14-5.
3. Heaton, J.H., and T.D. Gelehrter. 1989. Glucocorticoid induction of plasminogen activator and plasminogen activator-inhibitor messenger RNA in rat hepatoma cells. *Mol Endocrinol.* 3:349-55.
4. Andreasen, P.A., B. Georg, L.R. Lund, A. Riccio, and S.N. Stacey. 1990. Plasminogen activator inhibitors: hormonally regulated serpins. *Mol Cell Endocrinol.* 68:1-19.
5. Heaton, J.H., and T.D. Gelehrter. 1990. Cyclic nucleotide regulation of plasminogen activator and plasminogen activator-inhibitor messenger RNAs in rat hepatoma cells. *Mol Endocrinol.* 4:171-8.
6. Richards, J.S. 2001. New Signaling Pathways for Hormones and Cyclic Adenosine 3',5'- Monophosphate Action in Endocrine Cells. *Mol Endocrinol.* 15:209-218.
7. Tian, D., D. Huang, R.C. Brown, and R.A. Jungmann. 1998. Protein kinase A stimulates binding of multiple proteins to a U-rich domain in the 3'-untranslated region of lactate dehydrogenase A mRNA that is required for the regulation of mRNA stability. *J Biol Chem.* 273:28454-60.
8. Nachaliel, N., D. Jain, and Y. Hod. 1993. A cAMP-regulated RNA-binding protein that interacts with phosphoenolpyruvate carboxykinase (GTP) mRNA. *J Biol Chem.* 268:24203-9.
9. Themmen, A.P., L.J. Blok, M. Post, W.M. Baarends, J.W. Hoogerbrugge, M. Parmentier, G. Vassart, and J.A. Grootegoed. 1991. Follitropin receptor down-regulation involves a cAMP-dependent post-transcriptional decrease of receptor mRNA expression. *Mol Cell Endocrinol.* 78:R7-13.
10. Tholanikunnel, B.G., and C.C. Malbon. 1997. A 20-nucleotide (A + U)-rich element of b_2-adrenergic receptor (b_2AR) mRNA mediates binding to b_2AR-binding protein and is obligate for agonist-induced destabilization of receptor mRNA. *J Biol Chem.* 272:11471-8.
11. Tholanikunnel, B.G., J.R. Raymond, and C.C. Malbon. 1999. Analysis of the AU-rich elements in the 3'-untranslated region of b_2-adrenergic receptor mRNA by mutagenesis and identification of the homologous AU-rich region from different species. *Biochemistry.* 38:15564-72.
12. Wang, X., G. Nickenig, and T.J. Murphy. 1997. The vascular smooth muscle type I angiotensin II receptor mRNA is destabilized by cyclic AMP-elevating agents. *Mol Pharmacol.* 52:781-7.
13. Lu, D.L., H. Peegel, S.M. Mosier, and K.M. Menon. 1993. Loss of lutropin/human choriogonadotropin receptor messenger ribonucleic acid during ligand-induced down-regulation occurs post transcriptionally. *Endocrinology.* 132:235-40.
14. Danner, S., M. Frank, and M.J. Lohse. 1998. Agonist regulation of human b_2-adrenergic receptor mRNA stability occurs via a specific AU-rich element. *J Biol Chem.* 273:3223-9.
15. Huang, D., C.J. Hubbard, and R.A. Jungmann. 1995. Lactate dehydrogenase A subunit messenger RNA stability is synergistically regulated via the protein kinase A and C signal transduction pathways. *Mol Endocrinol.* 9:994-1004.
16. Lee, W.Y., P. Loflin, C.J. Clancey, H. Peng, and J.E. Lever. 2000. Cyclic nucleotide regulation of Na+/glucose cotransporter (SGLT1) mRNA stability. Interaction of a nucleocytoplasmic protein with a regulatory domain in the 3'-untranslated region critical for stabilization. *J Biol Chem.* 275:33998-34008.
17. Kash, J.C., and K.M. Menon. 1999. Sequence-specific binding of a hormonally regulated mRNA binding protein to cytidine-rich sequences in the lutropin receptor open reading frame. *Biochemistry.* 38:16889-16897.

18. Blaxall, B.C., A.C. Pellett, S.C. Wu, A. Pende, and J.D. Port. 2000. Purification and characterization of beta-adrenergic receptor mRNA-binding proteins. *J Biol Chem.* 275:4290-7.
19. Fan, X.C., and J.A. Steitz. 1998. Overexpression of HuR, a nuclear-cytoplasmic shuttling protein, increases the *in vivo* stability of ARE-containing mRNAs. *EMBO J.* 17:3448-60.
20. Wang, W., H. Furneaux, H. Cheng, M.C. Caldwell, D. Hutter, Y. Liu, N. Holbrook, and M. Gorospe. 2000. HuR regulates p21 mRNA stabilization by UV light. *Mol Cell Biol.* 20:760-9.
21. Pinol-Roma, S., and G. Dreyfuss. 1992. Shuttling of pre-mRNA binding proteins between nucleus and cytoplasm. *Nature.* 355:730-2.
22. Nargolwalla, C., D. McCabe, and I.B. Fritz. 1990. Modulation of levels of messenger RNA for tissue-type plasminogen activator in rat Sertoli cells, and levels of messenger RNA for plasminogen activator inhibitor in testis peritubular cells. *Mol Cell Endocrinol.* 70:73-80.
23. Tranque, P., R. Robbins, F. Naftolin, and P. Andrade-Gordon. 1992. Regulation of plasminogen activators and type-1 plasminogen activator inhibitor by cyclic AMP and phorbol ester in rat astrocytes. *GLIA.* 6:163-71.
24. Fukumoto, S., F.H. Allan, J.A. Yee, T.D. Gelehrter, and T.J. Martin. 1992. Plasminogen activator regulation in osteoblasts: parathyroid hormone inhibition of type-1 plasminogen activator inhibitor and its mRNA. *J Cell Physiol.* 152:346-55.
25. Thalacker, F.W., and M. Nilsen-Hamilton. 1992. Opposite and independent actions of cyclic AMP and transforming growth factor b in the regulation of type 1 plasminogen activator inhibitor expression. *Biochem J.* 287:855-62.
26. Georg, B., A. Riccio, and P. Andreasen. 1990. Forskolin down-regulates type-1 plasminogen activator inhibitor and tissue-type plasminogen activator and their mRNAs in human fibrosarcoma cells. *Mol Cell Endocrinol.* 72:103-10.
27. Konkle, B.A., P.R. Kollros, and M.D. Kelly. 1990. Heparin-binding growth factor-1 modulation of plasminogen activator inhibitor-1 expression. Interaction with cAMP and protein kinase C-mediated pathways. *J Biol Chem.* 265:21867-73.
28. DiBattista, J.A., J. Martel-Pelletier, N. Morin, F.C. Jolicoeur, and J.P. Pelletier. 1994. Transcriptional regulation of plasminogen activator inhibitor-1 expression in human synovial fibroblasts by prostaglandin E2: mediation by protein kinase A and role of interleukin-1. *Mol Cell Endocrinol.* 103:139-48.
29. Heaton, J.H., S. Kathju, and T.D. Gelehrter. 1992. Transcriptional and posttranscriptional regulation of type 1 plasminogen activator inhibitor and tissue-type plasminogen activator gene expression in HTC rat hepatoma cells by glucocorticoids and cyclic nucleotides. *Mol Endocrinol.* 6:53-60.
30. Kathju, S. 1994. Regulation of tissue type plasminogen activator gene expression by glucocorticoids and cyclic nucleotides in rat hepatoma cells. Ph.D. Thesis. *In* Department of Human Genetics. University of Michigan, Ann Arbor.
31. Heaton, J.H., M. Tillmann-Bogush, N.S. Leff, and T.D. Gelehrter. 1998. Cyclic nucleotide regulation of type-1 plasminogen activator-inhibitor mRNA stability in rat hepatoma cells. Identification of *cis*-acting sequences. *J Biol Chem.* 273:14261-8.
32. Tillmann-Bogush, M., J.H. Heaton, and T.D. Gelehrter. 1999. Cyclic nucleotide regulation of PAI-1 mRNA stability. Identification of cytosolic proteins that interact with an A-rich sequence. *J Biol Chem.* 274:1172-9.
33. Chen, C.Y., and A.B. Shyu. 1995. AU-rich elements: characterization and importance in mRNA degradation. *Trends Biochem Sci.* 20:465-470.
34. Kash, J.C., and K.M. Menon. 1998. Identification of a hormonally regulated luteinizing hormone/human chorionic gonadotropin receptor mRNA binding protein. Increased mRNA binding during receptor down-regulation. *J Biol Chem.* 273:10658-10664.
35. Heaton, J.H., W.M. Dlakic, M. Dlakic, and T.D. Gelehrter. 2001. Identification and cDNA cloning of a novel RNA-binding protein that interacts with the cyclic nucleotide

responsive sequence in the type-1 plasminogen activator inhibitor mRNA. *J Biol Chem.* 276:3341-3347.

36. Ranish, J.A., and S. Hahn. 1991. The yeast general transcription factor TFIIA is composed of two polypeptide subunits. *J Biol Chem.* 266:19320-7.
37. Ausubel, F.M., R. Brent, R.E. Kingston, D.D. Moore, J.G. Seidman, J.A. Smith, and K. Struhl. 1996. Current Protocols in Molecular Biology. Greene Publishing Associates and Wiley-Interscience, New York.
38. Loflin, P., C.Y. Chen, and A.B. Shyu. 1999. Unraveling a cytoplasmic role for hnRNP D in the in vivo mRNA destabilization directed by the AU-rich element. *Genes Dev.* 13:1884-97.
39. Montminy, M. 1997. Transcriptional regulation by cyclic AMP. *Annu Rev Biochem.* 66:807-22.
40. Gossen, M., and H. Bujard. 1992. Tight control of gene expression in mammalian cells by tetracycline-responsive promoters. *Proc Natl Acad Sci USA.* 89:5547-51.
41. Wang, Z., and M. Kiledjian. 2000. The poly(A)-binding protein and an mRNA stability protein jointly regulate an endoribonuclease activity. *Mol Cell Biol.* 20:6334-41.
42. Ford, L.P., J. Watson, J.D. Keene, and J. Wilusz. 1999. ELAV proteins stabilize deadenylated intermediates in a novel *in vitro* mRNA deadenylation/degradation system. *Genes Dev.* 13:188-201.
43. Brewer, G. 1998. Characterization of c-*myc* 3' to 5' mRNA decay activities in an *in vitro* system. *J Biol Chem.* 273:34770-4.
44. Albright, S.R., and R. Tjian. 2000. TAFs revisited: more data reveal new twists and confirm old ideas. *Gene.* 242:1-13.
45. Ceman, S., V. Brown, and S.T. Warren. 1999. Isolation of an FMRP-associated messenger ribonucleoprotein particle and identification of nucleolin and the fragile X-related proteins as components of the complex. *Mol Cell Biol.* 19:7925-32.
46. Wu, X.Q., S. Lefrancois, C.R. Morales, and N.B. Hecht. 1999. Protein-protein interactions between the testis brain RNA-binding protein and the transitional endoplasmic reticulum ATPase, a cytoskeletal g actin and Trax in male germ cells and the brain. *Biochemistry.* 38:11261-70.

9

POST-TRANSCRIPTIONAL CONTROL OF THE GAP-43 mRNA BY THE ELAV-LIKE PROTEIN HuD

Nora Perrone-Bizzozero and Rebecca Keller
University of New Mexico School of Medicine, Albuquerque, NM

Post-transcriptional control by mRNA-binding proteins is critical for shaping the temporal and spatial pattern of expression of a large number of developmentally regulated genes. Among these is the gene for GAP-43, a growth-associated protein expressed in neurons primarily during the initial establishment and remodeling of neural connections. Both transcriptional and post-transcriptional mechanisms control GAP-43 gene expression during development. While promoter activity determines the neural-specific expression of the gene, changes in mRNA stability modulate GAP-43 expression in neurons undergoing process outgrowth in response to growth factors and other signaling agents. For example, in PC12 cells induced to differentiate by nerve growth factor (NGF), GAP-43 mRNA levels are regulated primarily through selective changes in the rate of degradation of the mRNA. This process depends on the activation of protein kinase C (PKC) and is mediated by the interaction of highly conserved sequences in the 3' untranslated region (3'UTR) of the mRNA with neuronal-specific RNA-binding proteins. One of these proteins was recently identified as the ELAV-like protein HuD. ELAV is an RNA-binding protein that is critical for the development of the nervous system in Drosophila and HuD is one of four human homologs of this protein. This chapter discusses the evidence demonstrating a role for HuD in the control of GAP-43 mRNA stability, gene expression and neuronal differentiation, and presents recent findings on the molecular mechanisms underlying these effects.

BACKGROUND

The growth-associated protein GAP-43 (also called B-50, F1 and neuromodulin) is a phosphoprotein of the nerve terminal membrane that is expressed primarily during the initial establishment and remodeling of neural connections (1-3). Both transcriptional and post-transcriptional mechanisms contribute to the regulation of GAP-43 gene expression during nervous system development. Transcriptional factors of the bHLH family are known to control the neural-specific expression of the gene (4, 5). Yet, and most surprisingly, in several neuronal populations, the levels of gene transcription do not correlate well with the accumulation of the mature GAP-43 mRNA (6-9). In instances in which there is uncoupling between transcription and accumulation of the mRNA, GAP-43 gene expression was found to be controlled primarily by changes in the rate of turnover of the mRNA. As shown in the following sections, this process depends on the regulated interactions between *cis*-acting elements in the GAP-43 3' UTR and specific *trans*-acting factors.

Signals Controlling GAP-43 mRNA Stability

Rat pheochromocytoma PC12 cells respond to NGF by acquiring a neuronal-like phenotype. Multiple pathways are capable of signaling the expression of proteins required for PC12 cell differentiation. In the case of GAP-43, NGF induces the expression of this protein via PKC-dependent signaling (10). Phorbol esters such as TPA are potent activators of PKC and low levels of these agents are very effective at increasing GAP-43 mRNA levels and neurite outgrowth in PC12 cells. On the contrary, in the presence of PKC inhibitors or at high TPA concentrations that downregulate PKC, NGF is unable to induce GAP-43 expression and neuronal differentiation. Similar responses are observed in TPA-induced neuroblastoma cells, indicating that the actions of phorbol esters are independent of the neuronal cell type. Moreover, since these effects are not influenced by inhibition of protein synthesis, it is likely that PKC-dependent phosphorylation of pre-existing proteins is sufficient for controlling GAP-43 mRNA levels during neuronal differentiation.

Although phorbol esters influence transcription rates for a number of genes, the effect of these agents on GAP-43 expression appears to be solely at the post-transcriptional level. The rate of transcription of the GAP-43 gene in PC12 cells is not affected by either NGF (6, 7) or phorbol ester treatment. This observation is consistent with the absence of any typical TPA-responsive element (TRE) within the GAP-43 promoter (11, 12). Instead, phorbol esters induce GAP-43 expression in PC12 cells by selectively stabilizing the mRNA (10). As discussed below, the sequences required for the response to TPA are restricted to a small region in the 3'UTR of the mRNA that contains the recognition site for the RNA-binding protein, HuD.

Cis-Acting Determinants of GAP-43 mRNA Turnover

For several mRNAs of fast turnover, the presence of instability-conferring sequences within the 3'UTR greatly influences their rate of degradation (13, 14). One of such sequences is the AU-rich element (ARE) found in the 3'UTR of several proto-oncogene and cytokine mRNAs (15). Similarly, the main determinants of GAP-43 mRNA stability are localized in the 3'UTR of the mRNA (16, 17). These sequences are capable of destabilizing very stable messenger RNAs such as that for β-globin, indicating that they can act independently of the coding region to which they are attached. Moreover, these determinants are highly conserved in evolution (16). For example, the 3'UTR of GAP-43 mRNA shows 78% sequence identity between avian and rodent species. This level of conservation is similar to that of the coding region and is considerably higher than the 30% sequence conservation predicted to occur in the absence of selective pressure. Furthermore, portions of the GAP-43 3'UTR are conserved between more distantly related species, such as human and goldfish. Even though no typical AREs can be found in this region, there are several poly-pyrimidine stretches spanning the entire length of the 3'UTR (16). Among these sequences is a 200-nucleotide (nt) U-rich region that is sufficient for conferring TPA-responsiveness to the mRNA (see TPA-regulatory element in Figure 1A). Although there are sub-specializations within this element, the entire 200-nt region is required to display full responsiveness to TPA. Nucleotides in the 5' half of the element are destabilizing in nature and contain three GUUUG/C boxes (Fig. 1A). While these boxes are highly conserved, mutations of the central Us in all three of these motifs do not affect the half-life of the mRNA. Similar effects are seen for the GUUUG boxes in other mRNAs such as that for c-jun (18). The 3' half of the TPA-regulatory element contains the ARE-like sequence ACUUUCUCUCUAUUUCUCUCU. This sequence coincides with the recognition site for the RNA-binding protein HuD and is required for the HuD-dependent stabilization of the mRNA.

Trans-Acting Factors Controlling GAP-43 mRNA Stability

One likely reason for the conservation of sequences in the GAP-43 3'UTR is that they represent recognition sites for specific RNA-binding proteins. In fact, brain cytosolic proteins of 40, 65 and 95 kDa specifically interact with highly conserved sequences in the 3'UTR of the GAP-43 mRNA (16). Binding of these three proteins to the GAP-43 3'UTR is displaced by an excess of poly(U) RNA, while only the 65 kDa protein is displaced by high levels of poly(C) RNA. Proteins from either rat or chicken brain extracts bind efficiently to the GAP-43 3'UTR from either species and show the same sequence specificity and affinity for the RNA, demonstrating that both *cis-* and *trans*-acting factors of mRNA stability are highly conserved (16).

A

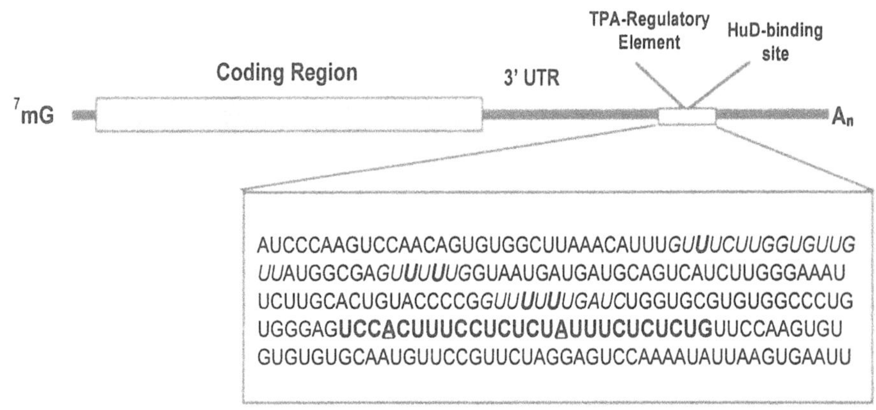

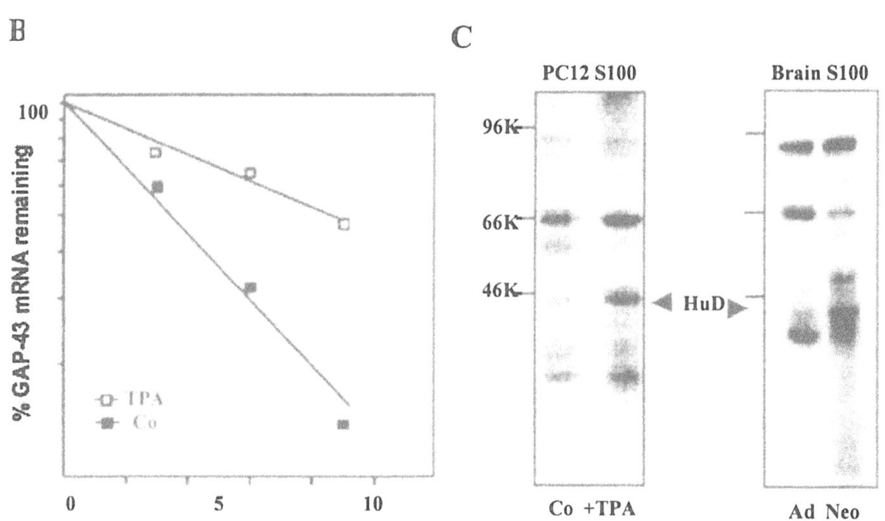

Figure 1. Sequences and proteins involved in the post-transcriptional regulation of the GAP-43 gene. A: *Diagram shows the localization of the TPA-regulatory element and the HuD-binding site in the GAP-43 mRNA. The HuD-binding site is shown in bold letters; ARE-like sequences within this region are underlined. GUUUG/C boxes are shown in Italics; bold letters indicate nucleotides whose mutation does not affect mRNA stability.* ***B:*** *TPA increases the half-life of the GAP-43 mRNA in PC12 cells.* ***C:*** *Binding of HuD to the GAP-43 mRNA is increased by TPA and during brain development. RNA binding assays used S100 extracts from control and TPA-treated PC12 cells or from adult (Ad) and neonate (Neo) rat brains.*

The properties of the 65 and 95 kDa GAP-43 mRNA-binding proteins suggest that they may be related to families of RNA-binding proteins described in other systems. These include other poly-pyrimidine-binding proteins identified because they interact with the regulatory element in the 3' UTR of the amyloid precursor protein (APP) mRNA (19, 20) or bind to a different region of the GAP-43 mRNA (21, 22). With regard to the 40 kDa protein, its neuronal specificity, apparent molecular weight, and ability to bind to U-rich sequences resembles that of the Hu family of nervous system-specific RNA-binding proteins. In fact, HuD binds to the GAP-43 mRNA with high affinity (K_d = 129 nM, 23) and, as shown in the following sections, is essential for controlling the stability of the mRNA both *in vitro* and *in vivo*.

As shown in Fig. 1, treatment of PC12 cells with phorbol esters not only increases the stability of the GAP-43 mRNA (Fig. 1B), but also enhances the binding of HuD to these sequences (Fig. 1 C). Since NGF does not change the expression of HuD in PC12 cells (24), these results suggest that the increased binding of HuD observed in the *in vitro* binding assays is due to an increased affinity of the protein for the mRNA. Given that HuD can be phosphorylated by PKC *in vitro*, it is likely that phosphorylation increases its affinity for the mRNA. Similarly, *in vitro* binding assays using adult and neonate brain extracts (Fig. 1C) show that the interaction of HuD with the GAP-43 mRNA is also developmentally regulated, with its highest affinity observed in the developing brain. Consistent with this finding, the neonatal rat brain contains 5-10 fold higher levels of GAP-43 mRNA than the adult brain and the mRNA itself is more stable than in mature neurons (Fig. 4C). Overall, there is an excellent correlation between the affinity of HuD for the GAP-43 3'UTR and the half-life of this mRNA in neurons differentiating both in tissue culture and *in vivo*.

FUNCTIONAL PROPERTIES OF HUD AND OTHER ELAV-LIKE PROTEINS

HuD is one of four mammalian ELAV/Hu proteins. As shown in the previous chapter, these proteins are homologous to the *Drosophila* protein ELAV (embryonic lethal abnormal visual system) and contain three RNA recognition motifs (RRM). In mammals, birds and *Xenopus*, three of the ELAV-like proteins, Hel-N1/HuB, HuC, and HuD, are developmentally regulated and neuronal specific. The fourth member, HuR (also called HuA), is also expressed in other tissues. The HuD mRNA is alternatively spliced and encodes three proteins, HuD, HuDpro, and HuDmex of molecular weight 35-40 kDa. Of these, HuDmex is the least abundant and is the only form that has an intact nuclear shuttling sequence (25). HuD is the major variant and has been shown to bind *in vitro* to sequences in the 3'UTRs of several mRNAs including c-fos (26), tau (27) p21 ^{waf1} (28), N-myc (29), and GAP-43 (23). Although there is no apparent consensus motif for the interaction of this protein, all the recognition sequences have in common a high U content. The first two RRMs of the HuD protein are necessary for this interaction (23) while the third RRM of HuD appears to bind very long poly(A) tracks with high specificity (30). It has also been shown that HuD is enriched in polysomes, the precise site where the mRNA degradation machinery is localized (14).

Role of ELAV-Like Proteins in Nervous System Development

ELAV was first described in *Drosophila melanogaster* where the gene is required for the normal development and maintenance of the nervous system. Deletion of the *elav* gene is embryonic lethal due to the failure of neurons to differentiate (31, 32). Temperature-sensitive mutations result in abnormal nervous system development, readily apparent in aberrant *Drosophila* eye formation. Mutant flies are incapacitated at non-permissive temperatures demonstrating that the continued expression of ELAV is essential for the function of mature neurons (33). The ELAV protein is primarily localized to the nucleus where it is known to participate in splicing mechanisms (34). Also, ELAV is capable of autoregulating its own expression by binding to the 3'UTR of its mRNA (35). Although additional targets of ELAV remain to be determined, it is apparent that this protein controls the expression of several genes that are essential for neuronal development and functioning.

In higher vertebrates and mammals, ELAV/Hu proteins constitute one of the earliest markers of neuronal differentiation (36, 37). In chicken embryos, HuR is the first protein to be expressed while neural precursors are still proliferating. The expression of HuD coincides with the earliest stages of neuronal differentiation and is maintained throughout the maturation of neurons. HuC is expressed at the end of differentiation and maintained during neuronal maturation (37). Similar types of expression patterns have been observed in the developing mouse (38) and rat (39) nervous systems and in the human brain (40). While all four ELAV-like proteins are important for neuronal differentiation, it is becoming apparent that their functional properties may be different. These proteins show a distinct developmental pattern of expression and map to different neuronal populations in the mature central (CNS) and peripheral nervous system (PNS)(38-40). Furthermore, they are localized to different subcellular compartments where they control the expression of different mRNAs (27, 41).

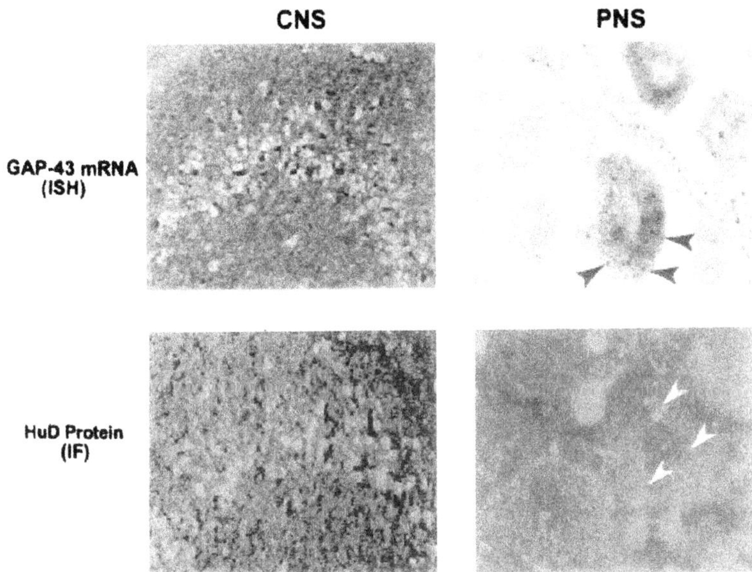

Figure 2. The HuD protein and the GAP-43 mRNA are localized in the same neurons in the central (CNS) and peripheral nervous systems (PNS)*. In situ hybridization (ISF) and immunofluorescence (IF) assays revealed the co-localization of HuD and the GAP-43 mRNA in the CA3 region of the hippocampal formation (CNS) and in dorsal root ganglion neurons (PNS). Arrows show that both molecules are localized to granules distributed throughout the cytoplasm but not in the nucleus.*

As shown in Fig. 2, the localization of HuD in the CNS and PNS coincides with that of the GAP-43 mRNA, further supporting the notion that HuD is involved in the control of this gene *in vivo*. Also, the subcellular localization of these molecules seems to be the same, as both HuD and the GAP-43 mRNA are localized to distinct granules in the cytoplasm of neurons (Fig 2, right panels). In agreement with these findings, other ELAV/Hu proteins are known to localize to granular structures, which presumably mark the location of polysomal aggregates in the cell (41).

Effect of HuD on GAP-43 Gene Expression and Neurite Outgrowth

Not only are the neuronal ELAV/Hu proteins expressed during development but also recent studies indicate that overexpression of these proteins is sufficient to induce neuronal differentiation both *in vitro* and *in vivo* (24,42-45). Overexpression of HuD increases neurite outgrowth in PC12 cells even in the absence of NGF (43). Similarly, as shown in Fig. 3, increased expression of HuD protein in primary cortical neurons *in vitro* leads to increased levels of the GAP-43 mRNA and protein and accelerated neurite outgrowth (46).

In contrast to neurons overexpressing HuD (Fig. 3), cells expressing HuD antisense RNA (24) or cells in which binding of HuD to the GAP-43 mRNA is competed out by excess GAP-43 3'UTR sequences (47) show a significant decrease in GAP-43 gene expression. Furthermore, in the absence of HuD, PC12 cells fail to differentiate in response to either NGF or phorbol esters, demonstrating that HuD is essential for PKC-dependent differentiation of these cells (24).

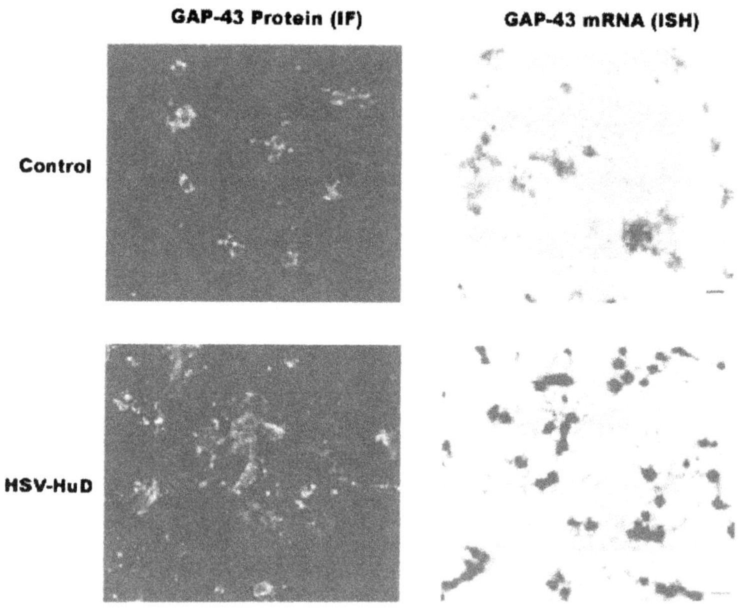

Figure 3. Overexpression of HuD increases GAP-43 mRNA, GAP-43 protein levels and neurite outgrowth in embryonic cortical cells in vitro. *E19 rat cortical neurons were cultured in the presence of control HSV-1 vector or vector containing HuD sequences (HSV-HuD) as described (47). Cells were fixed and processed for GAP-43 immunofluorescence (IF) or in situ hybridization (ISH). Panels show that overexpression of HuD increases GAP-43 mRNA levels in the cells leading to an increased expression of GAP-43 protein and accelerated neurite outgrowth. Scale bar represents 20 μm.*

Interestingly, a truncated form of the protein, HuD I+II, lacking its third RRM does not promote neuronal differentiation and is capable of blocking GAP-43 expression and neurite outgrowth even in the presence NGF (43). Furthermore, the truncated protein behaves in a dominant negative manner to disrupt the function of the endogenous HuD protein. The properties of the truncated HuD are similar to those of known C-terminal mutations of

Drosophila ELAV (33) and suggest that the third RRM of ELAV/Hu proteins is essential for their function.

ROLE OF HUD IN THE CONTROL OF GAP-43 mRNA DECAY: *IN VIVO* AND *IN VITRO* STUDIES

Overview of Decay

The first and often rate-limiting step of decay for many eukaryotic mRNAs is the exonucleolytic shortening of the poly(A) tail. In yeast, two distinct deadenylation-dependent degradation pathways have been described (48). The most prominent pathway is known as deadenylation-dependent decapping since deadenylation is coupled to the removal of the 5'cap (49, 50). In this process, the poly(A) tail is first shortened to approximately 10 nucleotides followed by the subsequent removal of the 5'cap. The remaining mRNA body is then exonucleolytically degraded from the 5'end. In addition to this mechanism, yeast mRNAs can be degraded by a slower pathway that is independent of decapping. In this process, deadenylation is followed by exonucleolytic degradation of the body from the 3'end, which is mediated by the exosome complex (51). In mammalian cells, the details of the degradation events are not well understood, however it appears that mRNA decay is still initiated with the removal of the poly(A) tail (52-54).

Control of mRNA degradation of select mRNAs is also mediated by AU-rich elements (AREs) found in the 3'UTR. These RNA-destabilizing elements vary in size and sequence and are associated with many highly labile mRNAs. These sequences have been shown to play a significant role in the stability of many different transcripts including those encoding certain proto-oncogenes and cytokines (13, 14, 55). There are at least ten different proteins that have been identified as having ARE-binding activity *in vitro*. These include AUF1, a.k.a. hnRNP D (56, 57, see also Chapter 9), hnRNP A1, hnRNP C (20, 58), and the ELAV family of RNA-binding proteins (59). The exact mechanism of mRNA stabilization conferred by these proteins is not known, however, a number of possibilities exist. It has been shown that HuR does not affect the deadenylation step of the decay pathway, and hence it is likely to stabilize the mRNA by binding to the 3'UTR only (60). The interaction of this protein upstream of the poly(A) tail may change the structure of the mRNA making it less sensitive to ribonuclease attack (61). On the other hand, HuD binds to GAP-43 mRNA molecules having long tails with higher affinity than those with short or no tail (62) and thus, it may stabilize the messenger RNA via its interactions with both the 3' UTR and the poly(A) tail. Supporting this idea, Ma et al, (30) found that HuD binds simultaneously to the poly(A) tail and to the ARE in the *c-myc* 3' UTR.

Stabilization of the GAP-43 mRNA by HuD: *In Vivo* Studies

Recent studies used transfection strategies to examine the function of HuD on mRNA stability in cultured cells. As shown in Fig. 4, increased expression of HuD protein in PC12 cells correlates with increased GAP-43 mRNA stability. In contrast, decreases in HuD expression caused by antisense treatment of the cells result in the destabilization of the mRNA. Overall, a strong correlation between HuD and GAP-43 mRNA turnover rates was demonstrated in cells containing levels of this protein that were either 3-fold lower to 3-fold higher than untransfected PC12 cells (24). Moreover, the changes in GAP-43 mRNA half-life lead to corresponding changes in GAP-43 mRNA and protein levels and neurite outgrowth (24). As indicated before, this process depends on the presence of the third RRM in HuD (43). Similarly, all three RRMs of HuR are required for stabilization of the c-fos mRNA (25). In contrast to mRNA stabilization by HuR, which is dependent on nuclear-cytosolic shuttling, HuD-dependent stabilization of the GAP-43 mRNA occurs at the cytoplasmic level. Like other ELAV/Hu proteins, HuD is enriched in polysomes, where it acts primarily on GAP-43 mRNA stability and is not involved in translational control (24).

Stabilization of the GAP-43 mRNA by HuD: *In Vitro* Studies

Cell-free systems have been widely used to investigate the degradation of numerous mRNAs. One of these decay assays uses polysomes as the source of the mRNA degradation machinery (14) and exogenous radiolabeled mRNAs as substrates. As shown in Fig. 4B, when *in vitro* synthesized, capped, and polyadenylated GAP-43 mRNA is incubated with polysomes obtained from adult rat brain, the mRNA decays with a half-life of about 15-20 min. This decay rate is similar to that observed for the endogenous mRNA (Fig.4C). Although the overall rate of decay in isolated polysomes is much faster than in cultured cells, the behavior of the mRNA and RNA-binding proteins in the *in vitro* system mimic that observed *in vivo.* For example, exchanging the instability-conferring sequences in the GAP-43 3'UTR with the 3'UTR of β-globin dramatically increases the stability of the transcript. Also, addition of exogenous HuD protein results in a marked increase in the overall stability of the GAP-43 mRNA (Fig. 4B).

The increase in GAP-43 mRNA stability by HuD occurs only in the presence of long poly(A) tails (> A_{150}). HuD protein does not have any effect on the half-life of mRNAs that contain short tails (A_{30}) or no tails (62). In correlation with these effects, HuD binds with the highest affinity to the 3'UTR of GAP-43 mRNAs that have long tails relative to those with short or no tails (62). Interestingly, messenger RNAs without tails are not degraded in isolated polysomes. These results are in agreement with data reported recently by Ford et al. (61) using a different *in vitro* decay system and suggest that the poly(A) tail is necessary for mRNA degradation to proceed.

As shown in Fig. 4 C, neonatal brain polysomes display increased stability of the GAP-43 mRNA, even in the absence of exogenous HuD. This is consistent with the higher binding of HuD to GAP-43 3' UTR sequences observed in neonatal brain extracts (Fig. 1C). Once again, there is an excellent correspondence between the affinity of HuD for the GAP-43 mRNA and its ability for protecting it against ribonuclease attack. Although the mechanism by which HuD exerts these effects *in vivo* is far from being elucidated, some features of the process are becoming apparent (Fig. 5).

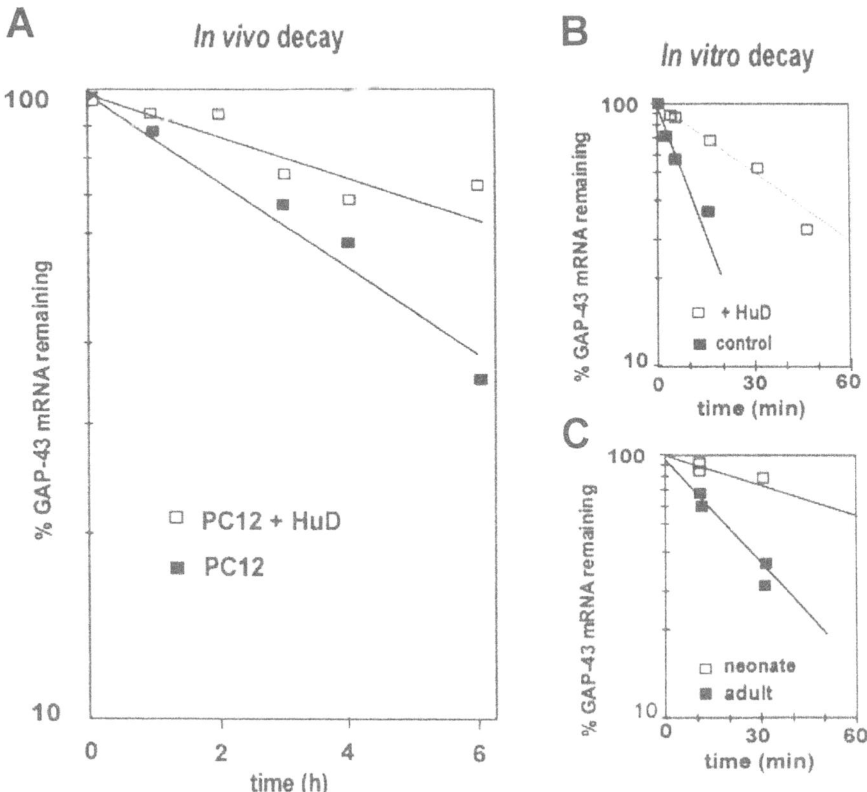

Figure 4. In vivo and in vitro effects of HuD on GAP-43 mRNA stability using PC12 cells (A) and isolated polysomes (B, C). *A: PC12 cells were transfected with a human HuD cDNA and mRNA decay assays were performed as described in reference 24. Rat brain polysomes were isolated and used for mRNA decay assays as described by Brewer and Ross (14). B: Addition of recombinant HuD protein to adult brain polysomes increases the stability of in vitro synthesized GAP-43 mRNA. C: Comparison of the GAP-43 mRNA half-life in adult and neonatal brain polysomes demonstrate the increased stability of this mRNA in the developing brain.*

Model for the Role of HuD in the Control of GAP-43 mRNA Stability

Based upon the results presented above, we propose that HuD controls GAP-43 mRNA stability by interacting with a highly conserved regulatory element in the 3'UTR of the GAP-43 mRNA. This process depends on the length of the poly(A) tail and on the developmental state of the neuron. As shown in Fig. 5, when neurons are stimulated to differentiate, HuD binds to the GAP-43 mRNA with higher affinity, leading to its stabilization and increased GAP-43 expression. This interaction takes place at the level of the polysome and is enhanced by signals that activate protein kinase C such as TPA.

The diagram in Fig. 5 also shows that binding of HuD to the GAP-43 mRNA can be increased by a simple mass action effect, by overexpression of this protein in the cell. In contrast, this interaction can be disrupted by blocking HuD expression via antisense RNA treatment (24) or by overexpression of specific sequences in GAP-43 3'UTR (47). These sequences compete with the endogenous mRNA for the binding of HuD and other *trans*-acting factors leaving the mRNA vulnerable to the attack of specific nucleases. As shown by Neve et al. (47), this treatment not only blocks GAP-43 expression but also disrupts neurite outgrowth in the cells. In addition, our model proposes that besides HuD, other polysomal-associated proteins may control GAP-43 mRNA stability via their direct interaction with the mRNA or indirectly, by binding to HuD.

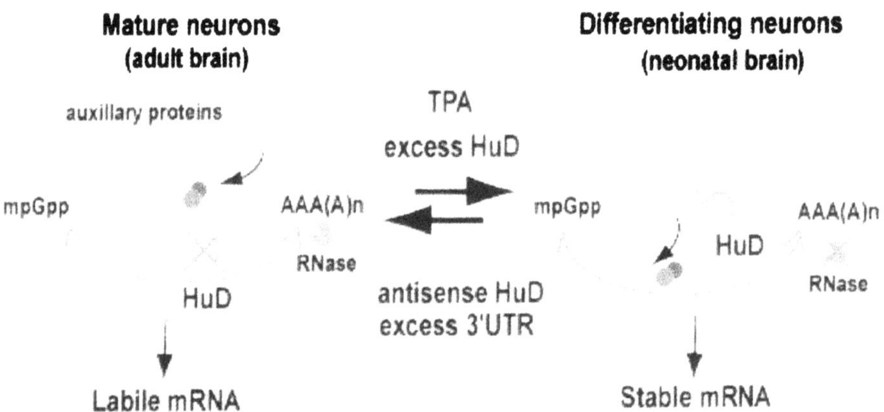

Figure 5. Model for the regulation of GAP-43 mRNA stability by HuD. *The diagram depicts some of the cellular factors that could affect the interaction of HuD with the GAP-43 3'UTR, thus controlling the stability of this mRNA. The structural/functional interaction of the 3' and 5' end of the mRNA is indicated by dotted lines. In addition to mRNA-specific factors, a number of proteins have been described for the general control of mRNA degradation (see ref. 63 and main text).*

Last, although not shown in the diagram, a number of proteins involved in the general control of mRNA stability and translation have been recently identified (63). These bind to either the 5' or 3' end of the mRNA and include the poly(A)-binding protein (PABP, 64), a poly(A)-specific ribonuclease PARN, a.k.a. DAN (65), AUF1 (57), HuR (25, 60), translation initiation factor eIF4G (66) and the cap-binding protein (eIF4E, 67). The potential interaction between HuD and these factors is currently under investigation.

CONCLUSIONS

It is becoming apparent that, together with transcriptional mechanisms, post-transcriptional regulation is critical for shaping the pattern of expression of several developmentally regulated genes. In the case of GAP-43, certain neurons of the developing and mature nervous system show dissociation between the levels of gene transcription and the stability of the mRNA (7,9). In some cells, the rate of transcription is high but the levels of the messenger RNA are quite low, in others, there is a low level of transcription but high accumulation of the GAP-43 mRNA, while in a third group of neurons, transcription and accumulation seem to go hand in hand. The uncoupling between the transcriptional and post-transcriptional regulation of the GAP-43 gene has obvious advantages, as it allows for a higher variation in the accumulation of the mRNA in different neuronal populations. It is also important to note that not all post-transcriptionally regulated genes show this uncoupling. Overall, it appears that the mechanisms controlling gene expression evolved according to the type of promoter found in the gene and to the time frame and levels in which the protein functions in the cell.

Our results indicate that HuD plays a critical role in the control of GAP-43 gene expression and neuronal differentiation and that it does so by stabilizing the mRNA. Although sketchy, the model proposed in Figure 5 provides a framework in which to ask more specific questions on the mechanism by which HuD controls the stability of the GAP-43 mRNA. Given the involvement of this protein in the development and repair of neural connections, this information has significant biological implications. Ultimately, understanding the mechanisms by which GAP-43 levels are regulated *in vivo* may be important for the treatment of several neurological conditions, ranging from neurodevelopmental disorders and brain trauma to spinal cord injury.

Summary of key concepts

- *Post-transcriptional mechanisms play an essential role in controlling the expression of a large number of developmentally regulated mRNAs.*
- *Expression of the GAP-43 gene is regulated by the PKC-dependent stabilization of the mRNA. This process is controlled by the interaction between regulatory sequences in the 3' UTR (cis-acting elements) and specific RNA-binding proteins (trans-acting factors).*
- *The regulatory elements in the GAP-43 3' UTR are U-rich and highly conserved in the evolution. Among these sequences is an ARE-like motif that contains the recognition site for the RNA-binding protein HuD.*
- *HuD is a member of the ELAV/Hu family of RNA-binding proteins. which contain three RRM motifs and are highly conserved.*
- *HuD stabilizes the GAP-43 mRNA by binding to its recognition site in the GAP-43 3' UTR. This interaction is increased during development and in response to signaling agents that activate PKC.*
- *The affinity of HuD for the GAP-43 mRNA depends on the size of the poly(A) tail, with the highest affinity for tails with more than 150 adenosines. These results are consistent with a model in which HuD's ability for stabilizing the GAP-43 mRNA is decreased as deadenylation proceeds and tails become shorter.*

Study Guide Questions

1) Decay plots in Fig. 4 show a linear relationship between the log of the remaining mRNA and decay time. What can you infer about the kinetics of GAP-43 mRNA decay?
2) Determine the % change in steady state levels of the GAP-43 mRNA in the following conditions: a) transcription stays the same but the mRNA half-life increases 3-fold and b) transcription decreases 2-fold and the half-life of the mRNA increases 2-fold.
3) The K_d of HuD for GAP-43 mRNAs with long tails is 10 nM while that for mRNAs with short tails is 100 nM. Assuming that half of the molecules of GAP-43 mRNA in the cell have short tails and half contain long tails, what is the proportion of HuD bound to each species?
4) For a large number of mRNAs, there is a good correlation between the structural requirements for translation and stability, with the cap and the tail being essential for both processes. Describe the significance of the coupling between translation and degradation of mRNAs in the economy of the cell.
5) Using the examples described in Question 3, what can you infer about the translation and stability of mRNAs containing short vs. long tails?

Acknowledgements
The authors wish to thank Dr. Andrea Beckel-Mitchener and Angel Miera for their contributions to this work. This research was supported by NIH grants NS30255 and GM52576 to NPB.

REFERENCES

1. Skene, J.H.P. 1989. Axonal growth associated proteins. *Ann. Rev. Neurosci.* **12**:127-156.
2. Benowitz, L.I., and Routtenberg, A. 1997. GAP-43: an intrinsic determinant of neuronal development and plasticity. *Trends Neurosci.* **20**:84-91.
3. Oestreicher, A.B., DeGraan, P.N.E., Gispen, W.H., Verhaagen, J., and Schrama, H. 1997. B-50, the growth-associated protein-43: modulation of cell morphology and communication in the nervous system. *Prog. Neurobiol.* **53**:627-686.
4. Chiaramello, A., Neuman, T., Reavy, D.R., and Zuber, M.X. 1996. The GAP-43 gene is a direct downstream target of the basic helix-loop-helix transcription factors. *J. Biol. Chem.* **271**:22035-22043.
5. Kinney, M., MacNamara, R.K., Valcourt, E., and Routtenberg, A. 1996. Prolonged alteration in E box binding after a single kainate injection: potential relation to F1/GAP-43 gene expression. *Brain Res. Mol. Brain Res.* **38**:25-36.
6. Federoff, H.J., Grabczyk, E., and Fishman, M.C. 1988. Dual regulation of GAP-43 gene expression by nerve growth factor and glucocorticoids. *J. Biol. Chem.* **263**:19290-19295.
7. Perrone-Bizzozero, N.I., Neve, R.L., Irwin, N., Lewis, S., Fischer, I., and Benowitz, L.I. 1991. Post-transcriptional regulation of GAP-43 mRNA levels during neuronal differentiation and nerve regeneration. *Mol. Cell Neurosci.* **2**:402-409.
8. Cantallops, I., and Routtenberg, A. 1999. Activity-dependent regulation of axonal growth: posttranscriptional control of the GAP-43 gene by the NMDA receptor in developing hippocampus. *J. Neurobiol.* **41**:208-20.
9. Namgung, U., and Routtenberg, A. 2000. Transcriptional and post-transcriptional regulation of a brain growth protein: regional differentiation and regeneration induction of GAP-43. *Eur. J. Neurosci.* **12**:3124-3136.
10. Perrone-Bizzozero, N.I., Cansino, V.V., and Kohn, D.T. 1993. Post-transcriptional regulation of GAP-43 gene expression in PC12 cells through PKC-dependent stabilization of the mRNA. *J. Cell Biol.* **120**:1263-1270.
11. Nedivi, E., Basi, G.S., Akey, I.V., and Skene, J.H.P. 1992. A neural-specific GAP-43 core promotor located between unusual DNA elements that regulate its activity. *J. Neurosci.* **12**:691-704.
12. Eggen, B.J.L., Nielander, H.B., Rensen-DeLeeuw, M.G.A., Schotman, P., Gispen, W.H., and Schrama L.H. 1994. Identification of two promotor regions in the rat B-50/GAP-43 gene. *Brain Res. Mol. Brain Res.* **23**:221-234.
13. Wilson, T., and Treisman, R. 1988. Removal of poly(A) and consequent degradation of *c-fos* mRNA facilitated by 3' AU-rich sequences. *Nature.* **336**:396-399.
14. Brewer, G., and Ross, J. 1990. Messenger RNA turnover in cell-free extracts. *Meth Enzymol.* **18**:203-209.
15. Shaw, G., and Kamen, R. 1986. A conserved AU sequence in the 3' untranslated region of GM-CSF mRNA mediates selective mRNA degradation. *Cell.* **46**:659-667.
16. Kohn, D.T., Tsai, K.-C., Cansino, V.V., Neve, R.L., and Perrone-Bizzozero, N.I. 1996. Role of highly conserved pyrimidine-rich sequences in the 3' untranslated region of the GAP-43 mRNA in mRNA stability and RNA-protein interactions. *Brain Res. Mol. Brain Res.* **36**:240-250.
17. Tsai, K., Cansino, V.V., Kohn, D.T., Neve, R.L., and Perrone-Bizzozero, N.I. 1997. Post-transcriptional regulation of the GAP-43 gene by specific sequences in the 3' untranslated region of the mRNA. *J. Neurosci.* **17**:1950-1958.

18. Peng, S.S., Chen, C.A., and Shyu, A. 1996. Functional characterization of a non-AUUUA AU-rich element from the c-*jun* proto-oncogene mRNA: evidence for a novel class of AU-rich elements. *Mol. Cell Biol.* **16**:1490-1499.
19. Zaidi, S.H., Denman, R., and Malter, J.S. 1994. Multiple proteins interact at a unique *cis*-element in the 3'-untranslated region of amyloid precursor protein mRNA. *J. Biol. Chem.* **269**:24000-24006.
20. Zaidi, S.H., and Malter, J.S. 1995. Nucleolin and heterogenous nuclear ribonucleoprotein C proteins specifically interact with the 3'-untranslated region of amyloid protein precursor mRNA. *J. Biol. Chem.* **270**:17292-17298.
21. Irwin, N., Baekelandt, V., Goritchenko, L., and Benowitz, L.I. 1997. Identification of two proteins that bind to a pyrimidine-rich sequence in the 3'-untranslated region of GAP-43 mRNA. *Nucleic Acids Res.* **25**:1281-1288.
22. DeFranco, C., Chicurel, M.E., and Potter, H. 1998. A general RNA-binding protein complex that includes the cytoskeleton-associated protein MAP 1A. *Mol. Biol. Cell.* **9**:1695-1708.
23. Chung, S.M., Eckrich, M., Perrone-Bizzozero, N.I., Kohn, D.T., and Furneaux, H.M. 1997. The Elav-like proteins bind to a conserved regulatory element in the 3' untranslated region of GAP-43 mRNA. *J. Biol. Chem.* **272**:6593-6598.
24. Mobarak, C.M., Anderson, K.D., Morin, M., Beckel-Mitchener, A., Rogers, S., Furneaux, H. M., King P., and Perrone-Bizzozero, N. I. 2000. The RNA-binding protein HuD is required GAP-43 mRNA stability, GAP-43 gene expression and PKC-dependent neurite outgrowth in PC12 cells. *Mol. Biol.Cell.* **11**:3191-3203.
25. Fan, X.C., and Steitz, J.A. 1998. Overexpression of HuR, a nuclear-cytoplasmic shuttling protein, increases the *in vivo* stability of ARE-containing mRNAs. *EMBO. J.* **17**:3448-3460.
26. Chung, S., Jiang, L., Cheng, S., and Furneaux, H.M. 1996. Purification and properties of HuD, a neuronal RNA binding protein. *J. Biol. Chem.* **271**:11518-11524.
27. Aranda-Abreu, G.E., Behar, L., Chung, S., Furneaux, H., and Ginzburg, I. 1999. Embryonic lethal abnormal vision-like RNA-binding proteins regulate neurite outgrowth and tau expression in PC12 cells. *J. Neurosci.* **19**:6907-6917.
28. Joseph, B., Orlian, M., and Furneaux, H. 1998. p21(waf1) mRNA contains a conserved element in its 3'-untranslated region that is bound by the ELAV-like mRNA-stabilizing proteins. *J. Biol.Chem.* **273**:20511-20516.
29. Lazarova, D.L., Spengler, B.A., Biedler, J.L., and Ross, R.A. 1999. HuD, a neuronal-specific RNA-binding protein, is a putative regulator of *n-myc* pre-mRNA processing/stability in malignant human neuroblasts. *Oncogene.* **18**:2703-2710.
30. Ma, W., Chung, S., and Furneaux, H. 1997. The Elav-like proteins bind to AU-rich elements and to the poly(A) tail of mRNA. *Nucleic Acid Res.* **25**:3564-3569.
31. Campos, A.R., Grossman, D., and White, K. 1985. Mutant alleles at the locus elav in *Drosophila melanogaster* lead to nervous system defects. A developmental-genetic analysis. *J. Neurogenet.* **2**:197-218.
32. Robinow, S., Campos, A.R., Yao, K.M., and White, K. 1988. The elav gene product of *Drosophila*, required in neurons, has three RNP consensus motifs. *Science.* **242**:1570-1572.
33. Homyk, T., Isono, K., and Pak, W.L. 1985. Developmental and physiological analysis of a conditional mutation affecting photoreceptor and optic lobe development in *Drosophila melanogaster*. *J. Neurogenet.* **2**:309-324.
34. Koushika, S.P., Lisbin, M.J., and White, K. 1996. ELAV, a *Drosophila* neuron-specific protein, mediates the generation of an alternatively spliced neural protein isoform. *Current Biology.* **6**:1643-1641.
35. Samson, M. 1998. Evidence for 3' untranslated region-dependent autoregulation of the *Drosophila* gene encoding the neuronal nuclear RNA-binding protein ELAV. *Genetics.* **150**:723-733.

36. Marusich, M.F., Furneaux, H.M., Henion, P., and Weston, J.A. 1994. Hu neuronal proteins are expressed in proliferating neurogenic cells. *J. Neurobiol.* **25**:143-155.
37. Wakamatsu, Y., and Weston, J.A. 1997. Sequential expression and role of Hu RNA binding proteins during neurogenesis. *Development.* **124**:3449-3460.
38. Okano, H.J., and Darnell, R.B. 1997. Heirarchy of Hu RNA binding proteins in developing and adult neurons. *J. Neurosci.* **17**:3024-3037.
39. Clayton, G.H., Perez, G.M., Smith, R.L., and Owens, G.C. 1998. Expression of mRNA for the ELAV-like neural-specific RNA binding protein, HuD, during nervous system development. *Brain Res. Dev. Brain Res.* **109**:271-80.
40. Perrone-Bizzozero, N. I., Thomas, D., Harji, F., Saland, L., Lynn, S., Hassinger, L., and Neve, R.L. 1999. Molecular and cytoarchitectural alterations in the hippocampus of patients with schizophrenia revealed by immunostaining against HuD, GAP-43 and other neuronal markers. *Schizophrenia Res.* (Suppl.) **36**:83.
41. Antic, D., and Keene, J.D. 1998. Messenger ribonucleoprotein complexes containing human Elav proteins: interactions with cytoskeleton and translational apparatus. *J. Cell Sci.* **111**:183-197.
42. Antic, D., Lu, N., and Keene, J.D. 1999. ELAV tumor antigen, Hel-N1, increases translation of neurofilament M mRNA and induces formation of neurites in human teratocarcinoma cells. *Gene Dev.* **13**: 449-461.
43. Anderson, K.D., Morin, M.A, Beckel-Mitchener, A., Mobarak, C.D., Neve, R. L., Furneaux, H. M., Burry, R.M., and Perrone-Bizzozero, N.I. 2000. Overexpression of HuD, but not its truncated form HuD I+II, promotes GAP-43 expression and neurite outgrowth in PC12 cells in the absence of nerve growth factor. *J. Neurochem.* **75**:1103-1114.
44. Akamatsu W. et al. 1999. Mammalian ELAV-like neuronal RNA-binding proteins HuB and HuC promote neuronal development in both the central and the peripheral nervous systems. *Proc. Natl. Acad. Sci. USA.* **96**:9885-9890.
45. Kasashima, K., Terashima, K., Yamamoto, K., Sakashita, E., and Sakamoto, H. 1999. Cytoplasmic localization is required for the mammalian ELAV-like protein HuD to induce neuronal differentiation. *Genes Cells.* **4**:667-683.
46. Anderson, K. D., Sengupta, J., Morin, M., Neve, R. L., Valenzuela, C. F., and Perrone-Bizzozero, N. I. 2001. Overexpression of HuD accelerates neurite outgrowth and increases GAP-43 mRNA expression in neurons and retinoic acid-induced embryonic stem cells *in vitro. Exp. Neurol.*, 168:250-258.
47. Neve, R.L., Ivins, K.J., Tsai, K.C., Rogers, S.L., and Perrone-Bizzozero, N.I. 1999. *Cis*-acting regulatory elements in the GAP-43 mRNA 3'-untranslated region can function in trans to suppress endogenous GAP-43 gene expression. *Brain Res. Mol. Brain Res.* **65**:52-60.
48. Beelman, C.A., and Parker, R. 1995. Degradation of mRNA in eukaryotes. *Cell.* **81**:179-183
49. Muhlrad, D., and Parker, R. 1994. Premature translational termination triggers mRNA decapping. *Nature.* **370**:578-581.
50. Tucker, M., and Parker, R. 2000. Mechanisms and control of mRNA decapping in *Saccharomyces cerevisiae. Ann. Rev. Biochem.* **69**: 571-595
51. Jacobs Anderson J.S., and Parker, R. 1998. The 3' to 5' degradation of yeast mRNAs is a general mechanism for mRNA turnover that requires the SK12 DEVH box protein and 3' to 5' exonuclease of the exosome complex. *EMBO J.* **17**:1497-1506.
52. Ross, J. 1995. mRNA stability in mammalian cells. *Microbiol. Rev.* **59**:423-450.
53. Shyu, A-B., Belasco, J., and Greenberg, M. 1991. Two distinct destabilizing elements in the *c-fos* message trigger deadenylation as a first step in rapid mRNA decay. *Genes Dev.* **5**:221-231.
54. Jacobson, A., and Peltz, S.W. 1996. Interrelationships between the pathways of mRNA decay and translation in eukaryotic cells. *Annu. Rev. Biochem.* **65**:693-740.
55. Chen, C., and Shyu, A-B. 1995. AU-rich elements: characterization and importance in mRNA degradation. *Trends Biochem. Sci.* **20**:465-470.

56. Kiledjian, M., DeMaria, C., Brewer, G., and Novick, K. 1997. Identification of AUF1 (heterogenous nuclear ribonucleoprotein D) as a component of the a-globin mRNA stability complex. *Mol. Cell. Biol.* **17**: 4870-4876.
57. Zhang, W., Wagner, B.J., Ehrenman, K., Schaefer, A.W., DeMaria, C.T., Crater, D., DeHaven, K., Long, L., and Brewer, G. 1993. Purification, characterization, and cDNA cloning of an AU-rich element RNA-binding protein, AUF1. *Mol. Cell Biol.* **13**:7652-7665.
58. Hamilton, B.J., Nagy, E., Malter, J.S., Arrick, B.A., and Rigby, W.F. 1993. Association of heterogenous nuclear ribonucleoprotein A1 and C proteins with reiterated AUUUA sequences. *J. Biol. Chem.* **268**:8881-8887.
59. Levine, T.D., Gao F.-B., King, P.H., Andrews, L.G., and Keene, J.D. 1993. Hel-N1: an autoimmune RNA-binding protein with specificity for 3' uridylate-rich untranslated regions of growth factor mRNAs. *Mol. Cell. Biol.* **13**:3494-3504.
60. Peng, S.S.-Y., Chen ,C.-Y.A., Xu, N., and Shyu, A.-B. 1998. RNA stabilization of the AU-rich element binding protein, HuR, an ELAV protein. *EMBO. J.* **17**:3461-3470.
61. Ford, L.P., Watson, J., Keene, J.D., and Wilusz, J. 1999. ELAV proteins stabilize deadenylated intermediates in a novel in vitro mRNA deadenylation/degradation system. *Genes Dev.* **13**:188-201.
62. Beckel-Mitchener, A., Keller, R., Miera, A., Mobarak, C.D., Kohn, D.T., Furneaux, H., and Perrone-Bizzozero, N.I. 2000 Poly(A) tail-dependent stabilization of GAP-43 mRNA by HuD. RNA Society meeting (Abstr.).
63. Gao, M., Fritz, D., Ford, L., and Willusz, J. 2000. Interaction between a poly(A)-specific ribonuclease and the 5' cap influences mRNA deadenylation in vitro. *Mol. Cell* **5**: 479-488
64. Bernstein, P., Peltz, S.W., and Ross, J. 1989. The poly(A)-poly(A) binding protein complex is a major determinant of mRNA stability. *Mol. Cell. Biol.* **9**:659-670.
65. Dehlin, E., Wormington, M., Körner, C., and Wahle, E. 2000. Cap-dependent deadenylation of mRNA. *EMBO J.* **19**:1079-1086.
66. Tarun, S.Z., Wells, S.E., Deardorff, J.A., and Sachs, A.B. 1997. Translation initiation factor eIF4G mediates *in vitro* poly(A) tail-dependent translation. *Proc. Natl Acad. Sci. USA.* **94**:9046-9051.
67. Morino, S, Imataka, H, Svitkin, YV, Pestova, TV, and Sonenberg, N 2000. Eukaryotic translation initiation factor 4E (eIF4E) binding site and the middle one-third of eIF4GI constitute the core domain for cap-dependent translation, and the C-terminal one-third functions as a modulatory region. *Mol. Cell. Biol.* **20:**468-477.

10

RNA-DEPENDENT PROTEIN KINASES

Raymond A. Petryshyn, Sergie Nekhai*, and Evelio D. Perez-Albuerne♦

*National Cancer Institute, Bethesda, MD, *Howard University, Washington, DC, and ♦Children's National Medical Center, Washington, DC 20010*

RNA molecules conduct various functions in living organisms by interacting with other biological molecules. The recognition of RNA molecules, usually by proteins, is often dependent on the shape into which the RNA folds, rather than on any specific nucleotide sequence (1). This review focuses on double-stranded RNA (dsRNA) dependent protein kinase (PKR), which phosphorylates the α subunit of eukaryotic initiation factor-2 (eIF-2α) (2). PKR contains two amino acid sequence motifs called dsRNA-binding motifs (DRBM) that allow binding to dsRNA and subsequently convert the protein from a latent to an active serine/threonine protein kinase (3). PKR is the only known kinase that depends on dsRNA for activation, although two closely related eIF-2α kinases, pancreatic eIF-2α kinase (PEK) (4) and PKR-like endoplasmic reticulum kinase (PERK) (5), have been described. PKR presents a unique paradigm for studying RNA:protein interaction because its activity depends on binding to dsRNA but not DNA, single-stranded RNA, or RNA:DNA hybrids. Well-known for mediating the antiviral effects of interferons (IFNs), PKR is also implicated in regulating cell differentiation, signal transduction, and in eliciting apoptosis in response to various stress induction agents (3,6). Although the protein is ubiquitous in cells, PKR activity is suppressed during cell proliferation and in tumor cells, suggesting a role for the kinase in the regulation of cell proliferation. This review summarizes the viral and cellular proteins and dsRNAs that activate and inhibit PKR, and the most recent findings in PKR knockout mice.

REGULATION OF PKR BY CELLULAR AND VIRAL PROTEINS

A number of cellular and viral proteins regulate the activity of PKR by direct interaction with PKR or by competing for dsRNA. The regulation of PKR by cellular proteins, by dsRNA-binding peptides and by viral proteins will be briefly discussed.

PKR Regulators Containing dsRNA-Binding Motifs (DRBMs)

Many proteins that regulate PKR contain DRBMs. Among the first of these, the TAR RNA binding protein (TRBP), was identified because of its ability to bind HIV-1 TAR RNA. TRBP contains three DRBMs (8). It is a member of the RNA-binding protein family that includes PKR, Drosophila Staufen (9) and Xenopus Xlrbpa (10). TRBP, however, appears to inhibit PKR activity (11,12) through an RNA-independent pathway (12). The mechanism of TRBP action may include formation of heterodimers with PKR (13).

Another TAR RNA-binding protein, NF90, was originally described as an interleukin (IL)-2 transcription activator (14). NF90 is retained by TAR-RNA affinity column and is implicated in the regulation of HIV-1 transcription (15). Recent studies indicate that NF90 also binds directly to a number of cellular mRNAs, inhibiting their translation by interfering with mRNA binding to polysomes (16,17). The large subunit of NF90 contains two DRBMs (18), allowing interaction with human adenovirus associated RNA (19). Further, NF90 is phosphorylated in a dsRNA-dependent manner (20). The NF90 homolog, DRBP76, was identified in a yeast two-hybrid screen with a mutant PKR as bait (21). DRBP76 interacts with PKR and with dsRNA *in vitro*, co-precipitates with PKR, and is phosphorylated by PKR *in vitro* (21). Since NF90 binds dsRNA with high affinity, it may inhibit PKR by sequestering limiting quantities of cellular dsRNAs. PKR is efficiently blocked by recombinant NF90 and overexpression of NF90 inhibits PKR (Nekhai, personal communication). Treatment of several human leukemic cell lines with interferon results in translocation of NF90 from the nucleus to the cytoplasm, which may explain the lack of PKR activity in these cells despite high levels of induced PKR protein (Nekhai, personal communication). These observations indicate several potentially important cellular roles for NF90, including the regulation of PKR. However, the true biological function of NF90 is yet to be fully uncovered.

The protein activator of PKR (PACT), identified in yeast-two hybrid screen with PKR as bait, is a dsRNA-binding protein (22). It forms heterodimers with PKR and activates the kinase in the absence of dsRNA (22,23). Overexpression of PACT results in increased phosphorylation of eIF-2α and inhibits protein synthesis(22). PACT-associated PKR activation occurs in response to diverse stress signals, and thus is postulated to function as a stress-modulated physiologic activator of PKR (23). The mouse homolog of PACT is a ubiquitously expressed PKR-associated protein (RAX) (24). Unlike PACT, overexpression of RAX does not induce PKR activation in cells, although RAX directly activates PKR *in vitro* (24). IL-3 deprivation of IL-3-dependent cells, as well as a number of other conditions that cause cell stress, stimulates the phosphorylation of RAX followed by RAX-PKR association and activation of PKR (24). It is likely that RAX may be a stress-activated, physiologic activator of PKR that couples transmembrane stress signals to protein synthesis (24).

The La antigen is an RNA-binding protein, which inhibits the activation of PKR by sequestering dsRNA (25). Both the N-terminal RNP motif and the C-terminal half of the La antigen are necessary for inhibition (25).

PKR is involved in regulating the growth of mouse embryonic 3T3-F442A fibroblasts prior to their subsequent differentiation to adipocytes (26). Extracts from 3T3-F442A cells cultured under conditions restrictive for differentiation contained elevated levels of an inhibitor of PKR called dRF. dRF inhibits the binding of dsRNA to PKR, preventing its phosphorylation and activation without a direct effect on the dsRNA (27). Phosphoprotein analysis shows reduced levels of PKR and eIF-2α phosphorylation that correlate with the expression of dRF (28).

Regulation of PKR Activity by dsRNA-binding Synthetic Peptides

PKR contains two dsRNA-binding motifs (DRBM1 and DRBM2) that independently interact with dsRNA (29-31). Synthetic peptides corresponding to the DRBM1 (residues 54-74 of murine PKR or residues 60-80 of human PKR) have been shown to inhibit the activation of PKR by disrupting RNA binding (32). Introduction of a cell permeable analog of the synthetic DRBM1 peptide into embryonic fibroblasts diminished PKR activity and eIF-2α phosphorylation, resulting in sustained cell proliferation even after cultures became confluent (33). These studies demonstrate the potential of using mimicking peptides targeted to DRBMs to explore the cellular function of PKR and other dsRNA-binding proteins and as reagents to enhance the proliferation of embryonic cells.

Viral Proteins that Regulate PKR

A number of diverse viruses (see Table 1) encode proteins that inhibit or overcome PKR activity to sustain efficient viral replication. Viral proteins abrogate PKR by one or more of the following mechanisms: sequestration of dsRNAs required for PKR activation; binding directly to PKR at critical sequences such as the DRBMs; and by behaving as pseudosubstrates for PKR by mimicking amino acid sequences within eIF-2α that are recognized by PKR. Viral proteins known to inhibit PKR and their mechanisms of action are summarized in Table 1.

Table 1. Viral proteins that inhibit PKR.

Virus	Protein	Effect on PKR	Ref.
Hepatitis C virus	E2 protein	Pseudosubstrate inhibition of PKR	(34)
Hepatitis C virus	Viral nonstructural 5A protein (NS5A).	Inhibition by direct binding to PKR	(35,36)
HIV-1	Tat	Inhibition by dsRNA-dependent and independent mechanism; pseudosubstrate inhibition of PKR	(37-39)
Parapox virus	Parapox virus orf OV20.0L gene product	Inhibition by binding to PKR	(40)
Porcine group C rotavirus	NSP3 protein	Binding to dsRNA and inhibition of PKR	(41)
Reovirus	Sigma3 virion outer shell protein	Binding to dsRNA and inhibition of PKR	(42)
Swinepox virus	C8L gene product	Pseudosubstrate inhibition of PKR	(43)
Vaccinia virus	E3L gene product, pE3	Inhibition by binding to dsRNA and to PKR	(44,45)
Vaccinia virus	K3L gene product, pK3	Pseudosubstrate inhibition of PKR	(46)

Viral and cellular dsRNAs that regulate PKR

Table 2. Viral dsRNAs that regulate PKR activity

Virus	RNA	Effect on PKR	Ref.
Adenovirus	VA RNA I	Inhibition of PKR	(47)
Epstein-Barr	EBER-1	Inhibition of PKR	(48)
HIV-1	TAR RNA	Activation and inhibition of PKR	(49,50)
Human δ hepatitis virus	Human δ hepatitis agent	Inhibition of PKR	(51)
Reovirus	Reovirus S1 gene, 1463 nucleotide mRNA	Activation of PKR	(52)
T-cell leukemia virus type 1 (HTLV-1)	Rex-RE RNA	Activation and inhibition of PKR	(53)

Several viruses also utilize their RNAs to regulate PKR and increase viral replication. The viral RNAs activate or inhibit PKR depending on the length of the double-stranded RNA region, the number of base-pair mismatches in the region, and the concentration of the viral RNA. The known viral RNAs that regulate PKR and their effects on PKR are summarized in Table 2.

While it is evident that PKR undergoes activation in cells and tissues in response to stress-inducing conditions in the absence of viral infection (6), it is not clear how much of the activation is due to cellular RNAs or to ancillary proteins such as PACT (22). However, a small number of cellular RNAs, including mRNAs, are observed to both activate (54-56) and inhibit (57) PKR through interactions at defined structural elements in the RNAs.

PERTURBATION OF dsRNA-BINDING KINASE ACTIVITY

The biological effects of increased or decreased activity of dsRNA-binding kinases (i.e. PKR and the family of PKR-like kinases) are beginning to be studied in culture and knockout mice. dsRNA-binding kinases appear to have important roles in: 1) cell proliferation, differentiation and apoptosis; 2) susceptibility to viral infection; 3) regulation of inflammatory signaling and modulating molecules. The majority of the studies have focused on PKR, using dominant-negative mutants of the protein and knockout mice.

PKR Knockout Mice

Yang *et al.* generated homozygous PKR knockout mice (PKR -/-(N) mice) by disrupting sequences that code for the N-terminal region of PKR, including the initiation codon (58). The PKR -/-(N) mice have a 129/SV(ev) x C57BL/6J background. Surprisingly, the disrupted gene is capable of directing the *in vitro* synthesis of a protein that could be immunoprecipitated by anti-PKR serum, probably corresponding to a C-terminal fragment of PKR. The protein has no detectable PKR-kinase activity *in vitro* and no immunoreactive band or PKR-kinase activity can be detected in cell extracts derived from the PKR -/-(N) mice.

The PKR -/-(N) mice have normal development and litters of normal size (58). The external appearance of the knockout is identical to the parental mouse strain and examination of their internal organs reveals no consistent abnormalities. There is no increase in spontaneous tumor frequency in PKR -/-(N) mice (58). Mouse embryo fibroblasts (MEFs) from wild-type and PKR -/-(N) mice were used to generate immortalized cell lines or infected with a retrovirus containing an oncogenic, mutated *ras* gene. Neither the parent MEFs, the immortalized cell lines, or the *ras* transfected MEFs are capable of forming colonies in soft agar or inducing tumors in nude mice (58).

These results contrast with the observation that mutant PKR proteins with absent or drastically reduced kinase function cause malignant transformation of NIH 3T3 cells in a dominant negative manner (59-62). The transformed cells are capable of growth in soft agar, and form tumors when injected into nude mice. Cells expressing a dominant negative PKR mutant have reduced levels of PKR activity (62) and lower levels of phosphorylated eIF-2α (59,62). The expression of a mutant form of eIF-2α that can not be phosphorylated at the major target site of PKR (serine 51) also leads to malignant transformation of NIH 3T3 cells (62). This suggests that decreased eIF-2α phosphorylation in cells expressing dominant-negative forms of PKR is sufficient to cause malignant transformation, and thus that decreased eIF-2α phosphorylation is downstream of dominant-negative PKR mutants in causing transformation.

Sequences in the catalytic domain were disrupted to generate a second type of homozygous PKR knockout mouse (the PKR -/-(C) mouse) (63). These mice have a 129/terSv x BALB/C background. No immunoreactive protein or PKR-kinase activity is detectable in PKR -/-(C) cell extracts. The PKR -/-(C) mice also have a normal appearance and litters of normal size (63). Upon treatment with erythropoietin, fewer erythroid colony-forming units (CFU) are seen in bone marrow from PKR -/-(C) mice, compared to marrow from wild-type mice. However, marrow from PKR -/-(C) mice forms the same number of myeloid CFU after treatment with IL-3, Kit ligand, or granulocyte-colony stimulating factor.

There is no difference in the level of eIF-2 αphosphorylation in MEFs derived from wild-type or PKR -/-(C) mice (63). Since the disrupted PKR genes can not direct the synthesis of a protein with catalytic function, the similar level of eIF-2α phosphorylation suggests that there are other redundant kinases that are active in the PKR -/-(C) mice.

The similar level of eIF-2α phosphorylation in wild-type and PKR -/-(C) cells provides some insight into why tissue culture cells with dominant-negative PKR mutants and those with PKR knockout behave differently in assays of malignant transformation. Since decreased eIF-2α phosphorylation appears to be downstream of dominant-negative PKR mutants during malignant transformation, it seems likely that PKR knockout cells do not undergo transformation (and PKR knockout mice do not develop tumors) because of redundant eIF-2α kinases. If this hypothesis is correct, then expression of either dominant-negative PKR mutants or mutant eIF-2α resistant to phosphorylation in PKR -/-(N) or PKR -/-(C) MEFs, or in their respective parent wild-type MEFs would be expected to generate a transformed phenotype. A complication that has yet to be adequately addressed is the modifying effect of genetic background differences in the mouse strains used for generating PKR knockouts. Such modifying effects may contribute to observed differences in malignant transformation between NIH 3T3 cells expressing dominant-negative mutants of PKR and PKR knockout MEFs.

Triply Deficient Mice

Other IFN-responsive systems in mice, such as the dsRNA-dependent 2',5'-oligoadenylate synthase/RNase L pathway and Mx1 protein, provide redundancy that may complicate the use of PKR knockout mice as tools to understand the role of PKR in resistance to viral infection. Zhou *et al.* generated mice (TD mice) that were triply deficient for PKR (containing the N-terminal disruption (58)), RNase L (64), and Mx1 (65). The TD mice appear normal and healthy. The response of TD mice to encephalomyocarditis virus (EMCV) infection is summarized in Table 3. The frequency of spontaneous tumors in TD mice is of interest, but is yet to be addressed. Further, treatment of TD or PKR knockout mice with known tumor inducers such as N-methyl-N-nitrosourea or 7,12-dimethylbenz[a]anthracene may prove to be a useful approach to defining and dissecting the role of PKR in tumorogenesis. This approach has been used in mice heterozygous for a deficiency in the DNA-mismatch repair gene, PMS2. PMS2 +/- mice do not have an increased frequency of spontaneous tumor formation, but are significantly more likely to develop intestinal tumors than wild-type mice after N-methyl-N-nitrosourea treatment (66).

Table 3. Outcomes of viral infection in knockout mice.

Knockout	Virus	Treatment	Outcome measure	Result	Ref.
PKR -/-(C)	Influenza	-	Lethal dose	Same	(63)
PKR -/-(C)	VSV	None or IFN-α/β	Survival	K died, WT survived	(67)
PKR -/-(N)	EMCV	-	Survival, blood IFN levels	Same	(58)
PKR -/-(N)	EMCV	Poly I:C or IFN-γ	Survival	WT twice as long	(58)
PKR -/-(N)	EMCV	IFN-α	Survival	Same	(58)
TD mice	EMCV	None or IFN-α	Survival time	K mice 3 days less	(65)

Abbreviations: VSV - vesicular stomatitis virus; poly I:C - polyinosinic-polycytidylic acid

Table 3 summarizes the effects of viral infection on wild-type (WT) and knockout (K) mice. Table 4 summarizes the effects of viral infection on MEFs derived from WT and K mice. The "Result" columns compare the outcome measure observed between WT and K.

Table 4. Outcomes of viral infection in knockout MEFs.

Knockout	Virus	Treat-ment	Outcome measure	Result	Ref.
PKR -/-(C)	Influenza	None or IFN-α/β	Apoptosis	Same	(63)
PKR -/-(C)	Vaccinia	-	Viral growth curve	Same	(63)
PKR -/-(C)	VSV	None, IFN-α or IFN-γ	Viral production	5-10x in K	(67)

There are significant differences in the effect of dsRNA-binding protein knockouts depending on the virus used for infection. Many of the outcomes of EMCV and VSV infection are different in knockout mice, while there is no change in the outcomes of influenza or vaccinia virus infections.

For some viral infections, the similarity in the course of infection in PKR knockout and wild-type animals may be the result of viral proteins that inactivate PKR even in wild-type cells. Influenza virus infection has this property (68). The influenza NS1 protein has been shown to inhibit PKR activation (69). Bergmann *et al.* generated an influenza virus that lacks the entire NS1 gene (delNS1 virus). After intranasal inoculation with either the delNS1 virus or the wild-type (A/PR/8/34) influenza virus, the viral titer in lung homogenates was determined. The wild-type virus titers were lower in wild-type mice during the initial phase of the infections, but were identical to the titers in PKR -/-(N) mice by day 6 after infection. In contrast, no virus could be recovered after delNS1 infection of wild-type mice, while titers of delNS1 were identical to the titers of wild-type virus in PKR -/-(N) animals.

A similar function was described in herpes simplex virus (HSV)-1 infection (70). Leib *et al.* used an HSV-1 mutated in the ICP34.5 gene, which directs the production of a protein that activates a cellular phosphatase that dephosphorylates eIF-2α (71). Since eIF-2α phosphorylation is a major mediator of PKR's antiviral activity, expression of the ICP34.5 gene antagonizes the antiviral activity of PKR. In mice infected by inoculation of the wild-type or ICP34.5-mutant virus into the cornea, the viral titers in cornea and trigeminal ganglion were measured 1 to 5 days after infection. Wild-type or PKR -/-(N) mice inoculated in the cornea with wild-type HSV-1 have similar viral titers (except for titers on day 3 of infection that are unexpectedly lower in PKR -/-(N) animals). The same inoculation using ICP34.5-mutant virus results in about 10-fold lower titers in wild-type animals, while there is no difference in titer in PKR -/-(N) animals. Mice with homozygous deletions in IFN-αβγ receptor (which is upstream of PKR in the IFN response pathway) have titers similar to PKR -/-(N) mice. Mice with homozygous deletions of RNase L, a component of the IFN-dependent antiviral pathway that does not include PKR and is not inactivated by ICP34.5 protein, had titers similar to wild-type mice. When infected with the wild-type virus intracerebrally, all the wild-type and PKR -/-(N) mice died (although the wild-type mice survived a median of 3 days longer). However, when infected with the ICP34.5-mutant virus intracerebrally, all the PKR -/-(N) mice died (with a time-course identical to infection with wild-type virus), but all the wild-type mice survived (70).

PKR Effects on Inflammatory Mediators

PKR -/-(N) mice challenged with lipopolysaccharide (LPS), a major component of the endotoxins expressed by pathogenic gram-negative bacteria, have peak serum levels of IL-6 and IL-12 that are about half the peak serum levels in wild-type mice (72). The concentration of IL-6 in supernatants from LPS-treated immortalized fibroblasts derived from PKR -/-(N) mice is more than three-fold lower than the concentration in similarly treated cells from wild-type mice. PKR -/-(N) fibroblasts also have decreased activation (compared with wild-type PKR fibroblasts) of the mitogen-activated protein kinases of the p38 and c-Jun N-terminal kinase subgroups in response to poly I:C, LPS, IFN-γ, IL-1β, and tumor necrosis factor (TNF)-α.

Bandyopadhyay *et al.* studied the expression of E-selectin in murine aortic endothelial (MuAE) cells (73). E-selectin is a cell surface adhesion molecule that mediates the initial step of leukocyte adherence to the endothelium. E-selectin is expressed on endothelial cells in acute and chronic inflammation, and can be induced by TNF-α or poly I:C (as well as IL-1). Compared to wild-type mice, MuAE cells derived from PKR -/-(N) mice have decreased induction of both E-selectin cell surface protein and mRNA when treated with TNF-α or poly I:C. The same pattern of decreased induction is seen in a transfected construct that fused the E-selectin promoter to a reporter gene. The E-selectin promoter contains both κB and ELAM-1 elements (that bind NF-κB and NF-ELAM-1, respectively). TNF-α or poly I:C-stimulated MuAE PKR -/-(N) cells have less NF-κB and NF-ELAM-1 DNA-binding activity, again compared with MuAE wild-type cells. These results suggest that decreased levels of E-selectin expression on endothelial cells exposed to inflammatory mediators could result in impaired leukocyte response to inflammation in PKR -/-(N) mice.

PKR also has a role in the regulation of nitric oxide (NO) in response to inflammation. NO is produced by NO synthetase 2 (NOS2) and has potent anti-viral and pro-inflammatory effects. The initial report of Uetani *et al.* (74), found that human epithelial airway cells expressed increased levels of NOS2 mRNA in response to infection with influenza virus or treatment with poly I:C or IFN-γ. The same cells have increased NF-κB and IFN regulatory factor-1 DNA-binding activity when treated with TNF-α, poly I:C or IFN-γ.

NOS2 mRNA induction by poly I:C or LPS is much less in PKR -/-(N) MEFs compared with wild-type MEFs. PKR -/-(N) MEFs also have decreased induction of NF-κB and IFN regulatory factor-1 DNA-binding activity after treatment with poly I:C or LPS. These results indicate that PKR is necessary for maximal NOS2 mRNA induction in the MEFs, and suggest that it may also mediate induction of NOS2 in human airway epithelial cells in response to influenza virus infection.

Maggi *et al.* studied the role of PKR in macrophage activation (75), which also involves expression of inducible nitric oxide synthase (iNOS). Treatment of a macrophage cell line (RAW264.7) with poly I:C stimulates iNOS protein expression and nitric oxide synthesis. Both responses are blocked in cells expressing dominant-negative mutants of PKR. IL-1 release in response to poly I:C is also blocked. Treatment of the cell line with a combination of poly I:C and IFN-γ overcomes the dominant-negative PKR block, resulting in expression of iNOS and nitric oxide and IL-1 release. This effect is mediated by an IFN-γ -dependent degradation of PKR, with > 90% loss of PKR immunoreactivity. Consistent with this finding, treatment of primary macrophages isolated from PKR -/-(N) mice with poly I:C and IFN-γ results in iNOS expression and nitric oxide and IL-1 release at levels identical to those in primary macrophages from wild-type mice treated similarly. These results, like those from transformation experiments discussed above, suggest that dominant-negative forms of PKR block the action of redundant kinases that are still present in PKR knockouts.

PERK and ER Stress Response

Harding *et al.* generated mouse embryonic stem cells homozygous for a disrupted PERK gene (PERK -/- cells) (76). Western blots of these PERK -/- cells confirm the absence of PERK protein. When WT cells are treated with agents that cause endoplasmic reticulum (ER) stress (thapsigarin, tunicamycin or dithiothreitol), the rate of protein synthesis is significantly reduced and mRNAs are released from polysomes. In contrast, in PERK -/- cells there is little change in the rate of protein synthesis or evidence of mRNA release. Wild-type cells have increased levels of the phosphorylated form of eIF-2α when treated with agents that cause ER stress, while no increase is seen in PERK -/- cells. PERK -/- cells have decreased survival, compared to wild-type cells, when treated with agents causing ER stress.

PKR Effects on Apoptosis

The apoptotic response of MEFs was also examined using actinomycin D with either TNF-α, LPS, or poly I:C. Der *et al.* examined apoptosis in wild-type and PKR -/-(N) MEFs (77), while Abraham *et al.* used PKR -/-(C) MEFs (63). Interestingly, different results were obtained in experiments using the two different knockout cell lines. Der *et al.* found that PKR -/-(N) MEFs have decreased viability (<40%) when treated with actinomycin D (50 ng/ml) plus either poly I:C, LPS, or TNF-α. Under identical conditions, wild-type MEFs have >80% viability. In contrast, under identical conditions, Abraham *et al.* found that both wild-type and PKR -/-(C) MEFs have >75% viability. Decreased cell viability is seen only with high concentrations (500 ng/ml) of actinomycin D plus TNF-α, and the decrease is similar in PKR -/-(C) and wild-type cells.

It is not clear whether the differences between the PKR -/-(N) and PKR -/-(C) MEFs are due to the methods that were used to disrupt the PKR gene or due to the different background mouse strains used. The genetic background alone makes a difference in wild-type cells treated with increasing concentrations of actinomycin D alone. The 129/SV(ev) x C57BL/6J background wild-type MEFs used by Der *et al.* (63) have 50% viability at a concentration of 100 ng/ml of actinomycin D, while129/terSv x BALB/C background wild-type MEFs have 65% viability at the maximum actinomycin D concentration (2500 ng/ml) used by Abraham *et al.* These studies suggest that the ability of genetic experiments to determine the effect of PKR on interesting biological phenotypes, such as infection resistance and tumor vulnerability, may vary depending on the background in which the genetic modification to PKR is made.

PKR and Differentiation

Salzberg *et al.* showed that PKR plays an important role in the myogenic differentiation of murine C2C12 cells (78). Under defined tissue culture conditions, C2C12 cells undergo myogenic differentiation, in a process that includes expression of MyoD and induction of the cyclin-dependent kinase inhibitor p21(WAF). Under conditions that induce myogenic differentiation, transfected clones of C2C12 cells expressing a dominant-negative form of PKR have reduced myotube formation, increased cell number, and increased thymidine uptake compared to control cells. The dominant-negative expressing cell lines also have reduced protein levels of MyoD, troponin T and creatine kinase activity, which are markers of muscle-specific differentiation. Delayed expression of p21(WAF) is also observed in the dominant-negative containing cells. These findings support a model in which the increased PKR activity arrests cell proliferation, which is necessary before the cells can express their differentiation program. A similar pattern of PKR activation and growth arrest prior to adipocyte differentiation is seen in 3T3-F442A cells (28).

CONCLUSIONS

Despite substantial biochemical and molecular information, and the use of PKR knockout mice, the full physiologic function of PKR, especially its role in tumor suppression, has yet to be resolved. Questions remain as to the scope and nature of the cellular RNAs that interact with the DRBMs of PKR, and the other cellular proteins involved in the regulation of PKR. Is the expression and activity of such agonists and antagonists of PKR altered under pathologic conditions? Do specific ancillary proteins function to activate cellular PKR in response to stress in the absence of dsRNA? Are there additional physiologic substrates for PKR? The answers to these and other questions will not only increase our understanding of the regulatory role of this clinically important kinase, but also establish approaches for studying the function of other dsRNA-binding proteins.

Summary of key concepts

- PKR is a serine/threonine protein kinase. Activation of PKR involves phosphorylation and is dependent on binding to dsRNA or dsRNA structures within single- stranded viral or cellular RNAs. The best characterized substrate of PKR is the α- subunit of the initiation factor eIF-2.
- PKR activation occurs in response to viral infection, cellular stress, and signals that oppose proliferation and promote differentiation. The activity of PKR is modulated by the interaction of protein or dsRNA with PKR directly, or by the interaction of proteins with cellular dsRNAs.
- Dominant-negative mutants of PKR can produce a transformed phenotype, but transformation is not observed in PKR-knockout mice or cell lines derived from them. Several lines of evidence suggest that this apparent paradox is the result of undiscovered proteins with PKR-like activity that provide redundant function in the knockouts, but which are inhibited by the dominant-negative mutant.
- PKR-knockout mice have abnormal responses to some viral infections and alterations in the production of inflammatory mediators.

Study Guide Questions

1) What evidence supports the hypothesis that the availability of cellular dsRNA limits the activity of PKR in cells not infected by a virus?
2) What other hypotheses, beside the presence of redundant proteins with PKR-like activity, could explain the difference in transformation phenotype seen between introduction of a dominant-negative PKR mutant and PKR knockout?
3) If virus X replicates to the same extent in wild type and PKR-knockout cells, what type of protein might you look for in the viral genome? How could you test for this protein genetically? Biochemically?

Acknowledgement

Supported by the Children's Cancer Foundation, National Childhood Cancer Foundation, Board of Lady Visitors of Children's National Medical Center, and by the Elaine H. Snyder Cancer Research Fund.

REFERENCES

1. Fierro-Monti I, Mathews MB. Proteins binding to duplexed RNA: one motif, multiple functions. Trends Biochem Sci 2000;25(5):241-6.
2. de Haro C, Mendez R, Santoyo J. The eIF-2alpha kinases and the control of protein synthesis. Faseb J 1996;10(12):1378-87.
3. Clemens MJ, Elia A. The double-stranded RNA-dependent protein kinase PKR: structure and function. J Interferon Cytokine Res 1997;17(9):503-24.
4. Shi Y, Vattem KM, Sood R, An J, Liang J, Stramm L, Wek RC. Identification and characterization of pancreatic eukaryotic initiation factor 2 alpha-subunit kinase, PEK, involved in translational control. Mol Cell Biol 1998;18(12):7499-509.
5. Harding HP, Zhang Y, Ron D. Protein translation and folding are coupled by an endoplasmic-reticulum- resident kinase [published erratum appears in Nature 1999 Mar 4;398(6722):90]. Nature 1999;397(6716):271-4.
6. Williams BR. PKR; a sentinel kinase for cellular stress. Oncogene 1999;18(45):6112-20.
7. Jaramillo ML, Abraham N, Bell JC. The interferon system: a review with emphasis on the role of PKR in growth control. Cancer Invest 1995;13(3):327-38.
8. Gatignol A, Buckler C, Jeang KT. Relatedness of an RNA-binding motif in human immunodeficiency virus type 1 TAR RNA-binding protein TRBP to human P1/dsI kinase and Drosophila staufen. Mol Cell Biol 1993;13(4):2193-202.
9. St Johnston D, Brown NH, Gall JG, Jantsch M. A conserved double-stranded RNA-binding domain. Proc Natl Acad Sci U S A 1992;89(22):10979-83.
10. Eckmann CR, Jantsch MF. Xlrbpa, a double-stranded RNA-binding protein associated with ribosomes and heterogeneous nuclear RNPs. J Cell Biol 1997;138(2):239-53.
11. Park H, Davies MV, Langland JO, Chang HW, Nam YS, Tartaglia J, Paoletti E, Jacobs BL, Kaufman RJ, Venkatesan S. TAR RNA-binding protein is an inhibitor of the interferon-induced protein kinase PKR. Proc Natl Acad Sci U S A 1994;91(11):4713-7.
12. Benkirane M, Neuveut C, Chun RF, Smith SM, Samuel CE, Gatignol A, Jeang KT. Oncogenic potential of TAR RNA binding protein TRBP and its regulatory interaction with RNA-dependent protein kinase PKR. Embo J 1997;16(3):611-24.
13. Cosentino GP, Venkatesan S, Serluca FC, Green SR, Mathews MB, Sonenberg N. Double-stranded-RNA-dependent protein kinase and TAR RNA-binding protein form homo- and heterodimers in vivo. Proc Natl Acad Sci U S A 1995;92(21):9445-9.

14. Kao PN, Chen L, Brock G, Ng J, Kenny J, Smith AJ, Corthesy B. Cloning and expression of cyclosporin A- and FK506-sensitive nuclear factor of activated T-cells: NF45 and NF90. J Biol Chem 1994;269(32):20691-9.
15. Traviss CE, McArdle JL, Marques SMP, Nekhai S, Burgess WH, Lamb NJC, A. K. Virology 2000, submitted.
16. Xu YH, Busald C, Grabowski GA. Reconstitution of TCP80/NF90 translation inhibition activity in insect cells. Mol Genet Metab 2000;70(2):106-15.
17. Xu YH, Grabowski GA. Molecular cloning and characterization of a translational inhibitory protein that binds to coding sequences of human acid beta-glucosidase and other mRNAs. Mol Genet Metab 1999;68(4):441-54.
18. Corthesy B, Kao PN. Purification by DNA affinity chromatography of two polypeptides that contact the NF-AT DNA binding site in the interleukin 2 promoter. J Biol Chem 1994;269(32):20682-90.
19. Liao HJ, Kobayashi R, Mathews MB. Activities of adenovirus virus-associated RNAs: purification and characterization of RNA binding proteins. Proc Natl Acad Sci U S A 1998;95(15):8514-9.
20. Langland JO, Kao PN, Jacobs BL. Nuclear factor-90 of activated T-cells: A double-stranded RNA-binding protein and substrate for the double-stranded RNA-dependent protein kinase, PKR. Biochemistry 1999;38(19):6361-8.
21. Patel RC, Vestal DJ, Xu Z, Bandyopadhyay S, Guo W, Erme SM, Williams BR, Sen GC. DRBP76, a double-stranded RNA-binding nuclear protein, is phosphorylated by the interferon-induced protein kinase, PKR. J Biol Chem 1999;274(29):20432-7.
22. Patel RC, Sen GC. PACT, a protein activator of the interferon-induced protein kinase, PKR. Embo J 1998;17(15):4379-90.
23. Patel CV, Handy I, Goldsmith T, Patel RC. PACT, a stress-modulated cellular activator of interferon-induced, double-stranded RNA activated protein kinase, PKR. J Biol Chem 2000, in press.
24. Ito T, Yang M, May WS. RAX, a cellular activator for double-stranded RNA-dependent protein kinase during stress signaling. J Biol Chem 1999;274(22):15427-32.
25. James MC, Jeffrey IW, Pruijn GJ, Thijssen JP, Clemens MJ. Translational control by the La antigen. Structure requirements for rescue of the double-stranded RNA-mediated inhibition of protein synthesis. Eur J Biochem 1999;266(1):151-62.
26. Petryshyn R, Chen JJ, Danley L, Matts RL. Effect of interferon on protein translation during growth stages of 3T3 cells. Arch Biochem Biophys 1996;326(2):290-7.
27. Judware R, Petryshyn R. Mechanism of action of a cellular inhibitor of the dsRNA-dependent protein kinase from 3T3-F442A cells. J Biol Chem 1992;267(30):21685-90.
28. Woldehawariat G, Nekhai S, Petryshyn R. Differential phosphorylation of PKR associates with deregulation of eIF- 2alpha phosphorylation and altered growth characteristics in 3T3-F442A fibroblasts. Mol Cell Biochem 1999;198(1-2):7-17.
29. Green SR, Mathews MB. Two RNA-binding motifs in the double-stranded RNA-activated protein kinase, DAI. Genes Dev 1992;6(12B):2478-90.
30. Bevilacqua PC, Cech TR. Minor-groove recognition of double-stranded RNA by the double-stranded RNA-binding domain from the RNA-activated protein kinase PKR. Biochemistry 1996;35(31):9983-94.
31. McCormack SJ, Thomis DC, Samuel CE. Mechanism of interferon action: identification of a RNA binding domain within the N-terminal region of the human RNA-dependent P1/eIF-2 alpha protein kinase. Virology 1992;188(1):47-56.
32. Nekhai S, Kumar A, Bottaro DP, Petryshyn R. Peptides derived from the interferon-induced PKR prevent activation by HIV-1 TAR RNA. Virology 1996;222(1):193-200.
33. Nekhai S, Bottaro DP, Woldehawariat G, Spellerberg A, Petryshyn R. A cell-permeable peptide inhibits activation of PKR and enhances cell proliferation. Peptides 2000;21(10):1449-56.

34. Taylor DR, Shi ST, Romano PR, Barber GN, Lai MM. Inhibition of the interferon-inducible protein kinase PKR by HCV E2 protein [see comments]. Science 1999;285(5424):107-10.
35. Gale M, Jr., Blakely CM, Kwieciszewski B, Tan SL, Dossett M, Tang NM, Korth MJ, Polyak SJ, Gretch DR, Katze MG. Control of PKR protein kinase by hepatitis C virus nonstructural 5A protein: molecular mechanisms of kinase regulation. Mol Cell Biol 1998;18(9):5208-18.
36. Gale MJ, Jr., Korth MJ, Tang NM, Tan SL, Hopkins DA, Dever TE, Polyak SJ, Gretch DR, Katze MG. Evidence that hepatitis C virus resistance to interferon is mediated through repression of the PKR protein kinase by the nonstructural 5A protein. Virology 1997;230(2):217-27.
37. Judware R, Li J, Petryshyn R. Inhibition of the dsRNA-dependent protein kinase by a peptide derived from the human immunodeficiency virus type 1 Tat protein. J Interferon Res 1993;13(2):153-60.
38. Brand SR, Kobayashi R, Mathews MB. The Tat protein of human immunodeficiency virus type 1 is a substrate and inhibitor of the interferon-induced, virally activated protein kinase, PKR. J Biol Chem 1997;272(13):8388-95.
39. Cai R, Carpick B, Chun RF, Jeang KT, Williams BR. HIV-I TAT inhibits PKR activity by both RNA-dependent and RNA- independent mechanisms. Arch Biochem Biophys 2000;373(2):361-7.
40. Haig DM, McInnes CJ, Thomson J, Wood A, Bunyan K, Mercer A. The orf virus OV20.0L gene product is involved in interferon resistance and inhibits an interferon-inducible, double-stranded RNA-dependent kinase. Immunology 1998;93(3):335-40.
41. Langland JO, Pettiford S, Jiang B, Jacobs BL. Products of the porcine group C rotavirus NSP3 gene bind specifically to double-stranded RNA and inhibit activation of the interferon-induced protein kinase PKR. J Virol 1994;68(6):3821-9.
42. Yue Z, Shatkin AJ. Double-stranded RNA-dependent protein kinase (PKR) is regulated by reovirus structural proteins. Virology 1997;234(2):364-71.
43. Kawagishi-Kobayashi M, Cao C, Lu J, Ozato K, Dever TE. Pseudosubstrate Inhibition of Protein Kinase PKR by Swine Pox Virus C8L Gene Product. Virology 2000;276(2):424-434.
44. Romano PR, Zhang F, Tan SL, Garcia-Barrio MT, Katze MG, Dever TE, Hinnebusch AG. Inhibition of double-stranded RNA-dependent protein kinase PKR by vaccinia virus E3: role of complex formation and the E3 N-terminal domain. Mol Cell Biol 1998;18(12):7304-16.
45. Sharp TV, Moonan F, Romashko A, Joshi B, Barber GN, Jagus R. The vaccinia virus E3L gene product interacts with both the regulatory and the substrate binding regions of PKR: implications for PKR autoregulation. Virology 1998;250(2):302-15.
46. Sharp TV, Witzel JE, Jagus R. Homologous regions of the alpha subunit of eukaryotic translational initiation factor 2 (eIF2alpha) and the vaccinia virus K3L gene product interact with the same domain within the dsRNA-activated protein kinase (PKR). Eur J Biochem 1997;250(1):85-91.
47. O'Malley RP, Duncan RF, Hershey JW, Mathews MB. Modification of protein synthesis initiation factors and the shut-off of host protein synthesis in adenovirus-infected cells. Virology 1989;168(1):112-8.
48. Elia A, Laing KG, Schofield A, Tilleray VJ, Clemens MJ. Regulation of the double-stranded RNA-dependent protein kinase PKR by RNAs encoded by a repeated sequence in the Epstein-Barr virus genome. Nucleic Acids Res 1996;24(22):4471-8.
49. Gunnery S, Green SR, Mathews MB. Tat-responsive region RNA of human immunodeficiency virus type 1 stimulates protein synthesis in vivo and in vitro: relationship between structure and function. Proc Natl Acad Sci U S A 1992;89(23):11557-61.
50. Edery I, Petryshyn R, Sonenberg N. Activation of double-stranded RNA-dependent kinase (dsl) by the TAR region of HIV-1 mRNA: a novel translational control mechanism. Cell 1989;56(2):303-12.

51. Robertson HD, Manche L, Mathews MB. Paradoxical interactions between human delta hepatitis agent RNA and the cellular protein kinase PKR. J Virol 1996;70(8):5611-7.
52. Henry GL, McCormack SJ, Thomis DC, Samuel CE. Mechanism of interferon action. Translational control and the RNA- dependent protein kinase (PKR): antagonists of PKR enhance the translational activity of mRNAs that include a 161 nucleotide region from reovirus S1 mRNA. J Biol Regul Homeost Agents 1994;8(1):15-24.
53. Mordechai E, Kon N, Henderson EE, Suhadolnik RJ. Activation of the interferon-inducible enzymes, 2',5'-oligoadenylate synthetase and PKR by human T-cell leukemia virus type I Rex-response element. Virology 1995;206(2):913-22.
54. Tian B, White RJ, Xia T, Welle S, Turner DH, Mathews MB, Thornton CA. Expanded CUG repeat RNAs form hairpins that activate the double- stranded RNA-dependent protein kinase PKR. Rna 2000;6(1):79-87.
55. Davis S, Watson JC. In vitro activation of the interferon-induced, double-stranded RNA-dependent protein kinase PKR by RNA from the 3' untranslated regions of human alpha-tropomyosin. Proc Natl Acad Sci U S A 1996;93(1):508-13.
56. Petryshyn RA, Ferrenz AG, Li J. Characterization and mapping of the double-stranded regions involved in activation of PKR within a cellular RNA from 3T3-F442A cells. Nucleic Acids Res 1997;25(13):2672-8.
57. Chu WM, Ballard R, Carpick BW, Williams BR, Schmid CW. Potential Alu function: regulation of the activity of double-stranded RNA-activated kinase PKR. Mol Cell Biol 1998;18(1):58-68.
58. Yang YL, Reis LF, Pavlovic J, Aguzzi A, Schafer R, Kumar A, Williams BR, Aguet M, Weissmann C. Deficient signaling in mice devoid of double-stranded RNA-dependent protein kinase. Embo J 1995;14(24):6095-106.
59. Barber GN, Wambach M, Thompson S, Jagus R, Katze MG. Mutants of the RNA-dependent protein kinase (PKR) lacking double- stranded RNA binding domain I can act as transdominant inhibitors and induce malignant transformation. Mol Cell Biol 1995;15(6):3138-46.
60. Koromilas AE, Roy S, Barber GN, Katze MG, Sonenberg N. Malignant transformation by a mutant of the IFN-inducible dsRNA- dependent protein kinase. Science 1992;257(5077):1685-9.
61. Meurs EF, Galabru J, Barber GN, Katze MG, Hovanessian AG. Tumor suppressor function of the interferon-induced double-stranded RNA- activated protein kinase. Proc Natl Acad Sci U S A 1993;90(1):232-6.
62. Donze O, Jagus R, Koromilas AE, Hershey JW, Sonenberg N. Abrogation of translation initiation factor eIF-2 phosphorylation causes malignant transformation of NIH 3T3 cells. Embo J 1995;14(15):3828-34.
63. Abraham N, Stojdl DF, Duncan PI, Methot N, Ishii T, Dube M, Vanderhyden BC, Atkins HL, Gray DA, McBurney MW and others. Characterization of transgenic mice with targeted disruption of the catalytic domain of the double-stranded RNA-dependent protein kinase, PKR. J Biol Chem 1999;274(9):5953-62.
64. Zhou A, Paranjape J, Brown TL, Nie H, Naik S, Dong B, Chang A, Trapp B, Fairchild R, Colmenares C and others. Interferon action and apoptosis are defective in mice devoid of 2',5'- oligoadenylate-dependent RNase L. Embo J 1997;16(21):6355-63.
65. Zhou A, Paranjape JM, Der SD, Williams BR, Silverman RH. Interferon action in triply deficient mice reveals the existence of alternative antiviral pathways. Virology 1999;258(2):435-40.
66. Qin X, Shibata D, Gerson SL. Heterozygous DNA mismatch repair gene PMS2-knockout mice are susceptible to intestinal tumor induction with N-methyl-N-nitrosourea. Carcinogenesis 2000;21(4):833-8.
67. Stojdl DF, Abraham N, Knowles S, Marius R, Brasey A, Lichty BD, Brown EG, Sonenberg N, Bell JC. The murine double-stranded RNA-dependent protein kinase PKR is required for resistance to vesicular stomatitis virus. J Virol 2000;74(20):9580-5.

68. Bergmann M, Garcia-Sastre A, Carnero E, Pehamberger H, Wolff K, Palese P, Muster T. Influenza virus NS1 protein counteracts PKR-mediated inhibition of replication. J Virol 2000;74(13):6203-6.
69. Lu Y, Wambach M, Katze MG, Krug RM. Binding of the influenza virus NS1 protein to double-stranded RNA inhibits the activation of the protein kinase that phosphorylates the elF-2 translation initiation factor. Virology 1995;214(1):222-8.
70. Leib DA, Machalek MA, Williams BR, Silverman RH, Virgin HW. Specific phenotypic restoration of an attenuated virus by knockout of a host resistance gene. Proc Natl Acad Sci U S A 2000;97(11):6097-101.
71. He B, Gross M, Roizman B. The gamma134.5 protein of herpes simplex virus 1 has the structural and functional attributes of a protein phosphatase 1 regulatory subunit and is present in a high molecular weight complex with the enzyme in infected cells. J Biol Chem 1998;273(33):20737-43.
72. Goh KC, deVeer MJ, Williams BR. The protein kinase PKR is required for p38 MAPK activation and the innate immune response to bacterial endotoxin. Embo J 2000;19(16):4292-7.
73. Bandyopadhyay SK, de La Motte CA, Williams BR. Induction of E-selectin expression by double-stranded RNA and TNF-alpha is attenuated in murine aortic endothelial cells derived from double- stranded RNA-activated kinase (PKR)-null mice. J Immunol 2000;164(4):2077-83.
74. Uetani K, Der SD, Zamanian-Daryoush M, de La Motte C, Lieberman BY, Williams BR, Erzurum SC. Central role of double-stranded RNA-activated protein kinase in microbial induction of nitric oxide synthase. J Immunol 2000;165(2):988-96.
75. Maggi LB, Jr., Heitmeier MR, Scheuner D, Kaufman RJ, Buller RM, Corbett JA. Potential role of PKR in double-stranded RNA-induced macrophage activation. Embo J 2000;19(14):3630-8.
76. Harding HP, Zhang Y, Bertolotti A, Zeng H, Ron D. Perk is essential for translational regulation and cell survival during the unfolded protein response. Mol Cell 2000;5(5):897-904.
77. Der SD, Yang YL, Weissmann C, Williams BR. A double-stranded RNA-activated protein kinase-dependent pathway mediating stress-induced apoptosis. Proc Natl Acad Sci U S A 1997;94(7):3279-83.
78. Salzberg S, Vilchik S, Cohen S, Heller A, Kronfeld-Kinar Y. Expression of a PKR dominant-negative mutant in myogenic cells interferes with the myogenic process. Exp Cell Res 2000;254(1):45-54.

11

REGULATION OF MESSENGER RNA-BINDING PROTEINS BY PROTEIN KINASES A AND C

Richard A. Jungmann

Northwestern University Medical School, Chicago, IL 60611 USA

The discovery in 1983 that catecholamines via the second messenger cAMP stabilize the mRNA of the lactate dehydrogenase-A (LDH-A) subunit gene indicated for the first time that intracellular mRNA levels are not only controlled at the level of transcription, but that post-transcriptional regulation of mRNA stability is an equally vital mechanism of gene control (1). This finding pointed to a major new mechanism controlling gene expression through binding of effector agents to plasma membrane receptors, a process that elevates intracellular levels of important second messengers such as diacyl glycerol, cAMP, inositol phosphates, ionized calcium and others. These second messengers in turn activate specific protein kinases leading to the phosphorylation and functional modification of putative trans- regulatory proteins that may participate in modulating the mRNA decay mechanism. Two of the major signal transduction pathways that control gene expression involve activation of protein kinase A (PKA) (2) and protein kinase C (PKC) (3). Indeed, evidence has accumulated over recent years showing that the stability of a number of mRNA species is changed by extracellular signals that activate PKA and/or PKC exerting both positive and negative effects on mRNA stability (1, 4-17). This implicates protein kinase target proteins in the process of mRNA stability regulation. Assuming the existence of phosphoproteins as trans-acting mRNA-binding factors, intriguing questions arise: How do these factors modulate mRNA turnover rates and what is their biochemical nature? Do factors interact with mRNA directly or through intermediates, and what is the molecular mechanism of the key interactions? The major goal of this chapter is to discuss the latest insights into the molecular mechanism of mRNA stabilization by two major signal transduction pathways represented by PKA and PKC.

BACKGROUND

Messenger RNAs have a characteristic half-life that determines, in part, their intracellular steady-state levels. However, in most cases half-lives are not fixed characteristics but may be modulated in response to various biological signals to maintain a precisely balanced homeostatic environment of gene expression (13,18,19). Key facets of mRNA turnover include: physiological signals initiating the stabilizing/destabilizing effects; structural element(s) within mRNA acting as stabilizing/destabilizing module(s); and *trans*-acting factors that modulate the decay process. While some of the *trans*-factors may be constitutively expressed in cells, others may be induced through controlled *de novo* synthesis and/or activation by effector agents *via* signal pathways. This chapter will discuss posttranscriptional regulation of lactate dehydrogenase A subunit mRNA (LDH-A mRNA) stability to illuminate this interesting and important aspect of gene regulation.

Characterization of LDH-A 3'-Untranslated Region

LDH-A mRNA is characterized by a relatively short half-life of about 45-55 min (1, 12). Untranslated sequences that are A- and U-rich are present in its 507-base 3'-UTR (bases 1103-1610). However, the 3'-UTR lacks previously identified consensus AU-rich elements, such as AUUUA, that is one of the recognition signals involved in mRNA stability/instability and as an attachment site for putative regulatory RNA-binding proteins. LDH-A 3'-UTR contains about 24% adenine, 28% uridine, 23% guanine, and 25% cytosine. Thus, it is not AU rich, however, a 99-nucleotide region (nt 1450-1549) has 70% AU content. This region does not share significant sequence similarity with some well-studied 3'-UTR instability elements, for instance those in c-fos, c-myc, GM-CSF, or histone. Common instability motifs such as UUAUUUU(A/U)(A/U) or AUUUA are not present, implying that LDH-A mRNA degradation may not be regulated in the same manner as mRNAs containing the above consensus motifs. The half-life of LDH-A mRNA ($t_{1/2} \approx$ 55 min) is increased in glioma cells after treatment with agents that activate the protein kinase (PK) A and C signal transduction pathways (1,12). These findings prompted us to ask: how is the mRNA structure involved in determining its relative stability; can *trans* factors be identified; and what is the mechanism that regulates LDH-A mRNA stability by PKA and C?

LDH-A 3'-UTR is Responsible for mRNA Instability

In order to investigate the functional role of LDH-A 3'-UTR in determining LDH-A mRNA half-life, expression vector pRc/FBB, containing the full-length rabbit β-globin sequence as a marker gene, was used for the construction of chimeric β-globin/ldh 3'-UTR expression vectors and for decay analysis of chimeric globin/ldh mRNAs. In this plasmid, the LDH-A 3'-UTR sequence as well as truncated 3'-UTR fragments had been inserted into the *Bgl* II site of pRc/FBB giving rise to chimeric β-globin/ldh 3'-UTR vectors (Fig. 1).

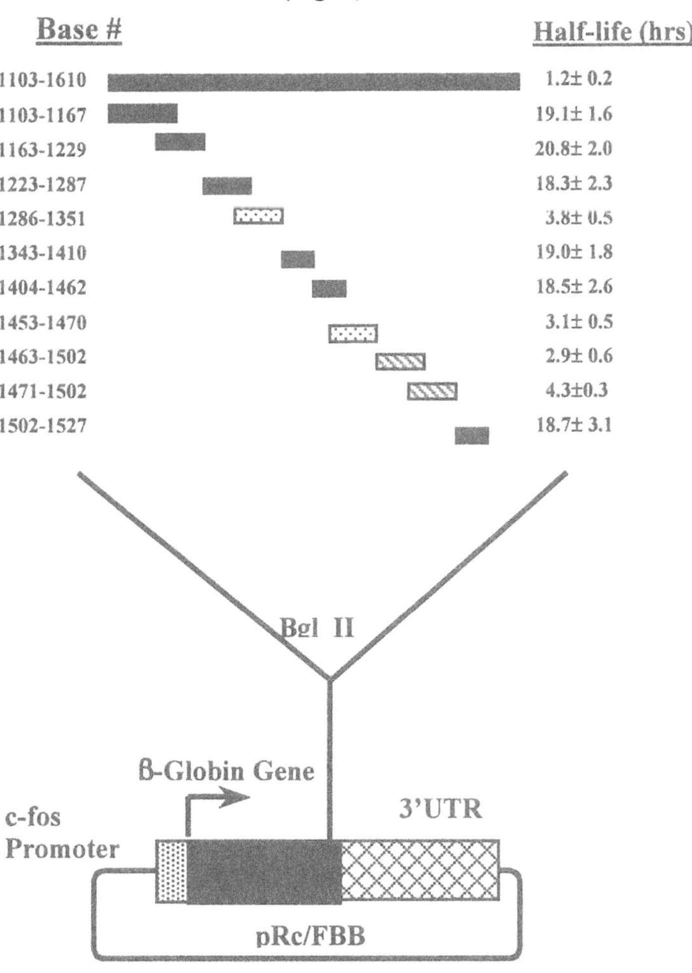

Figure 1. Rat C6 glioma cells were stably transfected with pRc/FBB in which the listed fragments of LDH-A 3'UTR (with 5' and 3' Bgl II ends) had been inserted into the Bgl II site of pRc/FBB. *After transfection and addition of serum, RNA was isolated at various time points up to 16 hours. Globin/ldh mRNA decay was quantitated by ribonuclease protection assay, and half-lives were calculated. Results are expressed as mean and SEM of four separate experiments. The half-life of pRc/FBB in which globin 3'-UTR had been replaced with the complete LDH 3'UTR (nt 1103-1610) is given for comparison. Dotted fragments, containing instability element(s) not regulated by either Sp-cAMPS or DG; diagonally hatched fragments, containing instability elements regulated by Sp-cAMPS and/or DG.*

After transfection and a transient pulse of transcription, decay analysis of chimeric globin/ldh mRNA with the complete LDH-A 3'-UTR (nucleotides 1103-1610) identified a half-life ($t_{1/2}$) of about 1.2 hours (15) similar to the half-life of wild-type LDH-A mRNA (1). This value is much lower than the wild-type β-globin mRNA (expressed from unmodified pRc/FBB) half-life of about 21 hours (Fig. 2).

Figure 2. Decay of chimeric globin/ldh mRNA as a function of time. Decay was assayed by ribonuclease protection assay (17). ▲, control; O, DG-treated cells; +, Sp-cAMPS-treated cells; • ; DG+Sp-cAMPS-treated cells; ■ ,wild-type β-globin mRNA.

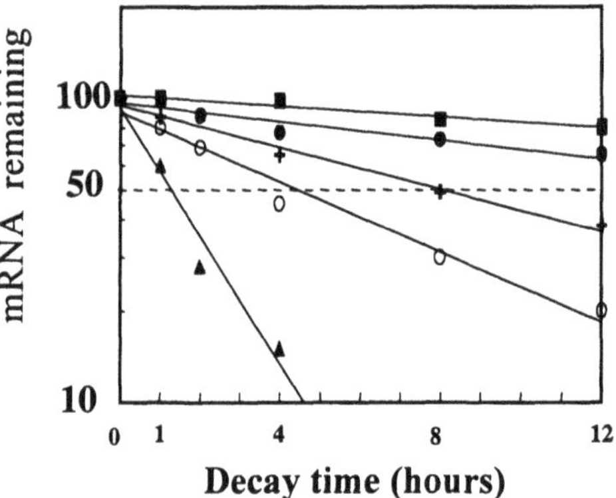

After truncation and insertion into pRc/FBB, several of the LDH-A 3'-UTR fragments did not significantly change wild-type globin mRNA half-life, and the chimeric mRNAs exhibited half-lives similar to wild-type globin. In contrast, insertion of several other fragments, consisting of nucleotides 1286-1351, 1453-1470, 1463-1502, and 1471-1502, markedly destabilized wild-type globin half-life and the corresponding chimeric globin/ldh mRNAs exhibited significantly reduced half-lives (between 2.9 and 4.3 hours). From these data, we conclude that LDH-A 3'-UTR contains at least three determinants of instability located in regions consisting of nucleotides 1286-1351, 1453-1470, and 1471-1502. Furthermore, LDH-A mRNA stability/instability is an inherent function of its 3'-UTR and is not affected by other regions of the mRNA.

The Half-Life of LDH-A mRNA is Affected by Activation of Protein Kinases A and C

Protein kinases A and C are members of the ser/thr/tyr protein kinase family that play pivotal roles in regulating metabolic processes, cellular differentiation and growth. Many of these cellular functions are regulated through phosphorylative and functional modification of target molecules such as enzymes and a myriad of other non-enzymatic regulatory nuclear and cytoplasmic proteins. The degree of phosphorylation is primarily determined by the enzymatic balance between the activity of protein kinases and phosphatases. Messenger RNA-binding proteins potentially represent key targets for protein kinases and phosphatases, since the reversible phosphorylation/dephosphorylation cycle may allow a fine-tuning of RNA-binding protein activity and, as consequence, mRNA stability. Disturbance of this balance may lead to pathological changes of gene expression caused by the altered homeostasis of intracellular mRNA levels.

Recently, some of the fundamental mechanisms mediating the effects of cAMP or phorbol ester on mRNA stability *via* activation of PKA and PKC have, in part, been elucidated (15, 17). DG, a specific activator of PKC, as well as Sp-cAMPS, a potent activator of PKA, achieved a marked increase of chimeric globin/ldh mRNA stability (Fig. 2). Activation of PKC stabilized the half-life from 65 minutes to about 4 hours, whereas activation of PKA led to a 7-fold increase of chimeric globin/ldh mRNA from 65 minutes to about 8 hours (Fig. 2). The most remarkable finding consisted of the identification of a synergistic action between PKA and PKC on chimeric globin/ldh mRNA stability. When a combination of DG + Sp-cAMPS was used as activator of the protein kinase pathways, the half-life of chimeric globin/ldh mRNA increased 18-fold to a half-life of about 21 hours. An identical synergistic effect is also observed for the decay of wild-type LDH-A mRNA in glioma cells (12).

Systematic Analysis of LDH-A 3'-UTR for the Presence of Protein Kinase A and C-Stabilizing Element(s)

The basal half-lives of chimeric globin/ldh mRNAs determined in unstimulated glioma cells are shown in Fig. 1. Using a similar experimental approach, the effects of protein PKA and/or PKC activation on chimeric mRNA decay were assessed. As shown in Table 1, several chimeric mRNAs were stable over the time course of the decay period. Others exhibited a relatively short half-life without being affected by Sp-cAMPS and/or DG. Chimeric β-globin/ldh mRNA containing the complete LDH-A 3'-UTR (nt 1103-1610) was stabilized by both DG and Sp-cAMPS. A combination of DG + Sp-cAMPS resulted in a synergistic 23.4-fold stabilization of the chimeric mRNA. The chimeric mRNA resulting from insertion of the truncated 31-nt fragment 1471-1502 (as well as fragment 1463-1502) into pRc/FBB similarly responded to DG as well as Sp-cAMPS and became stabilized. However, the degree of stabilization was only additive and not synergistic.

TABLE 1. Effect of activators of PKA and PKC on the half-life of chimeric globin/ldh mRNAs

				Half-life (hours)			
Frag. Insert (bases)	Control	Sp-cAMPS (ScA)	ScA/ Cont	DG	DG/ Cont	DG + ScA	DG+ ScA/ Cont
1103-1610	1.2±0.6	8.3±0.9	7.1	6.5±0.8	5.4	28.1±2.8	23.4
1103-1167	19.1±1.6	18.6±2.5	0.9	16.5±2.0	1.1	20.1±3.2	1.1
1163-1229	20.8±2.0	19.9±3.4	0.9	13.9±1.9	0.9	18.5±2.7	0.9
1223-1287	18.3±2.3	21.5±3.8	1.2	15.9±1.9	0.9	19.8±1.7	1.1
1286-1351	3.9±0.5	3.1± 0.7	0.8	2.9±0.8	0.7	3.0±0.4	0.8
1343-1410	19.0±1.8	20.7±1.9	1.1	21.5±2.9	1.1	18.7±1.1	0.9
1404-1462	18.5±2.6	19.5±2.9	1.1	20.1±2.4	1.1	17.9±2.1	0.9
1453-1470	2.9±0.9	3.6±1.2	1.2	3.1±0.7	0.7	2.1±0.9	0.7
1463-1502	4.1±0.6	17.5±2.1	4.3	13.9±2.2	3.4	24.6±1.9	6.0
1471-1502	3.9±0.8	17.8±2.5	4.6	13.7±1.	3.5	29.1±1.7	7.4

Rat C6 glioma cells were stably transfected with pRc/FBB in which the listed fragments of LDH-A 3'UTR (with 5' and 3' Bgl II ends) had been inserted into the Bgl II site of pRc/FBB (see Fig. 1). Cells were treated in serum-free medium for 6 hours with either 0.5 mM Sp-cAMPS, 100 nM DG, or a combination of both. After addition of serum to initiate a brief pulse of transcription, RNA was isolated at various time points up to 12 hours. Globin/ldh mRNA decay was assessed by ribonuclease protection assay and half-lives were determined. Results are expressed as mean and SEM of four separate experiments.

Table 1 shows that the fragment comprised of nucleotides 1471-1502 contains the regulatory site targeted by PKA and PKC and which is responsible, at least in part, for PKA- and PKC-mediated stabilization of LDH-A mRNA. It is equally clear that insertion of the entire LDH-A 3'-UTR into pRc/FBB is required to obtain synergistic stabilization. So far, no synergism has been observed with any of the truncated fragments.

The question remained whether fragment 1471-1502 contained only one unique element that responded to the activation of PKA and PKC, or two elements for protein kinase specific regulation. To resolve this question, detailed deletion and mutational analyses of fragment 1471-1502 were carried out (15,17). The analyses resulted in the identification of two critical determinants for PKA- and PKC-mediated LDH-A mRNA stabilization: (a) the PKA-stabilizing region (PASR) consisting of the 22-nucleotide fragment 1479-1500, and (b) the PKC-stabilizing region (PCSR) consisting of a 20-nucleotide region 1471-1490 (Fig. 3).

AUAUUUUCUGUAUUAUAUGUGU	PASR
CUACAGGAUAUUUUCUGUAU	PCSR

Figure 3. Homologous bases in PASR and PCSR. *The PASR (nucleotides 1478-1500) and PCSR (nucleotides 1471-1490) are lined up to show the 13-nucleotide homology. Homologous bases are underlined.*

Both elements exhibit multiple functions: (a) they are determinants of LDH-A instability, (b) they are high affinity binding sites for RNA-binding proteins and (c) they function specifically as modulators of LDH-A mRNA half-life depending on the state of activation of either PKA or PKC.

Perhaps the most surprising finding is that PASR and PCSR contain a 13-nt region (nt 1478-1491; -AUAUUUUCUGUAU-) that is common to both regulatory motifs (Fig. 3). Since the 13 nucleotides constitute 59% of the PASR and 65% of the PCSR, respectively, the question arises as to the significance of the overlapping arrangement of the two stabilizing regions. One functional property common to both PASR and PCSR, namely destabilization of LDH-A mRNA, is also exhibited by a class of short-lived mRNAs that share AU-rich motifs in their 3'-UTR (13). However, it is unlikely that relatively non-specific AU-rich sequences will provide for a selective regulation of mRNA stability, for mRNAs that are under posttranscriptional control by certain effector agents (e.g. cAMP and DG). Rather, additional determinants in PASR and PCSR, other than the 13-nt common sequence, seem to be required to modulate stability regulation in response to the PKA and PKC signal pathways. Thus, it can be hypothesized that the composite of the PKC- and PKA-regulated region of LDH-A 3'UTR consists of two discrete module(s): (a) the 13-nt overlapping non-specific AU-rich determinant in PASR and PCSR; and (b) specific site(s), e.g. the PASR-specific 3' end (nucleotides 1491-1500) and the PCSR-specific 5' region (nt 1471-1478), that in combination with phospho/dephosphoproteins are instrumental in determining the specificity of the stabilizing response.

Identification of LDH-A mRNA 3'-UTR-Binding Proteins

Several selectively acting LDH-A 3'-UTR-binding proteins have been identified and their molecular size, functional properties, and protein kinase regulation were elucidated (14). The identification was accomplished by using a radiolabeled RNA probe (fragment 1471-1502, containing both the PASR and PCSR) in RNA gel shift, Northern blot and UV-crosslinking assays. RNA gel shift assays identified several specific RNA/protein complexes using protein extracts from either Sp-cAMPS- or DG-stimulated cells but not from unstimulated cells (Fig. 4).

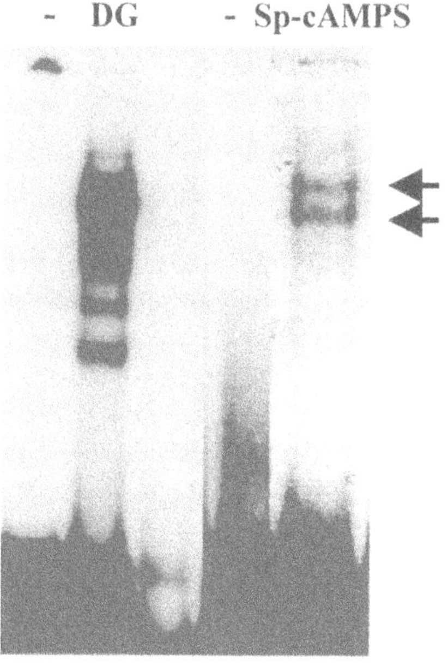

Figure 4. Electrophoretic band-shift analysis of ^{32}P-labeled fragment 1471-1502 with extracts of glioma cells. *Assays were performed with protein from unstimulated (-) and DG or Sp-cAMPS-stimulated cells.*

Figure 5. Electrophoretic band-shift and competition analysis (14). *Protein from Sp-cAMPS-treated glioma cells was tested for binding activity with ^{32}P-labeled fragment 1471-1502. The arrows indicate the position of two shifted RNA-protein complexes.*

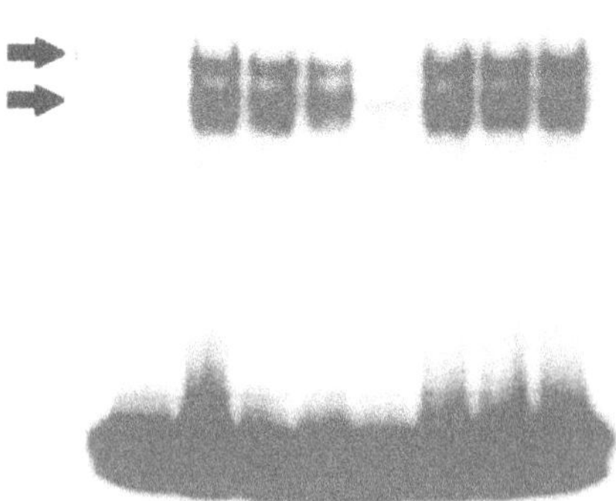

Competition analysis with a homologous (fragment nt 1471-1502) and non-homologous 3'-UTR fragment (fragment nt 1502-1527) showed the specificity of RNA-protein binding. Fragment 1471-1502 as competitor eliminated complex formation at a 25-fold molar excess, whereas fragment 1502-1527 failed to compete (Fig. 5). A similar result was obtained with protein extracts from DG-stimulated cells (not shown).

Northwestern blot (Fig. 6) as well as UV-crosslinking assays (Fig. 7) identified four proteins of about 95, 67, 52 and 50 kDa (named PASR/PCSR-binding proteins 1 through 4) in protein extracts that bind specifically to the PASR/PCSR as the result of Sp-cAMPS stimulation. The protein bands were competed by the PASR/PCSR fragment (nt 1471-1502) but not by fragment 1453-1471 indicating the specificity of binding. None of these proteins were observed in extracts from untreated cells. Comparable results were obtained after UV-crosslinking protein from Sp-cAMPS- or forskolin-stimulated cells to a ^{32}P-labeled PASR/PCSR fragment (Fig. 7).

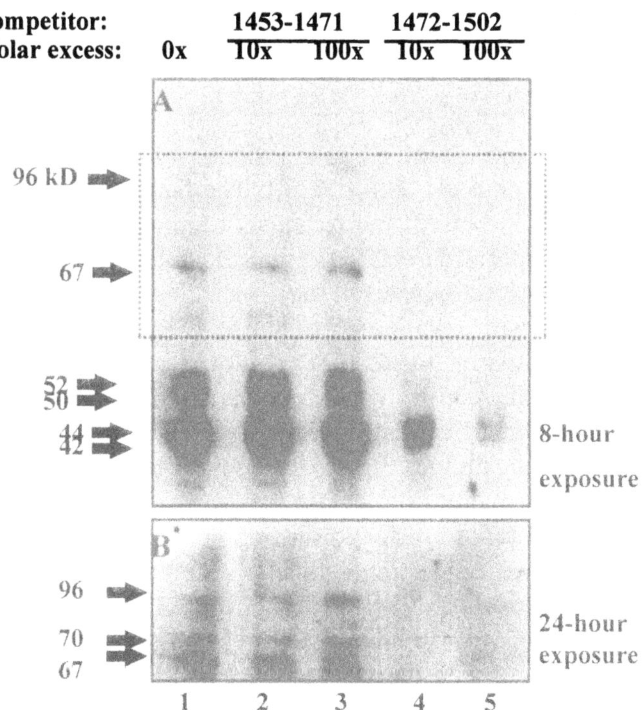

Figure 6. Northwestern blot analysis of PASR/PCSR binding activity of extracts from glioma cells treated with Sp-cAMPS (14). *The filter was probed with ^{32}P-labeled fragment 1471-1502 without competitor RNA (lane 1) and in the presence of the indicated amounts of unlabeled fragment 1453-1471 (lanes 2 and 3) and 1471-1502 (lanes 4 and 5). The positions of the PASR- PCSR-binding proteins are shown on the left side of the figure. The blot shown in A was exposed to film for an 8-hour period, whereas B shows an autoradiograph of the dotted rectangle in A after a 24-hour exposure.*

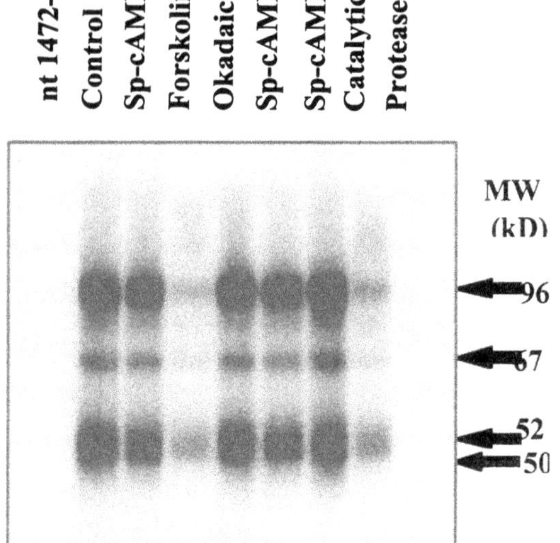

***Figure 7. UV-catalyzed PASR/PCSR-protein cross-linking assay.** The PASR/PCSR fragment was cross-linked by UV irradiation with protein extracts from glioma cells before (control) and after various treatments as indicated. Lane nt 1472-1502 contains ^{32}P-labeled RNA probe without protein.*

The results suggest that all four proteins may directly contact as yet unknown bases within the PASR/PCSR. The fact that binding of protein to PASR/PCSR was increased after treatment of glioma cells with activators of PKA, by the phosphatase inhibitor okadaic acid, and after over-expression of the catalytic subunit of PKA taken together with the finding that treatment of cellular extracts with alkaline phosphatase (Fig. 7; lane Sp-cAMPS+AP) prevented interaction of PASR/PCSR with protein, very strongly suggests a regulation of binding activity of all four RNA-binding proteins through their phosphorylative modification. Treatment of cells with Sp-cAMPS in the presence of cycloheximide (Fig. 7; lane Sp-cAMPS+CX) under conditions that inhibit protein synthesis indicated that *de novo* synthesis of active RNA-binding proteins was not a major consideration.

Further evidence suggesting a functional link between phosphorylation and subsequent functional modification of the PASR/PCSR-binding proteins is based on the following findings. The relatively rapid basal decay rate of LDH-A mRNA was considerably slowed in the presence of the protein phosphatase inhibitor okadaic acid (17) indicating a functional role for protein phosphorylation in the stabilization process. In glioma cells stably transformed with a PKA catalytic subunit expression vector, over-expression of the catalytic subunit stabilized LDH mRNA to the degree seen in forskolin- or Sp-cAMPS-treated cells (12). Dioctanoylglycerol (DG), a specific PKC activator, mimicked the effect of TPA, whereas staurosporin and bisindolylmaleimide, both inhibitors of PKC, prevented the TPA- or

DG-induced stabilization of LDH-A mRNA (12, 17). These data suggest a definite cause and effect relationship between phosphorylation, binding of protein to LDH 3'-UTR, and mRNA stabilization. It allows the conclusion that the PKA as well as PKC signal pathways play an active functional role in regulating LDH-A mRNA stability and act cooperatively in achieving stability regulation.

Role of Protein Kinases A and C in Regulation of mRNA Stability and mRNA-Binding Protein Activity

Protein kinase A: Virtually all of the evidence available to date supports the role of the catalytic subunit of PKA as the proximate intermediate in the effect of cAMP through phosphorylative and functional modification of key factors and enzymes (20). Based on this premise, it is conceivable that cAMP-mediated phosphorylative modification of RNA-binding proteins can lead to a change of RNA binding and stabilizing activities. Supporting evidence is given by a report demonstrating the activation of mRNA-binding protein AUF1 by the β-adrenergic receptor signal pathway suggesting the involvement of phosphorylative modification of AUF1 by PKA (21). Stephens and coworkers demonstrated increased binding activity of AUBP in 3T3-L1 preadipocytes treated with the cAMP analogue 8-bromo-cAMP (22). They suggested that the binding of AUBP with its AU-rich recognition sequence regulates the stability of GLUT1 mRNA half-life. Moreover, a 35-kDa β-adrenergic receptor mRNA-binding protein (β-ARB) is up-regulated by the catecholamine analogue isoproterenol and by cAMP (23). Peng and Lever reported that stabilization of the Na^{+} /glucose co-transporter (SGTL1) mRNA by cAMP is accompanied by binding of a 48-kDa protein to an U-rich domain within the 3'-UTR (24, 25). Considering these data, it is appropriate to postulate that regulation of the half-life of several of these mRNAs occurs through the mediation of RNA-binding phosphoprotein(s), and/or phosphoproteins that do not interact directly with RNA but with RNA-binding proteins through protein/protein interaction.

In this context, the mechanism of protein binding to phosphoenolpyruvate carboxykinase (PEPCK) 3'-UTR appears complex. In 1988, Hod and Hanson reported the stabilization of the PEPCK mRNA (6). Subsequently Nachaliel and coworkers reported the down-regulation of a 100-kDa PEPCK 3'-UTR-binding protein by cAMP under conditions that stabilized PEPCK mRNA (26) suggesting that the down-regulation was linked to PEPCK mRNA stabilization. In contrast, Christ and coworkers demonstrated an up-regulation of PEPCK 3'-UTR protein binding activity in glucagon-stimulated hepatocytes (8). To make the situation even more complex, Heise *et al.*

reported the purification of a 400- kDa cAMP-regulated 3'-UTR binding protein identified as the ferritin L chain (27).

Protein kinase C: Tumor promoters like 12-0-tetradecanoylphorbol-13-acetate (TPA) can modify the biological properties of cells by inducing an altered program of gene expression, through a process that includes activation of PKC (10, 28). A great number of examples of regulated mRNA stability (up- and down-regulation) in eukaryotic cells by phorbol ester or PKC activators are known (13, 29-41). Although these studies indicate that an increased level of phorbol ester is a sufficient signal for increased mRNA stability, the molecular mechanisms mediating the effects of phorbol ester have largely remained obscure. The fact that phorbol esters can activate PKC has led to the notion that mRNA stability regulation results from a cascade of events involving PKC isoenzymes and a regulatory system such as *cis*-and *trans*-acting factors. Phosphorylative modification of specific RNA-binding proteins by PKC would achieve the fine-tuning of mRNA half-life. In the case of LDH-A mRNA, the use of a specific inhibitor of PKC, bisindolylmaleimid, can abrogate the effect of TPA and diacylglycerol on PKC in the stabilization mechanism (12). The findings are in agreement with previous reports demonstrating that the stability of ribonucleotide reductase R1 mRNA is increased after phorbol ester treatment of cells. This phenomenon appears to be mediated through a 49-nt *cis*-elements within the 3'-UTR of ribonucleotide reductase mRNA and its interaction with specific binding proteins (42). Ribonuclease reductase R1 gene expression is elevated in BALB/c 3T3 fibroblasts treated with TPA. A 52-57-kDa R1 mRNA-binding protein (R1-BP) was identified that bound selectively to a 49-nt region of the ribonucleotide reductase R1 mRNA 3'-UTR. The R1-BP-RNA binding activity observed in unstimulated BALB/c 3T3 fibroblasts was rapidly and markedly down regulated after TPA treatment suggesting a role for R1BP in the mechanism of action of TPA-induced R1 message stabilization (43). Iron regulatory proteins (IRPs) are RNA-binding proteins that post-transcriptionally regulate iron uptake. The activity and expression of two IRPs is changed during phorbol ester-induced differentiation of HL-60 cells (44).

Potential Role of Nucleases in PKA/PKC-Mediated mRNA Stabilization

Conceivably, the combined action of degradative nucleases and factor(s) that modulate nuclease specificity and kinetics plays a central role in the decay process. Control of turnover of selective mRNAs through endonucleolytic action will require the involvement of highly specific nucleases that recognize unique sites on mRNA molecules. Given the assumption that the four PASR/PCSR-binding proteins identified in the LDH-A system are directly involved in the regulation of mRNA half-life, what might be their role? Do they possess specific nucleolytic activity that is functionally regulated through phosphorylation/dephosphorylation? Alternatively, do the proteins prevent RNA decay by binding to RNase cleavage sites safeguarding RNA against nucleolytic attack? Based on our present knowledge of mechanisms of mRNA stabilization and degradation pathways, it can be hypothesized that PASR/PCSR may function as a cleavage site for an endonuclease. In the absence of PKA or PKC activation, these sites are exposed due to the lack of phosphorylated PASR/PCSR-binding proteins and are, therefore, readily cleaved. In PKA/PKC-activated cells, on the other hand, PASR/PCSR-binding phosphoproteins protect the RNA from endonuclease attack by binding to the cleavage sites. This hypothesis is supported by the observation that PASR/PCSR contains two potential cleavage sites for a human homologue of bacterial RNase-E that is implicated in the regulation of many eukaryotic mRNAs (45, 46). Bacterial RNase-E has a consensus cleavage sequence, A/G-AUU*A/U (* denotes the cleavage point) present in bacterial mRNAs (45). However, cleavage is not restricted to this consensus sequence. For example, the human homologue of RNase E can also effectively catalyze the cleavage of sequence -UAUU*U- in the context of an AU-rich region of c-*myc* 3'-UTR (47). This implies that eukaryotic RNase-E may have a broader spectrum of target sequences. Alignment of PASR/PCSR with either consensus sequence or -UAUUU- reveals two potential RNase-E cleavage sites as shown below:

PASR/PCSR: 5'-CUACAGGAUAUU*UUCUGUAUU*AUAUGUGU - 3'
Consensus sequence: A/G-AUU*A/U A/G-AUU*A/U

PASR/PCSR has two sequences (underlined) similar to the RNase-E consensus motif. Close inspection of the two putative recognition sites of PASR/PCSR reveals that both of them match perfectly with the last four nucleotides of RNase-E consensus sequence, i.e. AUUU or AUUA, and that they differ with the consensus sequence only in the first nucleotide. This makes the two sequences potential candidates for recognition by a eukaryotic RNase-E like endonuclease.

Synergism Between the PKA and PKC Signal Pathways

The demonstration of a synergism involving PKA/PKC-mediated LDH-A mRNA stabilization is particularly intriguing (see Fig. 2) (12). It appears that the 3'-UTR contains discrete functional modules such as instability determinants that are not regulated by effector agents (see Fig. 1) and those that are regulated (PASR and PCSR) (see Table 1) containing recognition sites for mRNA stabilizing proteins. PASR/PCSR modules may function either independently or cooperatively in stabilizing mRNA. Synergism suggests a functional 'cross-talk' between the two signal transduction pathways involving *trans*-regulatory RNA-binding proteins that are substrates and targets either for PKA, PKC, or both (Fig. 8).

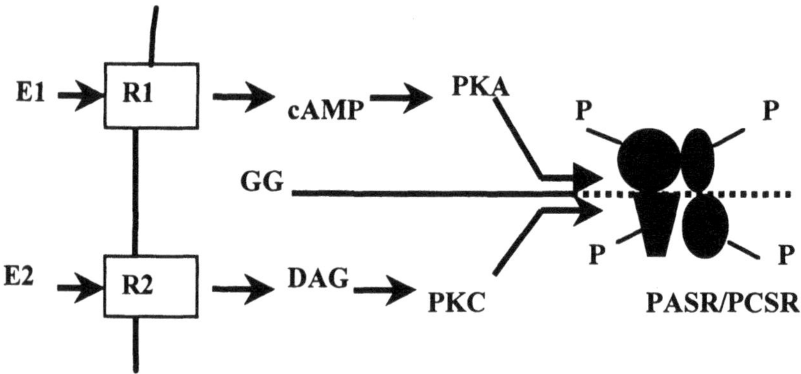

Figure 8. Hypothetical model illustrating the convergence of the PKA and PKC signal transduction pathways leading to cooperative mRNA stabilization*. Separate or simultaneous activation of two receptors (R1 , R2) by effector agents (E1, E2) leads to elevated second messenger levels (e.g. DG and cAMP) which in turn activate either the PKA and/or PKC signal transduction pathways. Activation of PASR/PCSR-binding proteins depends on the presence of serines and threonines that are specific sites of phosphorylation for the protein kinases. Depending on the selectivity of protein kinase action and phosphorylated sites, various combinations of binding of phosphorylated proteins to the PASR and PCSR sites can be envisioned leading to either synergistic or non-synergistic (mediated by either PKA or PKC) mRNA stabilization.*

Since PASR and PCSR possess a 13-nt region in common, the distinct possibility arises that PASR/PCSR-binding proteins are substrates for PKA and/or PKC, and selective phosphorylation of either one or more proteins may be mechanistically involved in the cooperative effect. Cross-talk between major signal transduction pathways is a well-recognized phenomenon known to occur in a number of systems (48-51).

Summary of key concepts

- *Posttranscriptional modulation of mRNA stability occurs through various effector agents and activation of their signal transduction pathways, such as the protein kinase A and C pathways. The modulation is an additional vital mechanism of gene control and is complementary to transcriptional regulation allowing a fine-tuning of intracellular mRNA levels and gene expression.*
- *In the case of LDH-A mRNA, several elements of instability can be identified within the 3'-UTR. However, stability regulation is achieved only through a single functional modular structure consisting of specific cis- and trans-elements. Functional interaction between these elements (PASR/PCSR and their binding proteins) is achieved through effector agent-mediated activation of the protein kinase A and C pathways, leading to phosphorylative and functional modification of the RNA-binding proteins. Subsequent stabilization of mRNA occurs as a consequence of these events by an, as yet, unknown mechanism.*
- *The interaction of the protein kinase A and C pathways leading to a synergistic mRNA stabilizing effect is an additional mechanism of gene control.*
- *From a biochemical standpoint, the cloning and identification of trans-acting RNA-binding proteins are needed to enable elucidation of the mechanism of mRNA stabilization. To this effect the development of cell-free systems to study mRNA decay and assay of RNA-binding proteins during purification is essential. Several methods of detection, such as RNA gel shift, UV crosslinking and Northwestern blotting are suitable to assay RNA-binding proteins during their purification. In addition to conventional purification, affinity chromatography using specific 3'-UTR target ligands should be helpful in the process.*

Study Guide Questions

1) Why is there a dual, seemingly redundant, regulation of gene expression by transcriptional as well as posttranscriptional controls, since both mechanisms seem to produce similar results, that is, they modulate intracellular mRNA levels?
2) Elucidation of the mechanisms whereby PASR- and PCSR-binding proteins achieve their mRNA-stabilizing effect ultimately requires cloning of the proteins. This could potentially be achieved through screening of a cDNA expression library with a radioactively labeled PASR/PCSR probe. However, this procedure will not be successful and will not identify expressed PASR/PCSR-binding proteins. Explain why this procedure cannot be used and suggest other procedure(s) leading to successful cloning.
3) It is important to recognize that LDH-A mRNA stability is regulated in a synergistic fashion by PKA and PKC. What is the physiological advantage for the cell exercising this type of mRNA stability regulation?
4) Messenger RNAs possess a specific secondary structure. Would changes of this structure have an effect on PASR/PCSR-binding protein function and how would one demonstrate this structure/function relationship experimentally?

Acknowledgements
Research performed in the author's laboratory was supported by NIH grant GM53115.

REFERENCES

1. Jungmann, R.A, Kelley, D.C, Miles, M.F, and Milkowski, D.M. 1983. Cyclic AMP regulation of lactate dehydrogenase. Isoproterenol and N6, 02'-dibutyryl cyclic AMP increase the rate of transcription and change the stability of lactate dehydrogenase messenger RNA in rat C6 glioma cells. *J. Biol. Chem.* **258**:5312-5318 .
2. Roesler, W.J, Vandenbark, G.R., and Hanson, R.W. 1988. Cyclic AMP and the induction of eukaryotic gene transcription. *J. Biol. Chem.* **263**:9063-9066 .
3. Nishizuka, Y. 1984.The role of protein kinase C in cell surface signal transduction and tumor promotion. *Nature* **308**:693-698 .
4. Mitchell, R.L, Zokas, L., Schreiber, R.D., and Verma, I.M. 1985. Rapid induction of the expression of proto-oncogene *fos* during human monocytic differentiation. *Cell* **40**:209-217 .
5. Levine, R.A, McCormack, J.E., Buckler, A., and Sonesheim, G.E. 1986. Transcriptional and posttranscriptional control of *c-myc* gene expression in WEHI 231 cells. *Mol. Cell. Biol.* **6**:4112-4116 .
6. Hod, Y., and Hanson, R.W. 1988. Cyclic AMP stabilizes the mRNA for phosphoenolpyruvate carboxykinase (GTP) against degradation *J. Biol. Chem.* **263**:7747-7752.

7. Sakaue, M., and Hoffman, B.B. 1991. cAMP regulates transcription of the a2A adrenergic receptor gene in HT-29 cells. *J. Biol. Chem.* **266**:5743-5749 .
8. Christ, B., Heise, T., and Jungermann, K. 1991.Binding of cytosolic protein from cultured rat hepatocytes to the 3'-end of phosphoenolpyruvate carboxykinase mRNA - significance for protein-mediated mRNA stabilization. *Biochem. Biophys. Res. Commun.* **177**:1273-1282 .
9. Chen, M., Schnermann, J., Smart, A.M., Brosius, F.C., Killen, P.D., and Briggs, J.P. 1993. Cyclic AMP selectively increases renin mRNA stability in cultured juxtaglomerular granular cells. *J. Biol. Chem.* **268**:24138-24144 .
10. Chen, F.Y., Amara, F.M., and Wright, J.A. 1994. Regulation of mammalian ribonucleotide reductase R1 mRNA stability is mediated by a ribonucleotide reductase R1 mRNA 3'-untranslated region cis-trans interaction through a protein kinase C-controlled pathway. *Biochem. J.* 302:125-132 .
11. Chen, F.Y., Amara, F.M., and Wright, J.A. 1994. Posttranscriptional regulation of ribonucleotide reductase R1 gene expression is linked to a protein kinase C pathway in mammalian cells. *Biochem. Cell Biol.* **72**:251-256 .
12. Huang, D., Hubbard, C.J., and Jungmann, R.A. 1995. Lactate dehydrogenase A subunit messenger RNA stability is synergistically regulated via the protein kinase A and C signal transduction pathways. *Molecular Endocrinol.* **9**:994-1004.
13. Ross J. 1995. mRNA stability in mammalian cells. *Microbiol. Rev.* **59**:423-450.
14. Tian, D., Huang, D., Brown, R.C., and Jungmann, R.A. 1998. Protein kinase A stimulates binding of multiple proteins to a U-rich domain in the 3'-untranslated region of lactate dehydrogenase-A mRNA which is required for mRNA stability regulation. *J. Biol. Chem.* **273**:28454-28460 .
15. Tian, D., Huang, D., Short, S., Short, M.L., and Jungmann, R.A. 1998. Protein kinase A-regulated instability site in the 3'-untranslated region of lactate dehydrogenase A subunit mRNA. *J. Biol. Chem.* **273**:24861-24866 .
16. Heaton, J.H., Tillmann-Bogush, M., Leff, N.S., and Gelehrter, T.D. 1998. Cyclic nucleotide regulation of type-1 plasminogen activator-inhibitor mRNA stability in rat hepatoma cells. Identification of cis-acting sequences. *J. Biol. Chem.* **273**:14261-14268 .
17. Short, S., Tian, D., Short, M.L., and Jungmann, R.A. 2000. Structural determinants for post-transcriptional stabilization of lactate dehydrogenase A mRNA by the protein kinase C signal pathway. *J. Biol. Chem.* **275**:12963-12969 .
18. Brawerman, G. 1987. Determinants of messenger RNA stability. *Cell* **48**:5-6 .
19. Ross, J. 1988. Messenger RNA turnover in eukaryotic cells. *Mol. Biol. Med.* **5**:1-14 .
20. Gonzalez, G.A., Yamamoto, K.K., Fischer, W.H., Karr, D., Menzel, P., Biggs III, W., Vale, W.W, and Montminy, M.R. 1989. A cluster of phosphorylation sites on the cyclic AMP-regulated nuclear factor CREB predicted by its sequence. *Nature* 337:749-752 .
21. Pende, A., Tremmel, K.D., DeMaria, C.T., Blaxall, B.C., Minobe, W.A., Sherman, J.A., Bisognano, J.D., Bristow, M.R., Brewer, G., and Port, J.D. 1996. Regulation of the mRNA-binding protein AUF1 by activation of the β-adrenergic receptor signal transduction pathway. *J. Biol. Chem.* **271**:8493-8501.
22. Stephens, J.M., Carter, B.Z., Pekala, P.H., and Malter, J.S. 1992. Tumor necrosis factor alpha-induced glucose transporter (GLUT-1) mRNA stabilization in 3T3-L1 preadipocytes. Regulation by the adenosine-uridine binding factor. *J. Biol. Chem* **267**:8336-8341.
23. Tholanikunnel, B.G., Granneman, J.G., and Malbon, C.C. 1995. The Mr 35,000 β-adrenergic receptor mRNA-binding protein binds transcripts of G-protein-linked receptors which undergo agonist-induced destabilization. *J. Biol. Chem.* **270**:12787-12793.
24. Peng, H., and Lever, J.E. 1995. Regulation of Na+-coupled glucose transport in LLC-PK1 cells. Message stabilization induced by cyclic AMP elevation is accompanied by binding of a Mr = 48,000 protein to a uridine-rich domain in the 3'-untranslated region. *J. Biol. Chem.* **270**:23996-24003.

25. Peng, H., and Lever, J.E. 1995. Post-transcriptional regulation of Na+/glucose cotransporter (SGTL1) gene expression in LLC-PK1 cells. Increased message stability after cyclic AMP elevation or differentiation inducer treatment. *J. Biol. Chem.* **270**:20536-20542.
26. Nachaliel, N., Jain, D., and Hod, Y. 1993A cAMP-regulated RNA-binding protein that interacts with phosphoenolpyruvate carboxykinase (GTP) mRNA. *J. Biol. Chem.* **268**:24203-24209.
27. Heise, T., Nath, A., Jungermann, K., and Christ, B. 1997. Purification of a RNA-binding protein from rat liver. Identification as ferritin L chain and determination of the RNA/protein binding characteristics. *J. Biol. Chem.* **272**:20222-20229.
28. Nishizuka, Y. 1986. Studies and perspectives of protein kinase C. *Science* **233**:305-312.
29. Akashi, M., Loussararian, A.H., Adelman, D.C., Saito, M., and Koeffler, H.P.1990. Role of lymphotoxin in expression of interleukin 6 in human fibibroblasts. Stimulation and regulation. *J. Clin. Invest.* **85**:121-129.
30. Bohjanen, P., Petryniak, B., June, C.H., Thompson, C.B., and Lindsten, T. 1991. An inducible cytoplasmic factor (AU-B) binds selectively to AUUUA multimers in the 3'-untranslated region of lymphokine mRNA. *Mol. Cell. Biol.* **11**:3288-3295.
31. Iwai, Y., Bickel, M., Pluznik, D.H, and Cohen, R.B. 1991. Identification of sequences within the murine granulocyte-macrophage colony-stimulating factor mRNA 3'-untranslated region that mediate mRNA stabilization induced by mitogen treatment of EL-4 thymoma cells. *J. Biol. Chem.* **266**:17595-17965.
32. Perrone-Bizzozero, N.I., Cansino, V.V., and Kohn, D.T. 1993. Posttranscriptional regulation of GAP-43 gene expression in PC12 cells through protein kinase C-dependent stabilization of the mRNA. *J. Cell Biol.* **120**:1263-1270.
33. Ahern, S.M., Miyata, T., and Sadler, J.E. 1993. Regulation of human tissue factor expression by mRNA turnover. *J. Biol. Chem.* **268**:2154-2159.
34. Sachs, A.B. 1993. Messenger RNA degradation in eukarykotes. *Cell* **74**:413-421.
35. Shih, S-C., Mullen, A., Abrams, K., Mukhopadhyay, D., and Claffey, K.P. 1999. Role of protein kinase C isoforms in phorbol ester-induced vascular endothelial growth factor expression in human glioblastoma cells. *J. Biol. Chem.* **274**:15407-15414.
36. Izzo, N.J., Tulenko, T.N., and Solucci, W.S. 1994. Phorbol esters and norepinephrine destabilize a1β-adrenergic receptor mRNA in vascular smooth muscle cells. *J. Biol. Chem.* **269**:1705-1710.
37. Lee, N.H., Earle-Hughes, J., and Fraser, C.M. 1994. Agonist-mediated destabilization of m1 muscarinic acetylcholine receptor mRNA. Elements involved in mRNA stability are localized in the 3'-untranslated region. *J. Biol. Chem.* **269**:4291-4298.
38. Ferry, R.C., Unsworth, C.D., Artymyshyn, R.P., and Molinoff, P.B. 1994. Regulation of mRNA encoding 5-HT2A receptors in P11 cells through a posttranscriptional mechanism requiring activation of protein kinase C. *J. Biol. Chem.* **269**:31850-31857.
39. Kijima, K., Matsubara, H., Murasawa, S., Maruyama, K., Ohkubo, N., Mori, Y., and Inada, M. 1996. Regulation of angiotensin II type 2 receptor gene by the protein kinase C-calcium pathway. *Hypertension* **27**:529-534.
40. Pang, J-HS., Wu, C-J., and Chau, L-Y. 1996. Post-transcriptional regulation of H-ferritin gene expression in human monocytic THP-1 cells by protein kinase C. *Biochem. J.* **319:**185-189.
41. Shin, KS., Park, J.Y., Kwon, H., Chung, C.H., and Kang, M.S. 1997. Opposite effect of intracellular Ca2+ and protein kinase C on the expression of inwardly rectifying K+ channel 1 in mouse skeletal muscle. *J. Biol. Chem.* **272**:21227-21232.
42. Amara, F.M., Hurta, R.R., Huang, A., and Wright, J.A. 1995. Altered regulation of message stability and tumor promoter-responsive cis-trans interactions of ribonucleotide reductase R1 and R2 messenger RNAs in hydroxyurea-resistant cells. *Cancer Research* **55**:4503-4505.

43. Chen, F.Y., Amara, F.M., and Wright, J.A. 1993. Mammalian ribonucleotide reductase R1 mNA stability under normal and phorbol ester stimulating conditions: involvement of a cis-trans interaction at the 3' untranslated region. *EMBO J.* **12**:3977-3986.
44. Schalinske, K.L., and Eisenstein, R.S. 1996. Phosphorylation and activation of both iron regulatory proteins 1 and 2 in HL-60 cells. *J. Biol. Chem.* **271**:7168-7176
45. Melefors, O., Lundberg, U. and Von Gabain, A.1993. In: Belasco, J., and Brawerman, G., editors. Control of messenger RNA stability. New York: Academic Press; p 53-70.
46. Lundberg, U., Melefors, L., Sohlberg, B., Georgellis, D., and Von Gabain, A.1995. RNase K: one less letter in the alphabet soup. *Mol. Microbiol.* **17**:595-596.
47. Wennborg, A., Sohlberg, B., Angerer, D., Klein, G., and Von Gabain, A. 1995. A human RNase E-like activity that cleaves RNA sequences involved in mRNA stability control. *Proc. Natl. Acad. Sci. U.S.A.* **92**:7322-7326.
48. Andersen, B., Milsted, A., Kennedy, G., and Nilson, J.H. 1988. Cyclic AMP and phorbol esters interact synergistically to regulate expression of the chorionic gonadotropin genes. *J. Biol. Chem.* **263**:15578-15583.
49. Houslay, M.D. 1991. 'Crosstalk': a pivotal role for protein kinase C in modulating relationships between signal transduction pathways. *Eur. J. Biochem.* **195**:9-27.
50. Morimoto, B.H., Koshland, Jr., D.E. 1994. Conditional activation of cAMP signal transduction by protein kinase C. The effect of phorbol esters on adenylyl cyclase in permeabilized and intact cells. *J. Biol. Chem.* **269**:4065-4069.
51. Li, Z., Vaidya, V.A., Alvaro, J.D., Iredale, P.A., Hsu, R., Hoffman, G., Fitzgerald, L., Curran, P.K, Machida, C.A., Fishman, P.H., and Dumas, R.S. 1998. Protein kinase C-mediated down-regulation of beta 1-adrenergic receptor gene expression in rat C6 glioma cells. *Mol. Pharmacol.* **54**:14-21.

12

POST-TRANSCRIPTIONAL REGULATION OF IRON METABOLISM

Tracey A. Rouault
NICHD, NIH, Bethesda, MD

Over the last fifteen years, much insight has been gained into the processes that determine how iron will be transported and utilized in cells and animals. Regulation of iron metabolism is important because iron is required for function of numerous proteins such as hemoglobin, but excess iron can react with oxygen species to generate free radicals and oxidative damage. Many human diseases are caused by insufficient or excess iron uptake, including iron deficiency anemia, a major health problem throughout the developing world, and hemochromatosis, an inherited iron overload syndrome that leads to serious disease in the Western world (1). *Because iron is both indispensable and potentially toxic, virtually all cells and organisms regulate uptake and utilization of iron. The regulation of many iron metabolism proteins, including ferritin and transferrin receptor depends upon binding of iron regulatory proteins to RNA stem-loops found within the transcripts that encode these proteins.*

OVERVIEW OF MAMMALIAN IRON METABOLISM

In mammals, individual cells in tissues throughout the body optimize iron status by regulating iron uptake and sequestration (2,3). Polarized epithelia regulate overall transport of iron into sites such as the systemic circulation, the central nervous system, and the testis by regulating expression of transmembrane iron transporters. The mammalian gut is an important tissue in mammalian iron homeostasis as it regulates dietary iron uptake (4). In iron deficient animals, expression levels of iron transporters increase markedly at the apical (5) and basolateral surfaces (6,7) of intestinal epithelial cells, mediating an influx of iron into blood that is distributed to tissues throughout the body by serum transferrin, a high molecular weight ferric (3+) iron binding protein (8). Cells in tissues throughout the body regulate iron uptake according to their needs by increasing or decreasing the amount of transferrin receptor (TfR) expressed at the plasma membrane (9). Once iron enters the cell, it poses a risk to the cellular constituents, including lipids, nucleic acids and proteins, because iron can react with oxygen species to produce highly reactive hydroxyl radicals (10). As a defense against iron excess, and also to create a reservoir of iron that can supply metabolic needs when external iron sources are insufficient, cells synthesize ferritin, a 500,000 MW multimeric protein composed of 24 subunits that assemble to form a sphere. There are two ferritin subunits, one of which, the ferritin H chain, has the ability to oxidize ferrous to the less soluble and toxic ferric form (11). The ferric iron is stored within the ferritin cavity as a hydroxyphosphate precipitate, and it is likely that iron sequestered by ferritin is unavailable to the cell until ferritin is degraded in lysosomes (12,13).

RNA-Protein Interactions Are Critical in Post-Transcriptional Regulation of Iron Metabolism

Much of the regulation of iron metabolism proteins is post-transcriptional, and the major components of the post-transcriptional regulatory machinery have been identified and characterized. In ferritin and TfR transcripts, RNA stem-loops found in the 5' and 3'UTRs function as binding sites for proteins known as iron regulatory proteins (reviewed in 23). The RNA stem-loop elements, (iron responsive elements, or IREs), have defined structural and sequence elements. IREs are stem-loop structures of between 28 and 35 nucleotides in length in which the stem consists of A-form helical regions interrupted by an unpaired cytidine positioned five base pairs 5' of a 6 nucleotide loop (see Fig. 1). In ferritin IREs, there is another potentially unpaired residue 5' of the bulge cytidine, and as a result, the ferritin IREs can be represented either with a three nucleotide bulge (14) or with base-pairs involving the two positions immediately 5' of the bulge C (15). Spontaneous human disease mutations and *in vitro* selection experiments indicate that the residue immediately 5' of the bulge C is most likely base-paired (15). The sequence of the loop is almost always CAGUGX, where X can be any base but G (reviewed in 9). The nucleotide sequences in the upper and lower stems of the IRE may vary as long as base-pairs maintain the structure of the upper and lower stems (16).

An unanticipated feature of the IRE structure is that the loop portion of the IRE is also highly structured. SELEX (Systematic Evaluation of Ligands by EXponential enrichment) studies indicate that base-pair formation between positions 1 and 5 of the loop is likely required for IRE function (17). Also, NMR solution structures have demonstrated that a Watson-Crick C-G base-pair exists between positions 1 and 5 of the loop, and that the adenine at position 2 stacks on position 5 (18,19). The residues in the loop at positions 3, 4 and 6 along with the bulge C of the stem are dynamic in solution. High sequence variability is tolerated at position 6 and moderate variability is tolerated at position 4 of the loop. Some evidence indicates that a guanine must be present at position 3 for high affinity binding (20).

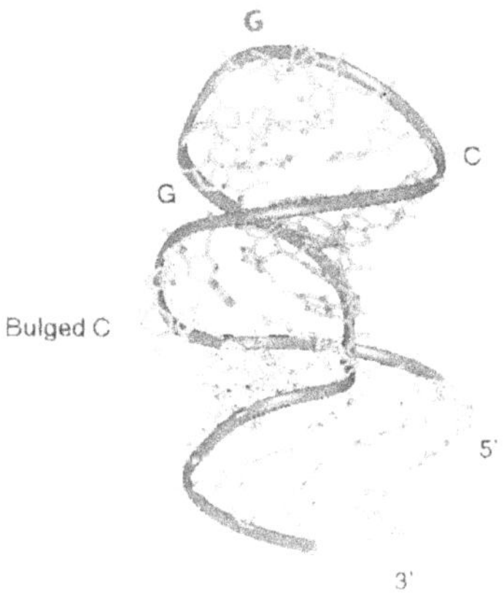

Figure 1. IREs contain a highly structured loop structure with a Watson-Crick base-pair between C1 and G5, stacking of A2 on G5, and a G at position 3 that is dynamic. *Upper and lower stems are composed of A-type RNA helices. Mutational analyses and SELEX studies indicate that the bulged C and the G at position 3 of the loop are residues that are extremely important in the high affinity binding of IRPs. (figure kindly provided by K. Addess, modified from Addess et al, 1997).*

Interestingly, the NMR solution structure of the IRE reveals that the guanine at position 3 adopts a somewhat unusual *syn* conformation that may be important in sequence specific recognition of the RNA by the protein (18). SELEX studies support the idea that the bulged C of the stem and G3 of the loop, residues that are highly dynamic in the NMR solution structure, are important in high-affinity binding of iron regulatory proteins (IRPs)(20) (see Fig. 1). In several other RNA-protein interactions, disordered residues in the free RNA that are critical to binding often become well-ordered upon RNA-protein complex formation (18). Thus, it is possible that the primary role of the fixed major structural features of the IRE is to create a "molecular ruler" that determines the spatial separation and orientation of the two residues that are most important for specific binding of IREs by IRPs, the bulged C of the stem and the G3 of the loop.

IREs Are Found in Numerous Transcripts of Iron Metabolism Proteins

Functionally important IREs have been found in the 5'UTR of ferritin H and L chain transcripts as well as in other 5'UTRs of iron metabolism genes, including the erythrocyte specific form of the heme biosynthetic enzyme 5-aminolevulinate synthase (21,22), mitochondrial aconitase (23,24), and succinate dehydrogenase subunit b of *Drosophila melanogaster* (24,25). In each of these examples, the IRE sequence is positioned near the cap site of the transcript and its presence confers the capacity for iron-dependent translational regulation upon the transcript in which it is found. IREs are also found in the 3'UTR of the TfR and a plasma membrane iron transporter known as the divalent metal transporter (DMT1) (4), DCT1 (5), or Nramp2 (26). The 3'UTR of the TfR transcript contains five IREs, and an endonucleolytic cleavage site (summarized in 27). One of the two alternative splice forms of DMT1 contains a single IRE that differs from consensus IREs because it contains an unpaired residue in the upper stem (5), but it is not yet clear how this IRE affects DMT1 expression.

Mutations in the IRE of ferritin L-chain are the cause of a genetic disease, hyperferritinemia and bilateral cataract syndrome

Over the past five years, the role of the IRE in regulation of ferritin translation has been confirmed by the discovery that mutations in the IRE of the ferritin L-chain cause the autosomal dominant disease, hereditary hyperferritinemia and bilateral cataract syndrome (HHCS) (28,29). Subsequently, numerous point mutations of the IRE have been identified as causes of the syndrome in unrelated families, and the severity of the syndrome is correlated with the degree of impairment of IRE-IRP binding (15). The major clinical findings are a marked increase in serum ferritin and tendency to form cataracts. The absence of other symptoms may be attributable to the fact that the ferritin L subunit has ferroxidase activity, and simple over-expression of ferritin L chain therefore, does not profoundly alter iron status. The discovery of this interesting human disease and elucidation of its molecular cause provides an entirely independent and convincing demonstration of the role of the IRE in regulation of ferritin gene expression.

Iron regulatory proteins are cytosolic iron-sensing proteins that bind IREs when cytosolic iron levels are low

In mammals, there are two cytosolic proteins that sense iron levels and modify binding to IREs accordingly. These proteins, known as iron regulatory proteins 1 and 2 (IRP1 and IRP2), bind to IREs when cells are depleted of iron. When they bind to an IRE near the 5' end of transcripts such as ferritin, they prevent initiation of translation, and new biosynthesis of ferritin is prevented (30). When they bind to IREs in the 3'UTR of the transcript, they protect the transcript from endonucleolytic cleavage, TfR mRNA levels increase, and accordingly, synthesis of TfR increases (27).

The key to translational regulation of ferritin is that IRPs generally bind to IREs only when cells are depleted of iron. IRE binding activity of IRPs can be assessed using an *in vitro* assay in which radiolabeled IRE is added to lysates and used to assess the amount of binding activity. Binding activity increases markedly in iron depleted cells, and the RNA protein complex formed by IRP1 can be distinguished from that of IRP2 because the RNA-protein complexes migrate to a different position in non-denaturing gels (15,31). The RNA–gel shift assay was important in allowing characterization, purification and cloning of IRPs (32-37).

IRP 1 AND 2 SENSE CYTOSOLIC IRON LEVELS BY DIFFERENT MECHANISMS

IRPs have been studied extensively to determine how they sense iron levels and modify their IRE binding activity. IRP1 and IRP2 transcript levels do not change when iron levels of cells in tissue culture are manipulated. Also, protein levels of IRP1 do not change when cellular iron status changes, but IRP2, which is 58% identical in amino acid sequence to IRP1, differs in its mode of regulation because IRP2 cannot be detected in iron-replete cells.

An Iron-Sulfur Cluster of IRP1 is the Key to Regulation of Its RNA Binding Activity

Cloning of the IRPs led to the discovery that IRP1 is actually a bifunctional protein. IRP1 and IRP2 are members of a larger gene family, the aconitase family (38). In iron-replete cells, IRP1 contains an iron-sulfur cluster and is a functional aconitase, which interconverts citrate and isocitrate in the cytosolic compartment of the cell. This enzymatic activity parallels that of mitochondrial aconitase, an enzyme that performs one of the core reactions of the mitochondrial citric acid cycle (39). Iron-sulfur clusters are notoriously sensitive to oxidative stress. For many iron-sulfur proteins, loss of an electron from the iron-sulfur cluster can lead to spontaneous cluster disassembly. A simple model for iron-sensing by IRP1 is as follows: the iron-sulfur prosthetic group that determines function is subject to constant turnover, and resynthesis depends upon iron-sulfur cluster assembly enzymes and the availability of elemental iron and a sulfur source (40). Two of the mammalian iron sulfur cluster assembly proteins have been cloned (41,42), and a multi-protein complex is involved in the process of synthesizing these prosthetic groups and inserting them into appropriate recipient proteins. Clearly, a deficiency in cytosolic iron would preclude reassembly of iron-sulfur clusters, and cytosolic iron deficiency would result in accumulation of the apoprotein (IRP1 that lacks an iron-sulfur cluster). In addition, molecules that promote oxidative disassembly of the iron-sulfur cluster could increase the proportion of IRP1 that is in the IRE- binding form (43), although oxidative stress may change IRP1 activity by a mechanism unrelated to iron-sulfur cluster disassembly (44).

The key to iron regulation of IRP1 is that IRP1 that lacks an iron-sulfur cluster has a high affinity for IREs. The crystal structure of mitochondrial aconitase (45) has been used to guide studies, because the structures of IRP1 and 2 and mitochondrial aconitase are likely to be quite similar (46). Evidence suggests that the iron-sulfur cluster prevents RNA binding by limiting access of RNA transcripts to critical RNA binding residues. In mitochondrial aconitase, the fourth domain of the protein is connected to domains 1-3 by a flexible hinge-linker peptide. In the holoprotein, the substrates citrate and isocitrate are bound by residues in all four domains. As active site residues of mitochondrial aconitase are derived from all four domains, it is likely that, similar to mitochondrial aconitase, the fourth domain of IRP1 is in close apposition to domains 1-3 in the functional enzyme (45). However, if the iron-sulfur cluster and substrate were no longer present, it is possible that the fourth domain could move on its flexible hinge-linker in relation to the rest of the protein, and the surfaces of the active site cleft could become accessible to large complex molecules. An important feature of this region of IRP1 is that the residues that line the active site cleft would thus be available for interactions with macromolecules only in cells in which IRP1 was in the apoprotein form. Evidence suggests that this unique region of IRP1 apoprotein is the region to which IREs bind. UV crosslinking of radiolabeled IRE to the RNA binding site has resulted in identification of several peptides, each of which maps to the active site cleft region of the protein (47,48). In addition, site-directed mutagenesis has resulted in identification of several arginines of the enzymatic active site that are important, or indispensable in IRE binding. The fact that arginines that are important in binding of IREs are also indispensable in aconitase function explains why IRP1 cannot function simultaneously as an IRE binding protein and a cytosolic aconitase (20). A further prediction of this model is that when the cysteines of IRP1 that are directly involved in ligation of the iron-sulfur cluster are mutagenized, an iron-sulfur cluster will no longer be bound, and the mutagenized protein will constitutively bind IREs, regardless of iron status. In fact, IRP1 with serine substitutions at the critical cysteines loses its ability to sense iron status and constitutively binds IREs (49,50). Thus the two activities of IRP1 are mutually exclusive, and the iron status of the cells is a major determinant of which of the two functions will predominate.

IRP2 is Iron-Dependently Degraded

The regulation of IRP2 differs from that of IRP1 in that IRP2 is physically absent in iron-replete cells. The key to regulation of IRP2 levels is that IRP2 is iron-dependently degraded in iron–replete cells (51,52). IRP2 contains an unique cysteine-rich domain that binds iron and subsequently undergoes iron-dependent oxidation, as indicated by acquisition of carbonyl modifications. The oxidized protein is a target for ubiquitination and the ubiquitinated protein is subsequently degraded by the proteasome (53). Oxidative stress may also enhance IRP2 turnover (54).

IRP1 and IRP2 are Potentially Highly Redundant in Function

Each of the two IRPs binds IREs with equal and high affinity with measured Kds of 20-50 picomolar (20). Also, each IRP is equally efficacious in translational regulation of ferritin (55). In addition, virtually all cell lines and tissue types appear to express both IRPs. Although the apparent functions of the IRPs appear to be partially or completely redundant in iron metabolism, genetic ablation of each of the IRPs in mice reveal partially non-overlapping roles of the two IRPs in normal iron physiology.

When mice with targeted deletions of each IRP were generated, it was expected that loss of IRP1 would result in misregulation of iron metabolism in the mouse, because in many tissues, including the liver and kidney, the mRNA for IRP1 is much more abundant than that of IRP2 (36). In addition, loss of cytosolic aconitase activity was expected to produce an unusual and interesting phenotype, because IRP2 does not have aconitase function (36,37). However, mice with genetic ablations of IRP1 (IRP1-/- mice) have a normal life-span, are fertile and do not show signs of misregulation of iron metabolism at autopsy (Rouault- unpublished observations). It is not uncommon that when duplicated gene pairs are evaluated, the redundant function of one protects against loss of function of the other. In such redundant gene pairs, genetic ablation of both of the genes often proves to be embryonic lethal, whereas the loss of either gene alone produces no phenotype.

Genetic Ablation of IRP2 Results in Adult-Onset Neurodegenerative Disease in Mice

Although IRP2-/- mice developed and reproduced normally, adult mice developed a movement disorder characterized by prominent tremor and loss of coordination. IRP2-/- mice abandoned normal grooming, and adopted an unusual posture. In addition, they moved very slowly, and showed a pronounced preference for walking backwards. At autopsy, significant accumulation of ferric iron was detected within intestinal epithelial cells (*unpublished observations*). To determine whether a similar maldistribution of iron was present in the brains of IRP2-/- mice with movement disorders, serial sections through the brain of IRP2-/- mice and wild-type littermates were stained for ferric iron. Accumulations of iron were observed in specific regions of the brain of IRP2-/- mice, including the caudate/putamen, the thalamus, and numerous white matter tracts throughout the brain. Integrity of axons was assessed using a stain known as the amino-cupric silver degeneration stain. The basis of the amino-cupric silver stain is that silver binds to negatively charged neurofilament proteins in degenerating axons that have lost membrane integrity and are therefore readily penetrated by the staining procedure. In the cerebellar white matter, axons that accumulate iron colocalize with those that are degenerating. In young animals, iron accumulation occurs before degenerative changes develop. Immunohistochemistry revealed that the pattern of distribution of ferritin matches that of ferric iron, implying that the increased ferric iron may well be sequestered within ferritin (56).

The increase in ferritin is most likely a result of sufficient IRE binding activity in cells that depend mostly on IRP2 for IRE-binding activity and translational repression. Over-expression of ferritin could lead to profound alterations in iron metabolism, because ferritin avidly sequesters iron at the expense of other proteins (11). In neurons, ferritin is apparently transported in axons, and thus, neurons could be subject to axonal iron overload when ferritin undergoes lysosomal degradation. Much remains to be learned about the mechanisms that underlie neurodegeneration in IRP2-/- mice.

Summary of key concepts

- *RNA binding proteins known as iron regulatory proteins (IRPs) are responsible for post-transcriptional regulation of ferritin, TfR, and several other iron metabolism proteins.*
- *A conserved RNA stem-loop is the binding site for IRPs.*
- *Mutations in the IRE of ferritin L chain cause human hyperferritinemia and bilateral cataract syndrome.*
- *Loss of IRP2 causes neurodegenerative disease in mice and therefore could be a cause of adult-onset neurodegeneration in humans.*

Study Guide Questions

1) What are the advantages of translational regulation of ferritin expression?
2) Why do you think that mutations in the ferritin H chain IRE are not found in HHCS?
3) How do you think the IRE-IRP system of regulation evolved?

Acknowledgements
Work in the authors' laboratory is supported by the National Institute of Child Health and Human Disease.

REFERENCES

1. Sheth, S. & Brittenham, G. M. (2000) Genetic disorders affecting proteins of iron metabolism: clinical implications *Annu Rev Med* **51**, 443-464.
2. Rouault, T. & Klausner, R. (1997) Regulation of iron metabolism in eukaryotes. *Curr Top Cell Regul* **35**, 1-19.
3. Hentze, M. W. & Kuhn, L. C. (1996) Molecular control of vertebrate iron metabolism: mRNA-based regulatory circuits operated by iron, nitric oxide, and oxidative stress. *Proc Natl Acad Sci U S A* **93**, 8175-8182.
4. Andrews, N. C., Fleming, M. D. & Gunshin, H. (1999) Iron transport across biologic membranes. *Nutr Rev* **57**, 114-123.
5. Gunshin, H., Mackenzie, B., Berger, U. V., Gunshin, Y., Romero, M. F., Boron, W. F., Nussberger, S., Gollan, J. L. & Hediger, M. A. (1997) Cloning and characterization of a mammalian proton-coupled metal-ion transporter. *Nature* **388**, 482-488.
6. Abboud, S. & Haile, D. J. (2000) A novel mammalian iron-regulated protein involved in intracellular iron metabolism. *J Biol Chem* **275**, 19906-19912
7. Donovan, A., Brownlie, A., Zhou, Y., Shepard, J., Pratt, S. J., Moynihan, J., Paw, B. H., Drejer, A., Barut, B., Zapata, A., Law, T. C., Brugnara, C., Lux, S. E., Pinkus, G. S., Pinkus, J. L., Kingsley, P. D., Palis, J., Fleming, M. D., Andrews, N. C. & Zon, L. I. (2000). Positional cloning of zebrafish ferroportin1 identifies a conserved vertebrate iron exporter. *Nature* **403**, 776-781
8. Ponka, P. (1999) Cellular iron metabolism. *Kidney Int Suppl* **69**, S2-11
9. Klausner, R. D., Rouault, T. A. & Harford, J. B. (1993) Regulating the fate of mRNA: the control of cellular iron metabolism *Cell* **72**, 19-28
10. Halliwell, B. & Gutteridge, J. M. (1992) Biologically relevant metal ion-dependent hydroxyl radical generation. An update. *FEBS Lett* **307**, 108-112
11. Cozzi, A., Corsi, B., Levi, S., Santambrogio, P., Albertini, A. & Arosio, P. (2000). Overexpression of wild type and mutated human ferritin H-chain in HeLa cells: *in vivo* role of ferritin ferroxidase activity. *J Biol Chem* **275**, 25122-25129
12. Theil, E. C. (1987). Ferritins *Ann Rev Biochem* **56**, 289-315
13. Radisky, D. C. & Kaplan, J (1998). Iron in cytosolic ferritin can be recycled through lysosomal degradation in human fibroblasts. *Biochem J* **336**, (Pt 1):201-5
14. Ke, Y., Wu, J., Leibold, E. A., Walden, W. E. & Theil, E. C. (1998). Loops and bulge/loops in iron-responsive element isoforms influence iron regulatory protein binding. Fine-tuning of mRNA regulation? *J Biol Chem* **273**, 23637-23640
15. Allerson, C. R., Cazzola, M. & Rouault, T. A. (1999). Clinical severity and thermodynamic effects of iron-responsive element mutations in hereditary hyperferritinemia-cataract syndrome. *J Biol Chem* **274**, 26439-26447

16. Leibold, E. A., Laudano, A. & Yu, Y. (1990). Structural requirements of iron-responsive elements for binding of the protein involved in both transferrin receptor and ferritin mRNApost-transcriptional regulation. *Nucleic Acids Res* **18**, 1819-1824
17. Henderson, B. R., Menotti, E., Bonnard, C. & Kuhn, L. C. (1994). Optimal sequence and structure of iron-responsive elements -selection of RNA stem-loops with high-affinity for iron regulatory factor *J Biol Chem* **269**, 17481-17489
18. Addess, K. J., Basilion, J. P., Klausner, R. D., Rouault, T. A. & Pardi, A. J. (1997). Structure and Dynamics of the Iron Responsive Element RNA: Implications for Binding of the RNA by Iron Regulatory Proteins. *J Mol Biol* **274**, 72-83
19. Gdaniec, Z., Sierzputowska-Gracz, H. & Theil, E. C. (1998). Iron regulatory element and internal loop/bulge structure for ferritin mRNA studied by cobalt(III) hexammine binding, molecular modeling, and NMR spectroscopy. *Biochemistry* **37**, 1505-1512
20. Butt, J., Kim, H. Y., Basilion, J. P., Cohen, S., Iwai, K., Philpott, C. C., Altschul, S., Klausner, R. D. & Rouault, T. A. (1996). Differences in the RNA binding sites of iron regulatory proteins and potential target diversity *Proc Natl Acad SciUSA* **93**, 4345-4349
21. Bhasker, C. R., Burgiel, G., Neupert, B., Emery-Goodman, A., Kuhn, L. C. & May, B. K. (1993). The putative iron-responsive element in the human erythroid 5-aminolevulinate synthase mRNA mediates translational control. *J Biol Chem* **268**, 12699-12705
22. Melefors, O., Goossen, B., Johansson, H. E., Stripecke, R., Gray, N. K. & Hentze, M. W. (1993). Translational control of 5-aminolevulinate synthase mRNA by iron-responsive elements in erythroid cells. *J Biol Chem* **268**, 5974-5978
23. Kim, H. Y., LaVaute, T., Iwai, K., Klausner, R. D. & Rouault, T. A. (1996). Identification of a conserved and functional iron-responsive element in the 5'UTR of mammalian mitochondrial aconitase. *J Biol Chem* **271**, 24226-24230
24. Gray, N. K., Pantopoulos, K., Dandekar, T., Ackrell, B. A. C. & Hentze, M. W. (1996). Translational regulation of mammalian and drosophila citric-acid cycle enzymes via iron-responsive elements. *Proc. Natl. Acad. Sci. U.S.A.* **93**, 4925-4930
25. Kohler, S. A., Henderson, B. R. & Kuhn, L. C. (1995). Succinate dehydrogenase b mRNA of Drosophila melanogaster has a functional iron-responsive element in its 5'-untranslated region. *J Biol Chem* **270**, 30781-30786
26. Fleming, M. D., Trenor, C. C. r., Su, M. A., Foernzler, D., Beier, D. R., Dietrich, W. F. & Andrews, N. C. (1997). Microcytic anaemia mice have a mutation in Nramp2, a candidate iron transporter gene. *Nat Genet* **16**, 383-386
27. Binder, R., Horowitz, J. A., Basilion, J. P., Koeller, D. M., Klausner, R. D. & Harford, J. B. (1994). Evidence that the pathway of transferrin receptor mRNA degradation involves an endonucleolytic cleavage within the 3'UTR and does not involve poly(A) tail shortening. *EMBO J* **13**, 1969-1980
28. Beaumont, C., Leneuve, P., Devaux, I., Scoazec, J. Y., Berthier, M., Loiseau, M. N., Grandchamp, B. & Bonneau, D. (1995). Mutation in the iron responsive element of the L ferritin mRNA in a family with dominant hyperferritinaemia and cataract. *Nat Genet* **11**, 444-46
29. Girelli, D., Corrocher, R., Bisceglia, L., Olivieri, O., De Franceschi, L., Zelante, L. & Gasparini, P. (1995). Molecular basis for the recently described hereditary hyperferritinemia-cataract syndrome: a mutation in the iron-responsive element of ferritin L-subunit gene (the Verona mutation). *Blood* **86**, 4050-4053
30. Muckenthaler, M., Gray, N. K. & Hentze, M. W. (1998). IRP-1 binding to ferritin mRNA prevents the recruitment of the small ribosomal subunit by the cap-binding complex eIF4F. *Mol Cell* **2**, 383-388
31. Leibold, E. A. & Munro, H. N (1988). Cytoplasmic protein binds *in vitro* to a highly conserved sequence in the 5' untranslated region of ferritin heavy- and light-subunit mRNAs. *Proc. Natl Acad Sci U S A* **85**, 2171-2175

32. Rouault, T. A., Tang, C. K., Kaptain, S., Burgess, W. H., Haile, D. J., Samaniego, F., McBride, O. W., Harford, J. B. & Klausner, R. D. (1990). Cloning of the cDNA encoding an RNA regulatory protein--the human iron-responsive element-binding protein. *Proc Natl Acad Sci U S A* **87**, 7958-7962
33. Hirling, H., Emery-Goodman, A., Thompson, N., Neupert, B., Seiser, C. & Kuhn, L. C. (1992). Expression of active iron regulatory factor from a full-length human cDNA by *in vitro* transcription/translation. *Nuc. Acids Res.* **20**, 33-39
34. Patino, M. M. & Walden, W. E. (1992). Cloning of a functional cDNA for the Rabbit Ferritin mRNA Repressor Protein: Demonstration of a Tissue Specific Pattern of Expression *J Biol Chem* **267**, 19011-19016
35. Yu, Y., Radisky, E. & Leibold, E. A. (1992). The Iron-Responsive Element Binding Protein: Purification, Cloning and Regulation in Rat Liver. *J Biol Chem* **267**, 19005-19010
36. Samaniego, F., Chin, J., Iwai, K., Rouault, T. A. & Klausner, R. D. (1994). Molecular Characterization of a Second Iron Responsive Element Binding Protein, Iron Regulatory Protein 2 (IRP2): Structure, Function and Post-translational Regulation. *J Biol Chem* **269**, 30904-30910
37. Guo, B., Yu, Y. & Leibold, E. A. (1994). Iron regulates cytoplasmic levels of a novel iron-responsive element-binding protein without aconitase activity. *J Biol Chem* **269**, 24252-24260
38. Gruer, M. J., Artymiuk, P. J. & Guest, J. R. (1997). The aconitase family: three structural variations on a common theme. *Trends Biochem Sci* **22**, 3-6
39. Kennedy, M. C., Mende-Mueller, L., Blondin, G. A. & Beinert, H. (1992). Purification and characterization of cytosolic aconitase from beef liver and its relationship to the iron-responsive element binding protein (IRE-BP). *Proc Natl Acad. Sci* **89**, 11730-11734
40. Rouault, T. A. & Klausner, R. D. (1996). Iron-sulfur clusters as biosensors of oxidants and iron. *Trends in Biochem Sci* **21**, 174-177
41. Land, T. & Rouault, T. A. (1998). Targeting of a human iron-sulfur cluster assembly enzyme, nifs, to different subcellular compartments is regulated through alternative AUG utilization. *Mol Cell* **2**, 807-815
42. Tong, W. H. & Rouault, T. (2000). Distinct iron-sulfur cluster assembly complexes exist in the cytosol and mitochondria of human cells. *EMBO J* **19**, 5692-5700
43. Pantopoulos, K. & Hentze, M. W. (1998). Activation of iron regulatory protein-1 by oxidative stress *in vitro*. *Proc Natl Acad Sci U S A* **95**, 10559-10563
44. Oliveira, L. & Drapier, J. C. (2000). Down-regulation of iron regulatory protein 1 gene expression by nitric oxide. *Proc Natl Acad Sci U S A* **97**, 6550-6555
45. Robbins, A. H. & Stout, C. D. (1989). The structure of aconitase. *Proteins* **5**, 289-312
46. Rouault, T. A., Stout, C. D., Kaptain, S., Harford, J. B. & Klausner, R. D (1991).. Structural relationship between an iron-regulated RNA-binding protein (IRE-BP) and aconitase: functional implications *Cell* **64**, 881-883
47. Swenson, G. R. & Walden, W. E. (1994). Localization of an RNA binding element of the iron responsive element binding protein within a proteolytic fragment containing iron coordination ligands. *Nuc Acids Res* **22**, 2627-2633
48. Basilion, J. P., Rouault, T. A., Massinople, C. M., Klausner, R. D. & Burgess, W. H. (1994). The iron-responsive element-binding protein: Localization of the RNA binding site to the aconitase active-site cleft. *Proc.Natl Acad Sci U.S.A.* **91**, 574-578
49. Philpott, C. C., Klausner, R. D. & Rouault, T. A. (1994). The bifunctional iron-responsive element binding protein/cytosolic aconitase: The role of active-site residues in ligand binding and regulation. *Proc Natl Acad Sci* **91**, 7321-7325
50. Hirling, H., Henderson, B. R. & Kuhn, L. C. (1994). Mutational analysis of the [4Fe-4S]-cluster converting iron regulatory factor from its RNA-binding form to cytoplasmic aconitase *EMBO Journal* **13**, 453-461
51. Guo, B., Phillips, J. D., Yu, Y. & Leibold, E. A. (1995). Iron regulates the intracellular degradation of iron regulatory protein 2 by the proteasome. *J Biol Chem* **270**, 21645-21651

52. Iwai, K., Klausner, R. D. & Rouault, T. A. (1995). Requirements for iron-regulated degradation of the RNA binding protein, iron regulatory protein 2. *EMBO J* **14**, 5350-5357
53. Iwai, K., Drake, S. K., Wehr, N. B., Weissman, A. M., LaVaute, T., Minato, N., Klausner, R. D., Levine, R. L. & Rouault, T. A. (1998). Iron-dependent oxidation, ubiquitination, and degradation of iron regulatory protein 2: implications for degradation of oxidized proteins. *Proc Natl Acad Sci U S A* **95**, 4924-4928
54. Hanson, E. S. & Leibold, E. A. (1999). Regulation of the iron regulatory proteins by reactive nitrogen and oxygen species. *Gene Expr* **7**, 367-376
55. Kim, H. Y., Klausner, R. D. & Rouault, T. A. (1995). Translational repressor activity is equivalent and is quantitatively predicted by *in vitro* RNA binding for two iron-responsive element binding proteins, IRP1 and IRP2. *J Biol Chem* **270**, 4983-4986
56. LaVaute, T., Smith, S., Cooperman, S., Iwai, K., Land, W., Meyron-Holtz, E., Drake, S. K., Miller, G., Abu-Asab, M., Tsokos, M., Switzer III, R., Grinberg, A., Love, P., Tresser, N. & Rouault, T. A. (2001) Targeted deletion of iron regulatory protein 2 causes misregulation of iron metabolism and neurodegenerative disease in mice. *Nature Genetics* **27**, 209-14.

13

CYTOCHROME P450 RNA–PROTEIN INTERACTIONS

Matti A. Lang and Françoise Raffalli-Mathieu

University of Uppsala, Uppsala, Sweden

Humans are constantly exposed to xenobiotics present in the environment, the food, or from clinical use. To counteract the toxic accumulation of foreign molecules, living organisms have evolved a defense system comprising a family of detoxifying enzymes (cytochromes P450, CYPs). Numerous xenobiotics induce CYP genes: the induction is an essential adaptive mechanism that allows organisms to adjust their detoxification capacity according to the needs. Only in a few cases are the induction mechanisms understood to some extent. While it is well admitted that both transcriptional and post-transcriptional (mRNA processing and turnover) control is important for cyp genes expression, very little is known about the post-transcriptional mechanisms of regulation. We have investigated the post-transcriptional regulation of two CYP genes, CYP1A2 and CYP2A5, and identified RNA-binding proteins interacting with their mRNA in an inducer-dependent manner. The strategy used to characterize the RNA-protein interaction and to identify the relevant RNA sequences as well as the regulatory factors are described in this chapter.

BACKGROUND

Cytochrome P450 enzymes (CYPs) form a superfamily of proteins involved in the metabolism of xenobiotics and endogenous molecules (1, 2). One of the most fascinating features of xenobiotics-metabolizing CYPs is that their expression can be increased to meet the detoxification needs of the organisms. Indeed, a large number of xenobiotics are inducers of the CYP isoenzymes that metabolize them. This adaptive process is essential for survival and is an important aspect of gene-environment interactions.

The expression of CYP genes is regulated at different stages, ranging from transcription to protein stabilization (3). The transcriptional activation of CYP genes has been widely studied and a number of intracellular receptors have been discovered that stimulate the transcription of specific CYP genes upon activation by xenobiotics (4). In addition, post-transcriptional regulation is important for many CYP genes (3). However, very little is known about the molecular mechanisms of the post-transcriptional regulation of these CYP genes, particularly the regulatory factors that affect mRNA metabolism during induction.

To stabilize or destabilize mRNAs is a widely used strategy to control the expression of transiently-expressed genes, such as those coding for cytokines, proto-oncogenes, and growth factors (5). Essential to this regulation is the existence of specific motifs within the mRNA (in the coding or non-coding regions) and the binding of specific proteins to these motifs. The RNA-protein interaction is the key to the regulation: RNA-binding proteins can stabilize or destabilize their target RNA, affect their maturation, transport or translation (6)

A large number of RNA-binding proteins are known. Amongst these, the heterogeneous nuclear ribonucleoproteins (hnRNPs) are particularly interesting. More than 20 hnRNPs have been identified and are known to be involved in transcriptional processes, alternative pre-mRNA splicing, pre-mRNA 3'-end processing and mRNA translation and turnover. Certain proteins in this family are multifunctional (7).

We have studied the regulation of the expression of 2 murine CYP genes: *CYP1A2* and *CYP2A5*, by 2 specific inducers: 3-methylcholanthrene (3MC) upregulates *CYP1A2* by partly post-transcriptional mechanisms (8) and pyrazole (pyr) induces *CYP2A5* post-transcriptionally (9). In this chapter we will present the characterization of specific CYP mRNA-protein complexes, the identification of the *cis*-elements (RNA sequences important for the interaction) and *trans*-acting factors (RNA-binding proteins), how xenobiotics affect the binding activity, and how the interaction can be relevant in gene regulation.

Detection of Proteins Binding to CYP1A2 and CYP2A5 mRNA

The binding of proteins to an mRNA of known sequence is typically investigated by using a radioactive *in vitro* synthesized RNA. In principle RNAs corresponding to the entire, or any fragment of the mRNA of interest can be synthesized.

Design of RNA probes: The synthesis of RNA probes is performed using combined PCR-*in vitro* transcription. The advantage of this method is its rapidity compared to cloning in T7 promoter-bearing vectors. In order for the amplified DNA to be subsequently transcribed, the sense primer in the PCR reaction must bear the T7 or T3 promoter sequence at its 5' end, so that the PCR product can be recognized by the T7 or T3 RNA polymerase in the transcription reaction.

PCR: The desired DNA fragments are synthesized using the relevant pairs of primers. For example the primers:
TAATACGACTCACTATAGGGAGAggccacgcttttccaagtga
(capital letters representing the T7 sequence) and gagacagccagagtaggca (nucleotide positions 1583-1602, and 1853-1872, according to the sequence from Kimura et al. (10)) allow the amplification of a DNA fragment comprising the T7 promoter sequence followed by the entire 3'UTR of CYP1A2.

In vitro transcription: *In vitro* transcription is rutinely performed using commercially available kits providing the RNA polymerase, ribonucleotides and the RNase inhibitor. The RNA is radioactively labeled with [^{32}P]-UTP, so that RNA-protein complexes can later be detected by autoradiography.

Protein extracts

From entire organ: To study the regulation of the murine *CYP1A2* and *CYP2A5*, which are predominantly hepatic isoforms, we have prepared crude, nuclear, cytosolic and polysomal extracts from mouse liver.

Crude extracts are prepared by taking the supernatant after low-speed centrifugation (12000*g*) of liver homogenates (11). Nuclei are prepared according to Dignam (12) and nuclear extracts are obtained following the method of Gorski (13), with slight modifications. Briefly, the liver is homogenized in a buffer containing 2M sucrose and the homogenate is centrifuged at 56000*g* through a cushion of the same buffer. The nuclei are found in the pellet. After lysis in a high-salt containing buffer and centrifugation at 2000*g*, the nuclear proteins are recovered in the supernatant, dialyzed, and frozen. The cytosolic (S130) and polyribosomal fractions are prepared according to the method of Brewer and Ross (14). Briefly, liver homogenates are centrifuged at 130000*g* through a cushion of 30% sucrose-containing buffer. The supernatant is the cytosolic fraction and the polyribosomes are contained in the pellet.

Electrophoretic mobility shift assay (EMSA): This assay allows detection of nucleic acid-protein complexes in non-denaturing conditions. The proteins are incubated in the presence of their target nucleic acid (radioactive probe) and the mixture is loaded onto an acrylamide non-denaturing gel. The conditions of the migration must preserve the electrostatic bonds between the nucleic acid and the bound proteins. The probe engaged in a complex with proteins migrates more slowly than the free probe. The free probe and shifted complex are detected by autoradiography.

When investigating the possible binding of proteins to CYP1A2 and CYP2A5 mRNA, we observed shifted bands in both cases (Fig. 1). Digestion of the extracts with proteinase K abolished the formation of retarded complexes (data not shown), which shows that the 3'UTR of these CYPs contain binding sites for RNA-binding proteins.

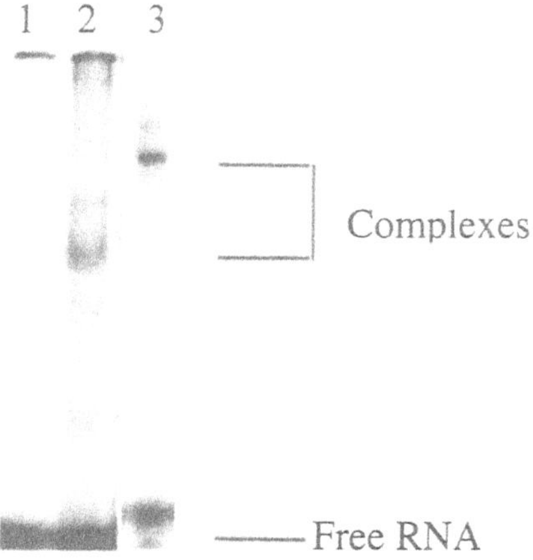

Figure 1. ***The EMSA was performed using the entire 3'UTR of CYP1A2 (lanes 1 and 2) and CYP2A5 (lane 3) and proteins from liver crude extracts (lanes 2 and 3) or no proteins (lane 1).*** *Data adapted from references 11 and 15.*

UV cross-linking assay (UVXL): By this technique the proteins in direct contact with the mRNA probes can be detected, and their size can be estimated. Similar incubation as for the EMSA are performed, and the RNA-protein complexes are made irreversible by UV light cross-linking. The unprotected RNA is digested with RNase A and the samples are denatured and loaded on SDS-PAGE (Fig. 2).

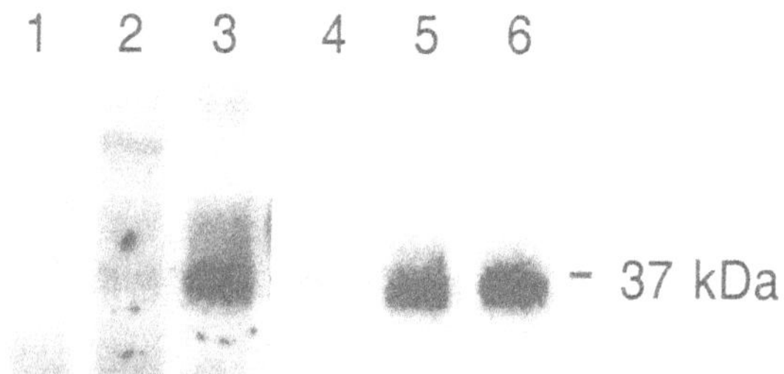

Figure 2. ***The UVXL was performed using the entire 3'UTR of CYP1A2 (lanes 1-3) and CYP2A5 (lane 4-6) and proteins from liver crude extracts (lanes 1 and 4), from polysomes (lanes 2 and 5) and from nuclei (lane 3 and 6).*** *Data adapted from references 11 and 15.*

Binding Site Mapping

Use of different RNA probes: The portion of RNA recognized by the RNA-binding proteins was studied using RNA probes, synthesized by PCR corresponding to various portions of the 3'UTR. Both in the case of CYP1A2 and CYP2A5, the use of partially overlapping probes allowed mapping the proteins binding sites to a relatively small fragment of the 3'UTR (approx. 70-120 nt) (Fig. 3).

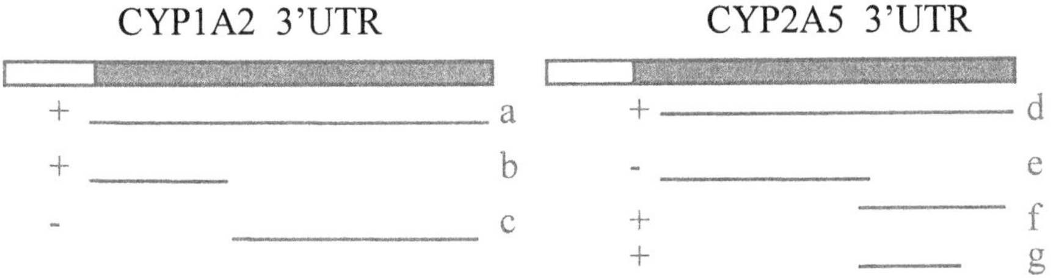

CYP1A2 3'UTR

+ a

+ b

- c

CYP2A5 3'UTR

+ d

- e

+ f

+ g

Figure 3: ***The straight lines correspond to different RNA probes (a-g) synthesized.*** *"+" and "-" indicate the formation or lack of formation of RNA-protein complex, respectively, as detected by UVXL. Data adapted from references 11 and 15.*

Secondary structure prediction and use of antisense oligonucleotides: Conformational analysis of probe g (see Fig. 4) shows that the 71 nt are arranged in a putative hairpin-loop structure.

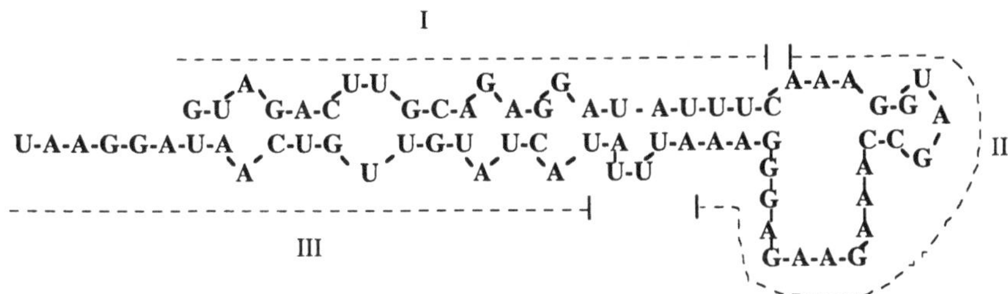

Figure 4: ***Predicted secondary structure of probe g.*** *The sequence complementary to the antisense oligonucleotides I, II and III are shown. Data adapted from reference 19.*

In order to more precisely map the protein binding site(s) on CYP2A5, we took advantage of three antisense phosphorothioate oligoribonucleotides (I, II and III, see Fig. 4), spanning the entire sequence of probe g. These were used as non-radioactive competitors in UVXL reactions containing the radioactive probe g.

The oligoribonucleotide complementary to region II exerted the strongest displacement of the radioactive probe (data not shown), suggesting that the primary binding site of the 37 kDa protein was situated at the loop of hairpin-loop structure. To further investigate the essential nucleotides for binding, various portions of the loop were deleted (see Fig. 5), using appropriate primers in PCR reactions and *in vitro* transcription.

Deletion studies

Figure 5: ***The sites of the deletions D1-D5 in probe g are indicated.***

UVXL assays were performed with the non-deleted probe and the 5 probes containing the deletions D1-D5. Probes containing the deletions D1-D3 bound to the 37 kDa protein with similar efficiency as the non-deleted probe, whereas the D5 probe showed reduced binding. No complex was formed with the D4 probe. These results show that the primary binding site of the 37 kDa protein to the 3'UTR of CYP2a5 mRNA is located at the tip of the loop and contains the nucleotides GGUAG.

Binding site mapping to CYP1a2 3'UTR

The nucleotide sequence of probe b (see Fig. 3) contains particular features: a 12U poly(U) stretch followed by a 16A poly(A) stretch separated by 4 nucleotides. This suggests that a hair-pin structure can potentially form at this site. Deletion of the U-stretch impaired the binding activity, as did the deletion of the A-stretch, although to a lesser extent (data not shown). These data suggested that the primary binding site of the CYP12 mRNA binding-protein lies within the U-stretch contained in the putative hairpin found in the b probe.

Activation of the RNA-Binding Activity by Xenobiotics

We have determined if the complex formation is affected during upregulation of the *CYP1A2* and *2A5* by xenobiotics. 3-methylcholanthrene and isosafrole were used to induce *CYP1A2* in C57/BL6 mice and pyrazole was used to induce *CYP2A5* in DBA/2N mice.

3-Methylcholanthrene: 3MC stimulates transcription of the CYP1a2 gene but it also acts at the post-transcriptional level (8, 16), possibly by modifying pre-mRNA processing in the nuclei. We have therefore, investigated whether 3MC alters the binding of the 37 kDa protein to the 3'UTR of CYP1A2 (Fig. 6).

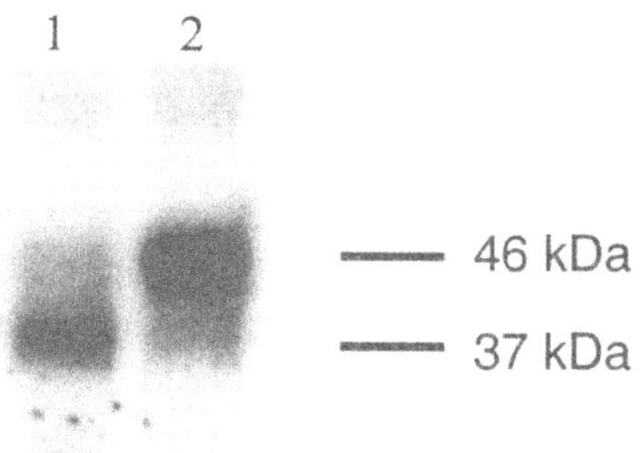

***Figure 6:** **UVXL was performed with protein extracts from untreated and 3MC-treated (200 mg/kg, one i.p. injection, 24 h) mice.** Data adapted from reference15.*

Only in nuclear extracts are RNA-protein complexes detected. 3MC dramatically affects protein binding to CYP1A2 3'UTR: the 37 kDa protein binding is decreased, while the binding of another 46 kDa protein strongly increases upon induction. The 46 kDa protein recognizes poly(U) sequences in a similar manner as the 37 kDa protein.

Isosafrole: This compound is a strong and rather selective inducer for CYP1A2. Its mechanism of action is not fully understood (17). We wanted to know if a CYP1A2 inducer, chemically unrelated to 3MC, affected RNA-protein interactions in a similar manner as 3MC.

UVXL with nuclear proteins from untreated and isosafrole-treated (150 mg/kg, one i.p. injection, 24 h) showed that isosafrole acts in a similar manner as 3MC: it increases the binding of the 46 kDa protein and simultaneously decreases that of the 37 kDa protein (our unpublished results).

Pyrazole: Pyrazole increases CYP2a5 mRNA half-life, without affecting the transcription rate of this gene (9). We therefore investigated its effect on protein binding to CYP2A5 3'UTR. Pyrazole strongly stimulates the binding of the 37 kDa protein to CYP2A5 3'UTR, as shown in Fig. 7.

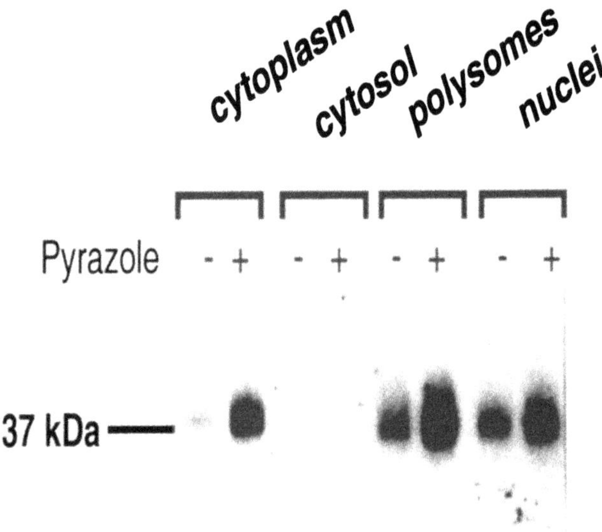

Figure 7: UVXL assays were performed using various subcellular extracts from mice untreated or treated with Pyrazole (200 mg/kg, 1 i.p. injection/day, during 3 days). *Data adapted from reference 11.*

Identification of the RNA-Binding Proteins

Based on several lines of evidence including biochemical properties, RNA binding characteristics as well as the subcellular localization, we hypothesized that the larger CYP1A2 mRNA binding protein is hnRNPC or a similar protein, and that the pyrazole-inducible CYP2A5 mRNA-binding 37 kDa protein, is the hnRNPA1 or a related protein. In order to test our hypothesis, we conducted a series of experiments as follows:

Partial proteolysis: Upon partial digestion with trypsin, hnRNPA1 releases a polypeptide of 25 kDa (20), detectable by UVXL. When our radioactive 37 kDa complex is cleaved with trypsin, a complex of 25 kDa is formed. The 37 kDa complex therefore exhibits the specific partial digestion pattern of hnRNPA1 (*our unpublished results*).

Competition Studies: A 20-mer RNA (known to be an excellent binding site for hnRNPA1 (18)) was used as a non-radioactive competitor in a UVXL assay with the radioactive CYP2A5 probe g. Figure 8 illustrates competition of the 20-mer RNA with the radioactive probe g.

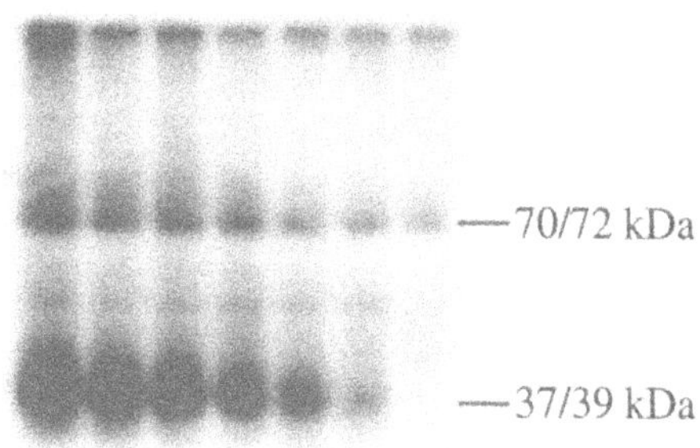

***Figure 8**: UVXL performed in the presence of increasing amounts of the "Winner" oligoribonucleotide (UAUGAUAGGGACUUAGGGUG).* *Data from ref.19.*

Immunoprecipitation: Specific monoclonal antibodies against the hnRNPA1 and hnRNPC (kindly provided by G. Dreyfuss, HHMI, University of Philadelphia, USA) were used to further identify the inducer-activable proteins (Fig. 9).

Typical UVXL assays were performed, but after the RNase digestion step, the antibodies against hnRNPA1 or hnRNPC proteins were added to the mixtures. Precipitation of the antibody-protein-RNA complexes was achieved by adding protein A-sepharose beads to the mixtures. After washing, the beads were denatured and the released complexes submitted to SDS-PAGE. Antibodies against CYP2E1 and CYP1A, as well as pre-immune serum were used as negative controls.

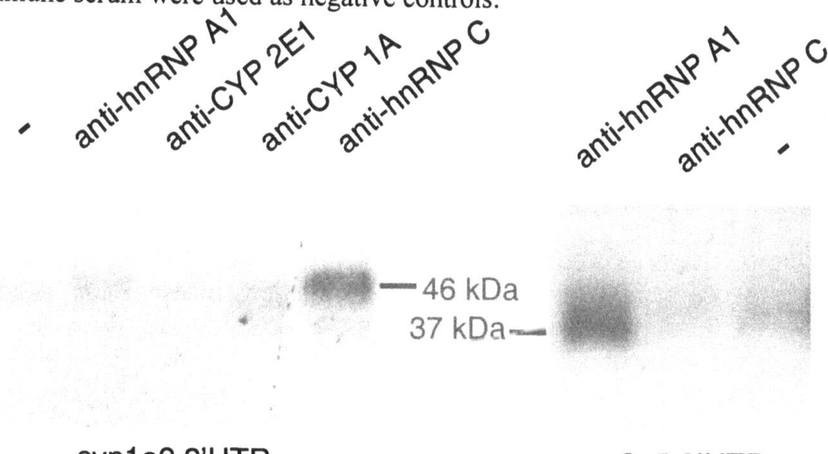

***Figure 9**: Immunoprecipitation of the CYP1A2 mRNA/46 kDa protein and CYP2A5 mRNA/37 kDa protein.* *Data adapted from references 11 and 15.*

The anti-hnRNPC antibody specifically reacted with the 46 kDa, 3MC-activable protein, and the anti hnRNPA1 antibody with the 37 kDa, pyrazole-activable protein. The anti hnRNPA1 antibody did not recognize the 37 kDa protein bound to CYP1A2 3'UTR in non-induced samples (15), showing that the 37 kDa proteins bound to CYP1A2 and CYP2A5 mRNA are different.

Future Investigations: Determining the Function of the CYP1A2 and CYP2A5 RNA-Binding Proteins.

CYP1A2 and differential splicing: Silver and Krauter have shown that the CYP1A2 mRNA is differentially spliced upon 3MC induction (8). Post-transcriptional regulation of the CYP1A2 gene could therefore, at least partially, be achieved by controlling the maturation of the pre-mRNAs in the nucleus. Our observation of 3MC-regulated binding of the hnRNPC to the 3'UTR of CYP1A2 mRNA is in accordance with this hypothesis. Other investigators have shown that the hnRNPC is a component of the splicing machinery (21). Therefore, 3MC could modify the maturation of the CYP1A2 pre-mRNAs by promoting the binding of the hnRNPC to the transcripts.

CYP2A5 and mRNA stability: CYP2A5 mRNA half-life is increased upon pyrazole induction *in vivo* (from 1.5 hours to 6 hours) (9). Following the method described by Brewer and Ross (14), we have shown that the poly(A) tails of pyrazole induced CYP2a5 mRNA are longer than those of non-induced CYP2A5 mRNA.

As increased poly(A) tail length are associated with increased mRNA stability, these results suggest that the pyrazole-inducible 37 kDa protein could be involved in the stabilization of CYP2A5 mRNA by controlling the length of poly(A) tails.

Other CYPs

It has been shown that post-transcriptional regulation is important for the regulation of several CYPs. In several investigations, transcription rates of specific isoforms have been compared to the nuclear and/or cytoplasmic levels of the corresponding mature transcripts. It has often been found in these studies that the inducers increase the amounts of mature transcripts to higher levels than can be expected from the level of transcriptional activation, leading to the hypothesis that message stabilization must occur. This is the case for the induction of CYP1A by polycyclic aromatic hydrocarbons (22, 23, 16), CYP2B by dexamethasone (24), CYP3A by triacetyloleandomycin (25), and CYP2E in diabetic rats (26). Only in a few cases have the regulatory mechanisms been further investigated, with respect to the identification of regulatory sequences within the mRNAs. Furthermore, no precise information is available to date on the possible RNA-protein interactions relevant for the regulation of the expression of these CYPs. As follows, a brief summary on investigations of other CYPs is presented.

CYP1A1: A polymorphism has been found in the 3'UTR of the cancer-prone Atlantic tomcod CYP1A1, which can affect the message stability. Since CYP1A1 is important in carcinogen metabolism, this might explain the susceptibility of this fish to hepatic neoplasia (27).

CYP2E1: The regulation of the expression the ethanol-inducible CYP2E1 gene is complex and occurs at different steps. Peng and Coon (28) have shown that the 3' UTR of the CYP2E1 gene contains information for the down-regulating effect of insulin. It has recently been proposed that the poly(A) tail and certain sequences within the 3' UTR are important for protecting CYP2E1 mRNA from RNase degradation (29).

CYP7 (Cholesterol 7α-hydroxylase): CYP7 initiates the conversion of cholesterol into bile acids in the liver. Agellon and Cheema (30) showed that the mouse cholesterol 7α-hydroxylase mRNA (which is relatively unstable *in vivo*) contains elements important for the post-transcriptional regulation of this gene by bile acids. These elements resemble the AU-rich elements found in many short-lived mRNAs. Some labile proteins could regulate the stability of the CYP7A1 gene (31).

CONCLUSIONS

We have shown that RNA-protein interactions occur at the 3'UTR of the transcripts of two xenobiotic inducible genes: the CYP1A2 and 2A5, and that these interactions are affected by specific CYP inducers. It is possible that RNA-protein interactions are important for the regulation of the expression of some other CYP genes as well. Beside transcriptional activation, this type of effect could be part of a mechanism by which xenobiotics affect gene expression.

Our observations are the first ones showing that the activity of certain mammalian RNA binding proteins is modulated by xenobiotics. Since relatively little is known regarding regulation of the activity and/or expression of RNA-binding proteins, it is relevant to investigate their mode of activation by xenobiotics, in the case of both CYP and non-CYP genes .

There is accumulating evidence that transcriptional and post-transcriptional processes are not independent from one another, but rather tightly linked (32). Our preliminary experiments show that modulation of the transcriptional activity of CYP2A5 affects its post-transcriptional regulation (unpublished observations). An interesting field of investigation will therefore be to study the links between transcriptional and post-transcriptional control of CYP genes expression, in particular the role of the regulatory proteins in this interplay.

We have shown that the proteins binding to the CYP1A2 and CYP2A5 transcripts belong to the hnRNP proteins. In the case of other genes, these proteins are known to be involved in transcript metabolism (maturation, translation, turnover). However, their exact role in CYP gene expression are unclear. To understand the signalling pathways controling the activity of CYP RNA-binding proteins as well as their precise function remain challenging issues and exciting areas of research for the coming years.

Summary of key concepts

- *Detection of CYP1A2 and CYP2A5 mRNA-binding proteins by EMSA and UVXL.*
- *Mapping of the protein binding sites by competition experiments and by assessing complex formation to truncated or deleted RNA probes.*
- *Identification of the RNA-binding proteins as pyrazole-activable hnRNPA1 and 3MC-activable hnRNPC, by partial proteolysis, competition studies, immunoprecipitation.*
- *Ongoing studies concerning physiologic action of the proteins: hnRNPA1 could modulate CYP2A5 mRNA stability by controlling the length of its poly(A) tail, and hnRNPC could be involved in the maturation of CYP1A2 pre-mRNAs.*

Study Guide Questions

1) Comparison of the EMSA and UVXL techniques: what type of information do they bring, what are their limitations, why are they complementary?
2) It is likely that transcriptional and post-transcriptional regulation of gene expression is tightly coupled. Why? Why are nucleic acid-binding proteins attractive candidates as linkers between these two types of regulation?
3) Describe several plausible mechanisms by which xenobiotics could activate RNA-binding proteins.
4) Why should drug-metabolizing enzymes be inducible? Why do several different mechanisms of induction occur?

Acknowledgements
We thank Doctor Gideon Dreyfuss for the kind gift of the anti-hnRNP antibodies.

REFERENCES

1. Nelson DR, Koymans L, Kamataki T, Stegeman JJ, Feyereisen R, Waxman DJ, Waterman MR, Gotoh O, CoonMJ, Estabrook RW, Gunsalus IC, Nebert DW. 1996 P450 superfamily: update on new sequences, gene mapping, accession numbers and nomenclature. Pharmacogenetics. 6(1):1-42
2. Rendic S, Di Carlo FJ. 1997 Human cytochrome P450 enzymes: a status report summarizing their reactions, substrates, inducers, and inhibitors. Drug Metab Rev. 29(1-2):413-580.
3. Porter TD, Coon MJ. 1991 Cytochrome P-450. Multiplicity of isoforms, substrates, and catalytic and regulatory mechanisms. J Biol Chem. 266(21):13469-72
4. Waxman DJ. 1999 P450 gene induction by structurally diverse xenochemicals: central role of nuclear receptors CAR, PXR, and PPAR. Arch Biochem Biophys. 369(1):11-23
5. Zubiaga AM, Belasco JG, Greenberg ME. 1995 The nonamer UUAUUUAUU is the key AU-rich sequence motif that mediates mRNA degradation. Mol Cell Biol. 15(4):2219-30.
6. Siomi H, Dreyfuss G. 1997 RNA-binding proteins as regulators of gene expression. Curr Opin Genet Dev. 7(3):345-53. Review.
7. Krecic AM, Swanson MS. 1999 hnRNP complexes: composition, structure, and function. Curr Opin Cell Biol. 11(3):363-71
8. Silver G, Krauter KS. 1990 Aryl hydrocarbon induction of rat cytochrome P-450d results from increased precursor RNA processing. Mol Cell Biol. 10(12):6765-8.
9. Aida K, Negishi M. 1991 Posttranscriptional regulation of coumarin 7-hydroxylase induction by xenobiotics in mouse liver: mRNA stabilization by pyrazole. Biochemistry. 30(32):8041-5.
10. Kimura S, Gonzalez FJ, Nebert DW. 1984 Mouse cytochrome P3-450: complete cDNA and amino acid sequence. Nucleic Acids Res. 12(6):2917-28
11. Geneste O, Raffalli F, Lang MA. 1996 Identification and characterization of a 44 kDa protein that binds specifically to the 3'-untranslated region of CYP2a5 mRNA: inducibility, subcellular distribution and possible role in mRNA stabilization. Biochem J. 313 (Pt 3):1029-37.
12. Dignam JD, Lebovitz RM, Roeder RG. 1983 Accurate transcription initiation by RNA polymerase II in a soluble extract from isolated mammalian nuclei. Nucleic Acids Res. 11(5):1475-89.
13. Gorski K, Carneiro M, Schibler U. 1986 Tissue-specific in vitro transcription from the mouse albumin promoter. Cell. 47(5):767-76.
14. Brewer G, Ross J. 1988 Poly(A) shortening and degradation of the 3' A+U-rich sequences of human c-myc mRNA in a cell-free system. Mol Cell Biol. 8(4):1697-708.
15. Raffalli-Mathieu F, Geneste O, Lang MA. 1997 Characterization of two nuclear proteins that interact with cytochrome P-450 1A2 mRNA. Regulation of RNA binding and possible role in the expression of the CYP1a2 gene. Eur J Biochem. 245(1):17-24.
16. Silver G, Reid LM, Krauter KS. 1990 Dexamethasone-mediated regulation of 3-methylcholanthrene-induced cytochrome P-450d mRNA accumulation in primary rat hepatocyte cultures. J Biol Chem. 265(6):3134-8.

17. Adams NH, Levi PE, Hodgson E. 1993 Regulation of cytochrome P-450 isozymes by methylenedioxyphenyl compounds. Chem Biol Interact. 86(3):255-74.
18. Burd CG, Dreyfuss G. 1994 RNA binding specificity of hnRNP A1: significance of hnRNP A1 high-affinity binding sites in pre-mRNA splicing. EMBO J. 13(5):1197-204.
19. Tilloy-Ellul A, Raffalli-Mathieu F, Lang MA. 1999 Analysis of RNA-protein interactions of mouse liver cytochrome P4502A5 mRNA. Biochem J. 339 (Pt 3):695-703.
20. Hamilton BJ, Nagy E, Malter JS, Arrick BA, Rigby WF. 1993 Association of heterogeneous nuclear ribonucleoprotein A1 and C proteins with reiterated AUUUA sequences. J Biol Chem. 268(12):8881-7.
21. Choi YD, Grabowski PJ, Sharp PA, Dreyfuss G. 1986 Heterogeneous nuclear ribonucleoproteins: role in RNA splicing. Science 231(4745):1534-9.
22. Pasco DS, Boyum KW, Merchant SN, Chalberg SC, Fagan JB. 1988 Transcriptional and post-transcriptional regulation of the genes encoding cytochromes P-450c and P-450d in vivo and in primary hepatocyte cultures. J Biol Chem. 263(18):8671-6.
23. Kimura S, Gonzalez FJ, Nebert DW. 1986 Tissue-specific expression of the mouse dioxin-inducible P(1)450 and P(3)450 genes: differential transcriptional activation and mRNA stability in liver and extrahepatic tissues. Mol Cell Biol. 6(5):1471-7.
24. Simmons DL, McQuiddy P, Kasper CB. 1987 Induction of the hepatic mixed-function oxidase system by synthetic glucocorticoids. Transcriptional and post-transcriptional regulation. J Biol Chem. 262(1):326-32.
25. Dalet C, Blanchard JM, Guzelian P, Barwick J, Hartle H, Maurel P. 1986 Cloning of a cDNA coding for P-450 LM3c from rabbit liver microsomes and regulation of its expression. Nucleic Acids Res. 14(15):5999-6015.
26. Song BJ, Matsunaga T, Hardwick JP, Park SS, Veech RL, Yang CS, Gelboin HV, Gonzalez FJ. 1987 Stabilization of cytochrome P450j messenger ribonucleic acid in the diabetic rat. Mol Endocrinol. 1(8):542-7.
27. Roy NK, Kreamer GL, Konkle B, Grunwald C, Wirgin I. C1995 haracterization and prevalence of a polymorphism in the 3' untranslated region of cytochrome P4501A1in cancer-prone Atlantic tomcod. Arch Biochem Biophys. 322(1):204-13.
28. Peng HM, Coon MJ. 1998 Regulation of rabbit cytochrome P450 2E1 expression in HepG2 cells by insulin and thyroid hormone. Mol Pharmacol. 54(4):740-7.
29. Kocarek TA, Zangar RC, Novak RF. 2000 Post-transcriptional regulation of rat CYP2E1 expression: role of CYP2E1 mRNA untranslated regions incontrol of translational efficiency and message stability. Arch Biochem Biophys. 376(1):180-90.
30. Agellon LB, Cheema SK. 1997 The 3'-untranslated region of the mouse cholesterol 7alpha-hydroxylase mRNA contains elements responsive to post-transcriptional regulation by bile acids. Biochem J. 328 (Pt 2):393-9.
31. Baker DM, Wang SL, Bell DJ, Drevon CA, Davis RA. 2000 One or more labile proteins regulate the stability of chimeric mRNAs containing the 3'-untranslated region of cholesterol-7alpha -hydroxylase mRNA. J Biol Chem. 275(26):19985-91.
32. Ladomery M. 1997 Multifunctional proteins suggest connections between transcriptional and post-transcriptional processes. Bioessays 19(10):903-9.

14

SITE-SPECIFIC CLEAVAGE OF INSULIN-LIKE GROWTH FACTOR II mRNAs

Erwin L. van Dijk and P. Elly Holthuizen*

*Centre de Genetique Moleculaire, Gif-sue-Yvette, Cedex, France, and University Medical Center Utrecht, Utrecht, The Netherlands**

Insulin-like growth factor II (IGF-II) is a protein involved in growth and differentiation of numerous cell types. Because of the importance of IGF-II in many different processes, various regulatory mechanisms are available in the cell to control IGF-II protein synthesis, reflecting the vital importance of fine tuning and rapid regulation of IGF-II mRNA levels and IGF-II protein amount. One such mechanism is the site-specific endonucleolytic cleavage of IGF-II mRNAs, which will be described in this chapter. The role of an RNA-binding protein that is associated with the highly structured RNA will be discussed as well as the requirements for recognition of the IGF-II mRNA by an endoribonuclease.

BACKGROUND

The insulin-like growth factors (IGFs) are potent mitogens for many different cell types and they play a central role in growth and development (1). The significance of the IGFs in these processes was demonstrated by the dramatic effects of mutations in IGF genes that result in the absence of these growth factors. Targeted disruption of the IGF-I or IGF-II gene leads to a 40% reduction in body weight at birth in otherwise normally proportioned mice (2,3). In addition to a function in general growth, locally produced IGFs have important roles in tissue regeneration and wound healing. All of these processes involve mitogenic stimulation and cell division and need to be controlled accurately. Various regulatory mechanisms can modulate the biological effects of IGFs. First, the number of receptors expressed on target cells determines the strength of the IGF signal. Both IGF-I and IGF-II act predominantly through the type I- IGF receptor resulting in a mitogenic response. Another receptor, the type II- IGF receptor, is involved in degradation of IGF-II. Second, the biological activities of the IGFs are controlled by their interaction with the IGF binding proteins (IGF-BPs) (4). All six IGF-BPs protect the IGFs against degradation and they facilitate transport to distinct body compartments. Third, the bioavailability of IGFs is regulated by the amounts of these growth factors produced and secreted by the cells capable of their synthesis. Serum concentrations of IGFs vary widely as a result of their individual expression profiles during prenatal and postnatal growth and development. In addition to the developmental stage-dependent synthesis of endocrine IGF, originating mainly from hepatocytes, IGF with local paracrine and autocrine functions is synthesized in a number of other cell types resulting in both proliferative and differentiating effects. IGFs are expressed in a wide variety of tissues by an intricate mechanism of regulating the levels of protein required in a particular tissue.

The human IGF-II gene is a complex transcription unit with many interesting regulatory aspects. It consists of 9 exons, of which exons 7, 8 and the first 237 nt of exon 9 encode the 180 aa long pre-pro-IGF-II, that is further processed into the 67 aa long mature IGF-II polypeptide. IGF-II expression is regulated in a tissue- and development-specific manner by differential activation of four promoters, resulting in a family of IGF-II mRNAs with four different leader sequences derived from the untranslated exons 1-6. The various leader sequences cause the transcripts to be translated with different efficiencies, where three of the four leaders allow efficient translation, but one leader strongly represses translation (5).

IGF-II gene expression is regulated both at the level of transcription and translation, and in addition, regulation also occurs at the level of post-transcriptional processing of the mRNAs and of processing of the IGF-II precursor protein. The transcription unit contains a very long 3'-UTR of 4 kb within exon 9 and this 3'-UTR region has been shown to contain a specific cleavage site. Cleavage destabilizes the full length IGF-II mRNAs and thus removes it from the pool of protein-producing mRNAs, thereby providing the cell with an additional way to control IGF-II protein synthesis. Thus, an elaborate set of regulatory mechanisms is available for the cell to control IGF-II protein synthesis, reflecting the vital importance of fine-tuning and rapidly regulating IGF-II mRNA and IGF-II protein amount.

In this chapter we will discuss the endonucleolytic cleavage of IGF-II mRNAs and we will address the following topics: (a) structure of IGF-II mRNA, (b) interaction of IGF-II mRNA with a specific binding protein, and (c) the function of IGF-II mRNA cleavage and/or the cleavage product.

Endonucleolytic Cleavage

The existence of a 1.8 kb long IGF-II RNA species that did not encode IGF-II protein but consisted solely of 3'-UTR sequences was first demonstrated by Northern blotting using a 3'-UTR specific IGF-II probe. Furthermore, it was shown that this 1.8 kb RNA was the result of cleavage of the full-length IGF-II mRNAs (6). Thus, in addition to the full-length IGF-II transcripts, a non IGF-II encoding RNA of 1.8 kb, derived from the IGF-II gene but not corresponding to any of the promoters, was detected. This 1.8 kb RNA species was present in human, rat and murine cells and was subsequently identified as a product generated by a site-specific endonucleolytic cleavage reaction in the 3'-UTR of full length IGF-II mRNAs (Fig. 1).

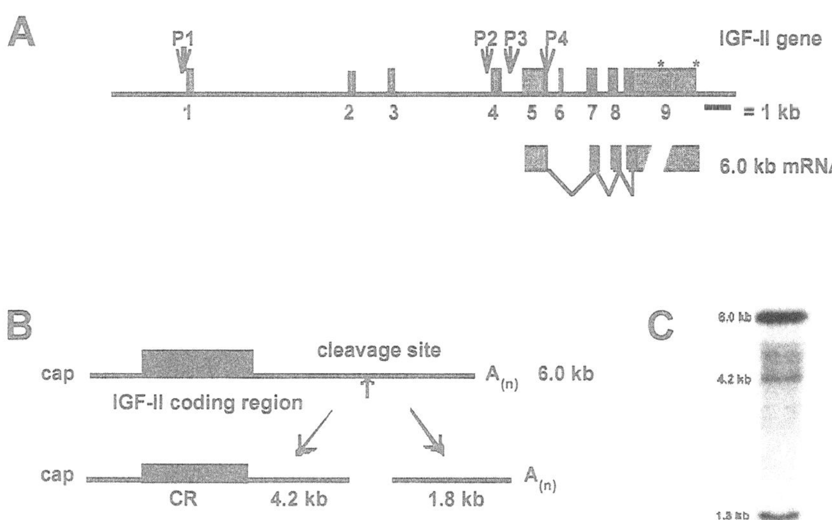

***Figure 1.** **A) Structure of the human IGF-II gene and the 6.0 kb transcript generated by transcription from the major fetal promoter P3.** Exon sequences are indicated by boxes and intron sequences are indicated by a line; the IGF-II coding exons 7, 8, and the first part of 9 are indicated in black. Promoters P1 to P4 are indicated by arrows. The two stars in exon 9 represent the two alternative polyadenylation sites in the 3'-UTR **B) Schematic representation of the endonucleolytic cleavage in the IGF-II 3'-UTR that results in an unstable capped 5'-cleavage product that contains the coding region and a stable polyadenylated 3'-cleavage product.** **C) Northern blot analysis of total RNA isolated from IGF-II-expressing human cells.** As a probe, a radiolabeled DNA fragment overlapping the cleavage site was used. Thus, the full length IGF-II transcript of 6.0 kb as well as both the 5' cleavage product of 4.2 kb and the 1.8 kb 3' cleavage product are detected. Note that the 5' cleavage product is less abundant than the 3' cleavage product due to its lower stability.*

Endonucleolytic cleavage of IGF-II mRNAs yields two degradation products: 1) a 5'-cleavage product that contains a cap structure, but lacks the protective poly(A) tail and is thus rapidly degraded and 2) a 3'-cleavage product of 1.8 kb that lacks the cap structure but is polyadenylated, and this product is surprisingly stable. Endonucleolytic cleavage of IGF-II mRNAs was shown to occur in the cytoplasm for all types of full-length IGF-II transcripts, irrespective of the promoter from which the transcripts are derived. The cleavage site in IGF-II mRNA has been mapped to the single nucleotide resolution by S1 mapping and primer extension experiments.

Elements Required for IGF-II mRNA Cleavage.

After mapping the exact location of the cleavage site in the 3'-UTR of IGF-II mRNAs, the next point to address was the identification of the regions that are necessary and sufficient for this cleavage reaction. In order to identify the RNA regions required for *in vivo* cleavage, an IGF-II minigene was constructed that contained a strong CMV promoter-enhancer and the IGF-II-encoding exons 7, 8, and 9.

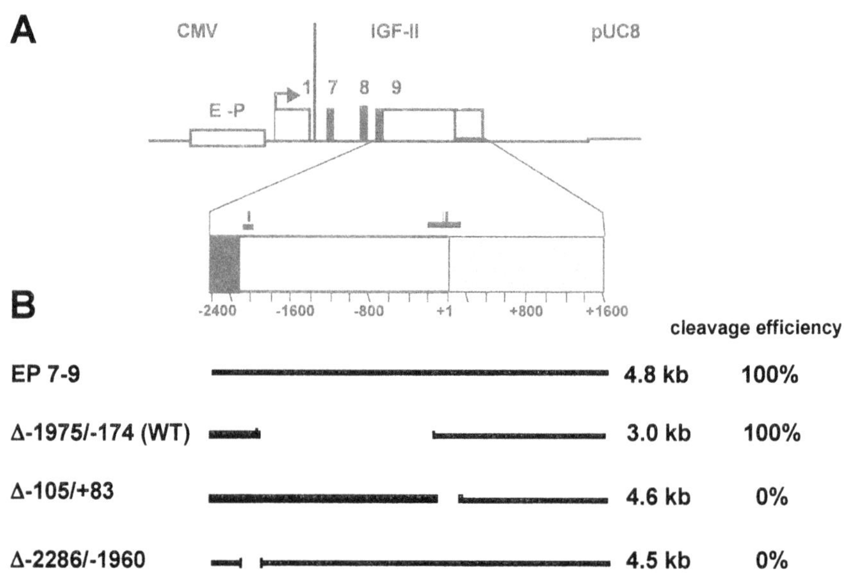

Figure 2. Schematic representation of the CMV-IGF-II minigenes. A) Construct EP7-9. *Construct EP7-9 contains the enhancer/promoter region and exon 1 of cytomegalovirus (CMV) fused to a 10.9 kb genomic fragment of the human IGF-II gene containing exons 7, 8, and 9 cloned in plasmid pUC8 (bold line). The CMV regions and the IGF-II and pUC regions are drawn at different scales. The arrow indicates the transcription start site. Translated and untranslated IGF-II regions are depicted by filled and open boxes, respectively. The gray area in the 3'-part of exon 9 represents the 1.8 kb RNA region.* ***B) Enlarged representation of human IGF-II exon 9.*** *Positions (bp) in exon 9 are relative to the cleavage site (+1). The two elements required for cleavage are indicated by bars (I, II). The black lines represent the wild type construct EP7-9 and its deletion derivatives. ΔWT lacks the region between elements I and II (-1955 to -174); Δ-105/+83 lacks the region overlapping the cleavage site; Δ-2286/-1960 lacks element I in the upstream region. The sizes of the transcripts and their relative cleavage efficiencies are indicated on the right. The transcript of ΔWT is cleaved as efficiently as EP7-9 (7).*

The effects on cleavage of various deletions in the 3'-UTR were tested in an *in vivo* system of cells transiently transfected with the IGF-II minigene constructs. This resulted in the identification of two widely separated elements in the 3'-UTR of IGF-II mRNAs that are essential for the site-specific endonucleolytic cleavage reaction to occur. It was shown that a 350 nt region (element II) surrounding the cleavage site is necessary but not sufficient for cleavage, and that an additional 150 nt long element (element I) located 2 kb upstream of element II is also required to confer cleavage (Fig. 2). When the two elements were transferred into the 3'-UTR of a heterologous gene, β-globin, it was shown that the presence of both elements I and II in the 3'-UTR of β-globin mRNA was sufficient for cleavage (7). In summary, two elements are necessary and sufficient for endonucleolytic cleavage of IGF-II mRNAs: element I located between positions –2286 and –1960 relative to the clevage site; and element II located around the cleavage site from positions –173 to +150 (Fig. 2).

The Structure of the 3'-UTR of IGF-II mRNAs.

After establishing that the two elements in the 3'-UTR of IGF-II mRNAs were sufficient to confer cleavage, RNA folding algorithms were used to predict a secondary structure for the 3'-UTR of IGF-II mRNAs. Structural analysis using RNA folding algorithms suggested that the cleavage site in element II is situated between two highly structured domains, a region immediately upstream of the cleavage site that can form two stem-loop structures and a region downstream of the cleavage site that is very G-rich (Fig. 3A). Furthermore, a stable 80 nt long double-stranded (ds) RNA stem structure could be formed between the G-rich region in element II and the very C-rich element I (8).

Based on these results a model was tested suggesting that elements I and II, that are distant in the primary sequence, are brought into proximity by RNA folding, thereby forming the recognition determinant for endonucleolytic cleavage. Using the *in vivo* IGF-II minigene expression system, it was confirmed that the formation of this stable stem-structure between element I and the G-rich part of element II is necessary for cleavage of IGF-II mRNAs. In addition, the cleavage site itself is located in an open conformation, flanked by two stem-loop structures. After the full length IGF-II mRNA has been cleaved, the 3'-product remains very stable, although it lacks a cap-structure. The most likely explanation for this is that the G-rich stretch directly downstream of the cleavage site region, when no longer forming the long double stranded RNA structure, can now adopt another highly folded structure, the so-called guanosine-quadruplex structure (9) (Fig. 3B).

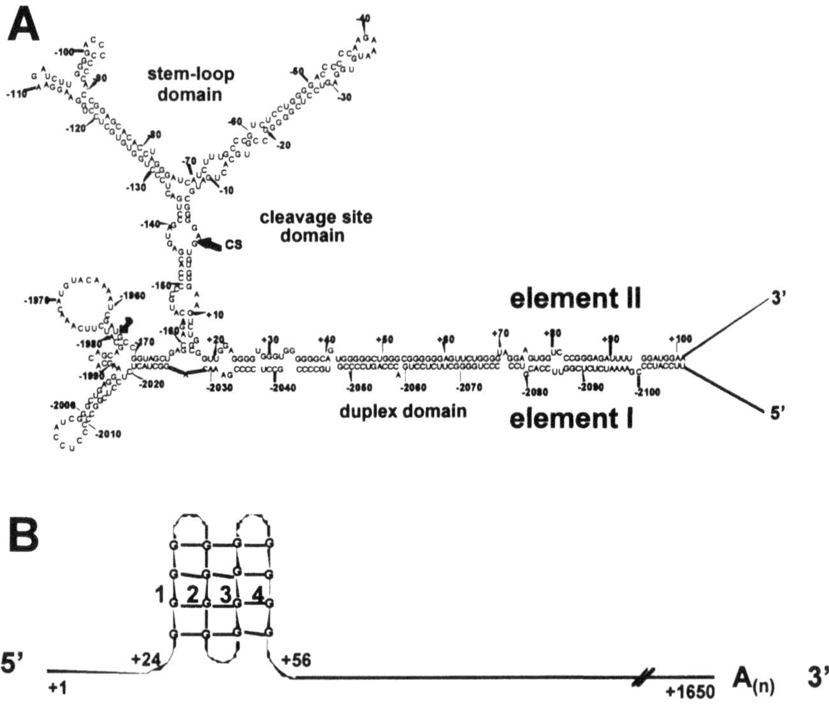

Figure 3. A) Secondary structure prediction of the RNA structural domains formed by elements I (positions -2116 to -2031) and II (positions -173 to +150) in the 3' UTR of the IGF-II mRNAs. *Shown is a folding where the spacing between the elements is identical to construct ΔWT, that lacks the region between elements I and II from -1955 to -174 and that cleaves with wild type efficiency. The site of the deletion is indicated by a bent arrow; the cleavage site is indicated by a straight arrow.* ***B) Schematic representation of the G-quadruplex structure formed by the G-rich region in element II after cleavage.*** *This compact structure, composed of four stretches of four G-residues each, is thought to protect the 3' cleavage product against exonucleolytic degradation, thus accounting for its unusual stability.*

The model is that the stem-structure is the predominant structure in full-length IGF-II mRNA, while subsequently, after cleavage of the full length IGF-II mRNA has taken place, the guanosine-quadruplex structure can form at the 5' end of the 1.8 kb RNA in such a way that the intramolecular guanosine-quadruplex structure protects the 1.8 kb RNA from rapid 5'-exonucleolytic degradation. This would account for the unusual stability of the 3'-terminal cleavage product.

RNA-Protein Complex Formation at the Stem-Structure.

In search for *trans*-acting factors that specifically bind to the RNA elements required for cleavage, electrophoretic mobility shift assays (EMSAs) were carried out using *in vitro* synthesized IGF-II RNAs containing elements I and II and cytoplasmic extracts from human cells that endogenously express and cleave IGF-II mRNAs. A distinct complex, specific for IGF-II mRNA, was observed only when the synthesized RNAs contained element II. In addition, two non-specific complexes were present (Fig. 4B)(10).

To determine the binding site of the protein more precisely, RNAs with various deletions in element II were used. Initial binding experiments with RNAs containing 5'- and 3'-truncations in element II roughly mapped the binding site of the protein to the stem-structure between positions –133 to –73 (10). Deletion of a side-loop (Δ-103/-92) had no effect on protein binding, but mutation of a part of the stem from position -89 to -85 completely disrupts formation of the complex (Fig. 4). Even more subtle mutations of only 2 nucleotides in the region from –88 to –81, such as mutant M3 (-84/-83) completely abolished binding of the protein (Fig. 4B). This confirms that the protein specifically binds to the stem and not to the loop. Specificity of binding was further assessed in competition experiments. Effective competition of the specific RNA-protein complex was only observed with unlabeled wild type element II RNA and with mutant Δ-103/-92. In summary, a protein that specifically interacts with the RNA region from positions -89 to -81 pairing with nucleotides -126 to -117 in the stem-structure 133 to 73 nucleotides upstream of the cleavage site was detected.

To examine whether this RNA-protein complex actually plays a role in cleavage, the same mutations in the stem structure that abolish formation of the specific complex *in vitro* were introduced in the IGF-II minigene and the *in vivo* cleavage efficiency of the different mutants was determined. The mutant with the deleted side-loop, Δ-103/-92, which did not disrupt binding of the protein *in vitro,* was also tested for its effect on the cleavage efficiency *in vivo*. No reduction in the cleavage efficiency was observed in this mutant (Fig. 4C). However, for mutants –89/-85 and M3 (-84/-83), where RNA-protein complex formation is absent, a reduction of approximately 40% in cleavage efficiency was observed. From this we conclude that although RNA-protein complex formation is not absolutely essential for cleavage, its absence does have a reducing effect on cleavage efficiency. The most likely explanation for this observation is that the structure of the stem-loop to which the protein is binding serves as a structural determinant to keep the cleavage site in its proper conformation.

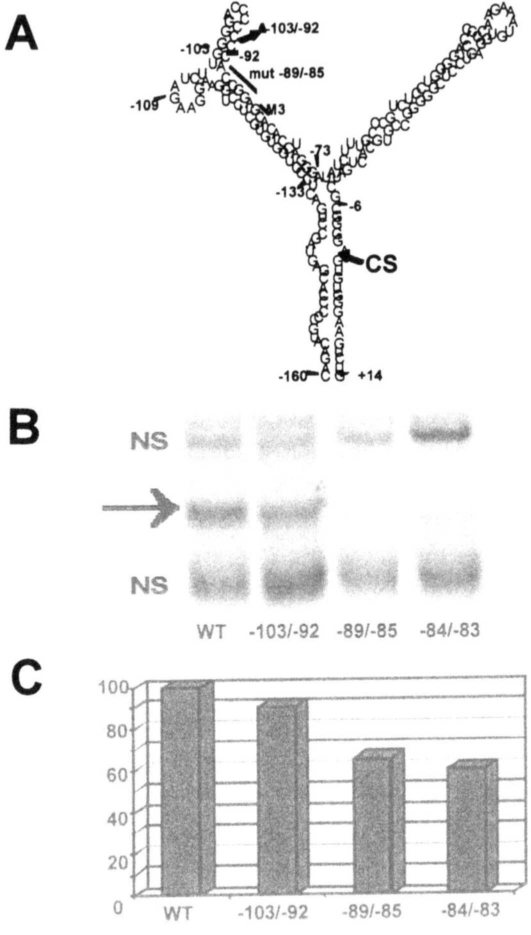

Figure 4. Identification of a specific RNA-protein complex in the proximity of the cleavage site. A) Secondary structure representation of the region around the cleavage site. Mutations introduced in order to map the binding site of an RNA-binding protein that *specifically interacts with RNAs containing element II. The region from -103 to -92 was deleted (Δ-103/-92), generating a truncated stem-loop; nucleotides -89 to -85 (5'-CGGAG-3') in the stem were mutated to 5'-AUAUA-3' (mut -89/-85); and nucleotides -84C and -83A were mutated to -84G/-83U (M3).* ***B) In vitro binding assay to test the effect of the mutations on formation of the specific complex.*** *Radiolabeled in vitro synthesized RNAs containing the various mutations were incubated with cytoplasmic extract from human IGF-II-expressing cells and RNA-protein complexes were separated from unbound RNA by native polyacrylamide gel electrophoresis. Two nonspecific complexes, that are also formed with non-related RNAs (not shown), are indicated by 'NS'. The arrow on the left indicates the element II-specific RNA-protein complex.* ***C) Effects on in vivo IGF-II mRNA cleavage of the various mutations that either have no effect on formation of the specific complex or totally disrupt it (see (B)).***

THE ROLE OF THE VARIOUS IGF-II RNA STRUCTURAL DOMAINS IN CLEAVAGE

The RNA regions found to be required for cleavage can be subdivided into three separate folding domains: (1) the upstream stem-loop domain consisting of a 5'- and a 3'-stem-loop; (2) the duplex domain consisting of the downstream region of element II that folds into an extended duplex of 83 nucleotides with element I, and (3) the cleavage site itself, located in an internal loop flanked by two helical regions and a second internal loop. We will refer to this latter region as the 'cleavage site domain'. Several experiments were performed to dissect the functions of the various RNA regions in cleavage (Fig. 5).

In order to identify recognition determinants in the cleavage site domain, several nucleotide substitutions around the cleavage site were introduced (10,11)(Fig. 5A). First, a G-residue at position +1 relative to the cleavage site was changed into a U; this caused a reduction in cleavage efficiency to 30% of that of the wild type construct (Fig. 5B). However, this substitution also caused a small change in the predicted structure of the loop, which might be responsible for the observed effect. In order to restore the predicted wild type structure of the loop, a compensatory A to C substitution was introduced in the loop at position -144. Rather than restoring the wild type cleavage efficiency, this mutation caused a further decrease in cleavage efficiency to 5% of the wild type, indicating that the specific identity of the nucleotides at these positions is important. To obtain insight into the importance of nucleotide identity in the flanking helices, two GC base pairs in the downstream duplex were changed to AU in construct mut[-149,-148,+5.+6]. These mutations resulted in a complete loss of cleavage, indicating that also in the flanking region of the cleavage site, specific nucleotides are recognized.

To test whether the upstream stem-loops are required for cleavage, specific deletions were introduced in the stem-loop domain, removing all or part of the stem-loops. The deletions were designed in such a way that the predicted wild type secondary structure around the cleavage site would remain intact (Fig. 5A). First, the region from nucleotides –104 to –78 was deleted, leading to a truncated 5' stem-loop and, more importantly, to the loss of the binding site for the identified binding protein. Deletion of this protein-binding region caused a reduction to 50% of the wild type cleavage efficiency, comparable to that observed with the point mutations in the protein-binding domain. This suggests that loss of protein binding may be responsible for the effect. To examine the influence of the remaining part of the first stem-loop as well as the second stem-loop structure, another mutant Δ-135/-18 was constructed (Fig. 5A).

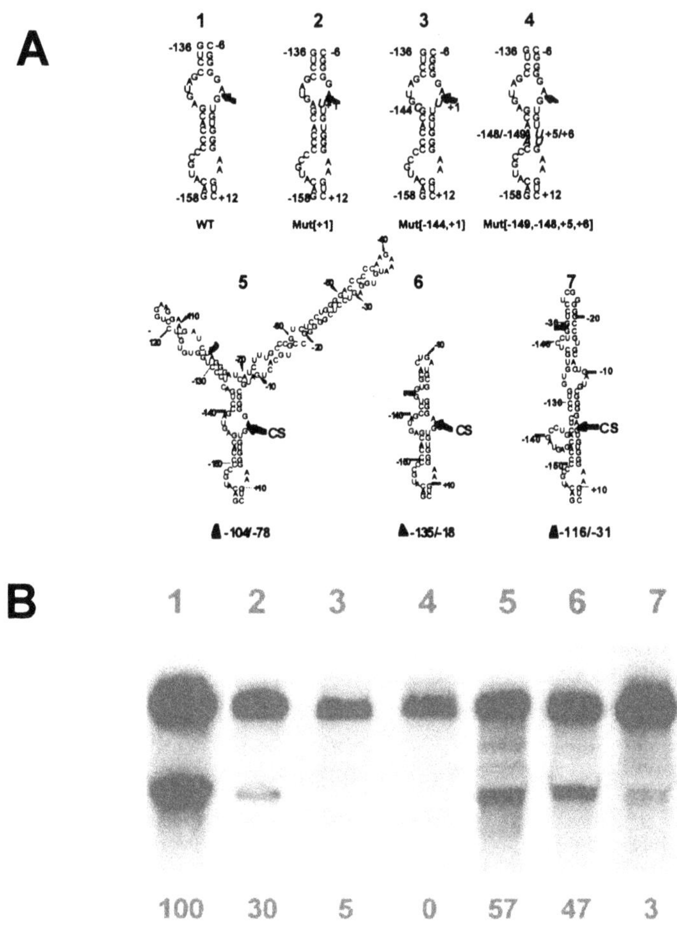

Figure 5. ***A) Predicted local secondary structures around the cleavage site in ΔWT and the various mutant minigene constructs.*** *In each structure, the cleavage site is indicated by a straight arrow; altered nucleotides are indicated in bold italics and deletions are indicated by a bent arrow.* ***B) Northern blot analysis of total RNA isolated from human cells transiently transfected with the various minigene constructs indicated in (A).*** *The blot was hybridized with a 3'-UTR specific probe that detects both full-length IGF-II mRNA as well as the 3'-cleavage product. The numbers of the lanes correspond with the numbers of the constructs shown in (B). Mean cleavage efficiencies of 3 separate experiments are given as a percentage relative to the efficiency of cleavage of ΔWT, which was set at 100%. Variation between the calculated cleavage efficiencies in independent experiments was less than 10%.*

Although the complete stem-loop domain from position -135 to -18 was deleted in this mutant, it was cleaved with approximately the same efficiency as Δ-104/-78, indicating that cleavage can occur even in the complete absence of both stem-loops. This suggests that as long as the cleavage site domain is intact and in the proper folding, cleavage can occur, albeit with a two-fold reduced activity due to the absence of the RNA-binding protein. To evaluate the importance of the structure around the cleavage site more specifically, mutant Δ-116/-31 was constructed. Deletion of this region results in a construct with a predicted altered structure formation around the cleavage site. Cleavage was completely absent in this RNA with altered conformation. This result shows that the maintenance of a native conformation around the cleavage site is important.

In summary, the results indicate that: (1) the identity of specific nucleotides in the cleavage site loop as well as in the flanking helix is important for recognition by an endonuclease; (2) provided that the predicted structure around the cleavage site is maintained, even complete deletion of the stem-loop domain results in only a two-fold reduction of cleavage efficiency, most likely due to the loss of the protein binding site; (3) disruption of the structure around the cleavage site completely abolishes cleavage.

On the basis of the above-mentioned data, we suggest that the main function of the stem-loop domain and the RNA-binding protein is to stabilize the cleavage site domain. The 80 nt long duplex domain between element I and the G-rich region of element II may have a similar role, since its disruption directly alters the structure in the cleavage site domain thereby abolishing cleavage. Specific nucleotides in the cleavage site domain are important for recognition by an endoribonuclease, either because the enzyme directly contacts them, or because they are involved in providing a specific RNA conformation in the cleavage site loop that creates a suitable binding surface. This latter option was suggested by chemical probing data that revealed subtle structural changes due to point mutations, that could not be predicted by computer folding algorithms. Thus, the stem-loop and the duplex domains may act in a cooperative manner to maintain a correct structure in the cleavage site domain. A general implication of these results is that a specific RNA recognition site can be heavily dependent on complex surrounding RNA regions for the maintenance of an appropriate structural context.

Functional Aspects of IGF-II mRNA Cleavage.

An interesting question is what the physiological function of the cleavage reaction is. The most likely explanation is that it functions as an additional regulatory step to rapid adaptation of altered growth conditions, and thus is a growth factor requirement, in the cell. Since the full-length IGF-II mRNAs were shown to be very stable, an additional mechanism would be required to down-regulate protein synthesis and thus IGF-II bioavailability. After cleavage of the full length IGF-II mRNAs takes place, the 5'-cleavage product that contains the IGF-II coding region is rapidly degraded by exonucleases due to the absence of a poly(A) tail (6). In this respect it is noteworthy that the stability of IGF-II mRNAs seems to depend on the growth conditions of the cells. Human hepatoma Hep3B cells endogenously express IGF-II when grown in serum free medium. However, when after serum deprivation the medium is supplemented with 10% fetal calf serum containing IGF-II, full-length IGF-II mRNAs are rapidly degraded while the amount of 1.8 kb RNA remains high. This suggests that the cleavage reaction is a regulated event.

Another interesting option is that the extremely stable 1.8 kb 3'-cleavage product has an intrinsic function as a non-coding RNA species. In this respect it is noteworthy to mention that 3'-UTRs of genes expressed in a myogenic cell line can promote myogenic differentiation or function as tumor suppressors (12). Whether or not the 1.8 kb IGF-II cleavage product can also fulfill such an unconventional role in cellular processes remains to be established.

Comparison of IGF-II mRNA Cleavage with Related Systems.

Regulation of mRNA stability is a common mechanism of controlling gene expression in eukaryotic cells. In general, degradation of mRNAs proceeds through the action of exo- and endonucleases in cooperation with regulatory factors. *Cis*-acting sequences involved in stability regulation are often found in the 3'-UTR, although there are also reports of stability determinants present in the coding region or 5'-UTR. A number of mRNAs in different organisms have been described to be targets for specific endonucleolytic cleavage, much like the IGF-II mRNAs (13).

In *Xenopus* oocytes, the transcript from the homeobox gene *Xlhbox2* is specifically destabilized during development, suggesting that this mRNA is a target for a specific nuclease (14). It has been demonstrated that the Xlhbox2 mRNA is endonucleolytically cleaved at 4 sites in a 90 nucleotide region in the 3'-UTR close to the translation stop codon; the cleavage sites are flanked by 5'-ACCT-3' repeats and must be in a single-stranded configuration to allow cleavage. Subsequently, it was shown that a single copy of the sequence 5'-ACCUACCUACCCACCUA-3' is sufficient for cleavage of a heterologous β-globin mRNA and reduces its half-life from 57 to 15 minutes. Thus, the recognition determinants appear to be much less complex than in the case of the IGF-II mRNAs, requiring only a short sequence that needs to be single-stranded. Furthermore, it was shown that *Xenopus* oocytes contain a specific RNA-binding protein that protects the Xlhbox2 mRNA from cleavage. The protein is regulated in a development-specific manner and may thus account for the developmental regulation of Xlhbox2 mRNA stability.

The *9E3* (or CEF-4) gene is expressed in chicken embryo fibroblasts (CEF) and encodes a secreted 6 kDa protein that is related to platelet growth factor 4 and to a number of other inducible inflammatory mediator proteins (15). In CEF infected with a temperature-sensitive strain of Rous sarcoma virus (RSV), 9E3 mRNA is more stable at a permissive temperature than at a non-permissive temperature, suggesting that the 9E3 mRNA is stabilized in RSV-transformed CEF. Evidence that the 9E3 mRNA stability is regulated by endonucleolytic cleavage in the 3'-UTR was provided by the presence of small RNA species that were identified as degradation intermediates of 9E3 mRNA. The relative amount of the small RNAs was increased under conditions in which 9E3 mRNA was less stable, suggesting that increased cleavage activity may account for the destabilization of the 9E3 mRNA. The cleavage site was mapped to a region with the sequence 5'-GUAAUUCCUCCUGCUCCCUGG-3'.

A third system that shows interesting similarities with the IGF-II system is the *transferrin receptor (TfR)* mRNA. The half-life of TfR mRNA is mediated through a region in the 3'-UTR that contains a rapid turnover determinant of about 250 nt and 5 RNA motifs named iron responsive elements (IREs) which are binding sites for a cytoplasmic IRE binding protein (IRE-BP) (16,17). Binding of the IRE-BP to the IREs inhibits degradation of the transcript, suggesting that the function of the rapid turnover determinant is masked. Subsequently, this rapid turnover determinant was shown to be a target site for endonucleolytic cleavage. A shorter TfR RNA species appeared upon treatment of cells with an iron source. The cleavage site was mapped to a single-stranded region near one of the iron-responsive elements in the 3'-UTR with the sequence 5'-AUAAAG/AACAAG-3'. Nucleotide substitutions around the cleavage site render the RNA refractory to cleavage, indicating that the exact identity of specific nucleotides is important for recognition by an endoribonuclease. Furthermore, cleavage was found to be independent of poly(A) tail shortening since the 3'-cleavage product appears to be polyadenylated to the same extent as the full-length RNA. Interestingly, cleavage of the TfR mRNA requires rather large regions of RNA structure, including a segment located about 300 nucleotides upstream of the cleavage site. This provides an interesting similarity with the IGF-II system and suggests that the endonuclease recognition site in the TfR mRNA is relatively complex in comparison with various other systems where short sequences are sufficient for recognition.

Summary of key concepts

- *A growing number of mRNAs appear to be targets for specific endoribonucleases. The various systems differ among each other with respect to the cleavage site requirements.*
- *Specific RNA-binding proteins may act as so-called protective factors that mask the endonucleolytic cleavage site. In fact, regulation of cleavage through protective binding proteins appears to be a common mechanism for most of the systems described.*
- *Alternatively, specific RNA-binding proteins may act as modulators of RNA structure, resulting in exposure of the cleavage site. These functions of RNA-binding proteins would allow the specific regulation of a number of mRNAs by a limited set of endoribonucleolytic enzymes.*
- *The identification of such RNA-binding proteins as well as the endonucleases recognizing the specific RNA targets will lead to a better understanding of the function of these highly structured mRNAs.*

Study Guide Questions

1. When IGF-II mRNA is cleaved in its 3'-UTR, which part contains the coding region? What are the implications for IGF-II protein synthesis?
2. Regulation of IGF-II expression takes place at the transcriptional as well as the post-transcriptional level. What is the advantage of this dual regulation?
3. Formation of an IGF-II mRNA-protein complex takes place at a position approximately 90 nt upstream of the actual cleavage site. What is the function of complex formation at this position?
4. The IGF-II mRNA-protein complex formed at the first stem-loop affects cleavage efficiency. How could one test if specific structural determinants are involved in recognition of the target RNA binding site of the protein?
5. It has not been possible to identify a complex between the endonuclease and the cleavage site region. What could be an explanation for this fact?

Acknowledgements

This work was supported bu a grant form the Netherlands Organization for the Advancement of Pure Research (NWO).

REFERENCES

1. Daughaday,W.H. and Rotwein,P. 1989. Insulin-like growth factors I and II. Peptide, messenger ribonucleic acid and gene structures, serum, and tissue concentrations. *Endocr.Rev.* 10:68-91.
2. DeChiara,T.M., Efstratiadis,A., and Robertson,E.J. 1990. A growth-deficiency phenotype in heterozygous mice carrying an insulin-like growth factor II gene disrupted by targeting. *Nature* 345:78-80.
3. Liu,J.P., Baker,J., Perkins,A.S., Robertson,E.J., and Efstratiadis,A. 1993. Mice carrying null mutations of the genes encoding insulin-like growth factor I (Igf-1) and type 1 IGF receptor (Igf1r). *Cell* 75:59-72.
4. Rajaram,S., Baylink,D.J., and Mohan,S. 1997. Insulin-like growth factor-binding proteins in serum and other biological fluids: regulation and functions. *Endocr.Rev.* 18:801-831.
5. Holthuizen,P.E., Steenbergh,P.H., and Sussenbach,J.S. 1999. Regulation of IGF gene expression. In The IGF System. Molecular Biology, Physiology, and Clinical Applications. R.G.Rosenfeld and Roberts,C.T., editors. Humana Press., Totowa, New Jersey, USA. 37-61.
6. Meinsma,D., Holthuizen,P., Van den Brande,J.L., and Sussenbach,J.S. 1991. Specific endonucleolytic cleavage of IGF-II mRNAs. *Biochem.Biophys.Res.Commun.* 179:1509-1516.
7. Meinsma,D., Scheper,W., Holthuizen,P., Van den Brande,J.L., and Sussenbach,J.S. 1992. Site-specific cleavage of IGF-II mRNAs requires sequence elements from two distinct regions of the IGF-II gene. *Nucleic Acids Res.* 20:5003-5009.
8. Scheper,W., Meinsma,D., Holthuizen,P.E., and Sussenbach,J.S. 1995. Long-range RNA interaction of two sequence elements required for endonucleolytic cleavage of human insulin-like growth factor II mRNAs. *Mol.Cell.Biol.* 15:235-245.
9. Christiansen,J., Kofod,M., and Nielsen,F.C. 1994. A guanosine quadruplex and two stable hairpins flank a major cleavage site in insulin-like growth factor II mRNA. *Nucleic Acids Res.* 22:5709-5716.
10. van Dijk,E.L., Sussenbach,J.S., and Holthuizen,P.E. 1998. Identification of RNA sequences and structures involved in site-specific cleavage of IGF-II mRNAs. *RNA.* 4:1623-1635.
11. van Dijk,E.L., Sussenbach,J.S., and Holthuizen,P.E. 2000. Distinct RNA structural domains cooperate to maintain a specific cleavage site in the 3'-UTR of IGF-II mRNAs. *J Mol Biol* 300:449-467.
12. Rastinejad,F., Conboy,M.J., Rando,T.A., and Blau,H.M. 1993. Tumor suppression by RNA from the 3' untranslated region of a- tropomyosin. *Cell* 75:1107-1117.
13. Ross,J. 1997. A hypothesis to explain why translation inhibitors stabilize mRNAs in mammalian cells: mRNA stability and mitosis. *BioEssays* 19:527-529.
14. Brown,B.D., Zipkin,I.D., and Harland,R.M. 1993. Sequence-specific endonucleolytic cleavage and protection of mRNA in *Xenopus* and *Drosophila*. *Genes Dev.* 7:1620-1631.
15. Stoeckle,M.Y. and Hanafusa,H. 1989. Processing of 9E3 mRNA and regulation of its stability in normal and Rous sarcoma virus-transformed cells. *Mol.Cell Biol.* 9:4738-4745.
16. Binder,R., Horowitz,J.A., Basilion,J.P., Koeller,D.M., Klausner,R.D., and Harford,J.B. 1994. Evidence that the pathway of transferrin receptor mRNA degradation involves an endonucleolytic cleavage within the 3' UTR and does not involve poly(A) tail shortening. *EMBO J.* 13:1969-1980.

17. Schlegl,J., Gegout,V., Schläger,B., Hentze,M.W., Westhof,E., Ehresmann,C., Ehresmann,B., and Romby,P. 1997. Probing the structure of the regulatory region of human transferrin receptor messenger RNA and its interaction with iron regulatory protein-1. *Endocrinol.Metabol.Clin.North Am.* 3:1159-1172.

15

HORMONAL REGULATION OF THE EGF/RECEPTOR SYSTEM

Stephen W. Spaulding, and Lowell G. Sheflin

VA Western New York and SUNY at Buffalo, Buffalo NY

The EGF/receptor system is involved in the development, maintenance and repair of various organs, particularly those involving ductal systems. Disruptions of the EGF/receptor system have been implicated in a variety of clinical disorders including cancer and aging. Some hormones alter the expression of EGF ligands, EGF receptors and of downstream factors that mediate EGF receptor action during development. These hormonal effects vary according to the species, organ and stage of development under study. Androgens alter EGF mRNA levels in several rodent tissues, yet no androgen-responsiveness is detectable within 7 kB upstream of the promoter of the EGF gene, which supports earlier evidence that androgens have post-transcriptional actions on EGF expression. AU-rich elements that can influence transcript stability and subcellular localization are present in the 3' UTR of EGF mRNA. Androgens regulate EGF mRNA polyadenylation site usage, poly-A tail length and mRNA stability in the mouse submaxillary salivary gland (SMG) but not in the kidney, although androgens do regulate other genes in the kidney. Androgens also regulate the levels of several AU-rich RNA binding proteins known to affect the subcellular localization, translation and stability of AU-rich mRNAs. Androgen-dependent changes in the subcellular levels of these RNA binding proteins also indicate a causal connection with the androgen-dependent changes in EGF mRNA and its translation.

THE PHYSIOLOGY OF EGF

EGF was discovered in mouse submaxillary salivary gland (SMG) extracts by its ability to open the eyes of newborn mice prematurely. A complex system of EGF-like ligands and EGF receptors is now known to regulate the development and repair of many organs. EGF receptor expression, various ligands (including EGF and TGF-α), as well as downstream elements that transmit their signals are all modulated during development.

Specific hormones modulate the EGF system to produce appropriate tissue-specific development. Disruptions in the EGF system can cause hormones to promote inappropriate cell apoptosis or proliferation. In rodent SMGs, androgen modulates several features of EGF transcripts at the post-transcriptional level, but in other models, androgen acts on different elements in the EGF system. Conversely, EGF can modulate the androgen responsiveness of some systems.

The full-length transcript of the EGF gene encodes a 130 kDa precursor protein with a potential transmembrane domain, but EGF transcripts can be variously processed to produce endocrine, paracrine or juxtacrine factors. In the rodent SMG, most of the prepro-EGF molecule is eliminated, and only the proximal extracellular domain is stored bound to a protease/binding protein in granules in the granular convoluted tubule cells. The SMG then releases the mature 6 kDa EGF peptide whose characteristic motif of three intra-molecular disulfide bonds is crucial for ligand-binding to EGF receptors. A larger fragment of prepro-EGF is released into the urine from the kidney, the other major site of EGF synthesis in the rodent. If full-length prepro-EGF molecule is expressed on the surface of a cell, it can interact with EGF receptors on neighboring cells. EGF - and a score of similar ligands - bind to EGF receptor family members, causing the receptors to homodimerize and activate tyrosine kinase, promoting downstream protein-protein interactions and activating phosphorylation csacades. EGF receptors can also heterodimerize with other members of the EGF receptor family, Her/neu (c-erbB-2), c-erbB-3 and c-erbB-4.

The EGF System in Embryogenesis

In pig ovarian follicles, EGF mRNA and protein are detected in oocytes, whereas granulosa and thecal cells in the follicles express EGF receptor mRNA (1). In the developing blastocyst TGF-α transcripts are detectable from days 8-12, while EGF mRNA appears around day 15, and then EGF immunoreactivity appears in developing lung, gut and amnion (2).

A truncated form of the EGF receptor is expressed in some tissues. This alternatively spliced transcript, lacking the transmembrane and intracellular domains, can bind to full-length receptors to modulate their function. Truncated EGF receptor is detected in mouse and pig uteri, whereas the full-length transcript is detected in preimplantation blastocysts (3,4).

Knocking out the EGF receptor gene in CF-1 mice causes preimplantation death due to degeneration of the inner cell mass of the embryo. In 129/Sv mice, death of EGF receptor knock-outs occurs around mid-gestation, while in CD-1 mice, death in homozygotes occurs neonatally due to abnormalities in skin, kidney, brain, liver and gastrointestinal tract (5). Thus other genetic factors modulate lethality, once the EGF receptor is knocked out.

The EGF System in Kidney Development

In developing human fetal kidney cells, EGF is detectable in mesonephric vesicles and metanephric caps, and rises again after birth. TGF-α is expressed in the mesonephros but not in the metanephros. The EGF receptor is expressed at high levels in condensing nephrogenic cells, in newly-induced epithelium and cortical branches of ureter and at low levels in nephronic cells, but is nearly absent from renal stroma and medullary collecting ducts (6). As the kidney begins to develop, EGR receptor immunoreactivity, the binding of EGF to renal membranes, and the phosphorylation of the receptor all increase during late gestation, then they fall toward adult levels after birth (7). Thyroid hormone regulates the level of EGF and its mRNA in the loop of Henle in adult rats. If the thyroid hormone level in newborn rats is elevated, it reduces renal EGF receptor and TGF-α mRNA levels, whereas EGF mRNA levels rise (8).

The EGF System in Submaxillary Gland (SMG) Development

The level of androgen to which a mouse embryo is exposed *in utero* affects the level of EGF in the adult SMG. Females that were located between male fetuses *in utero* have higher levels of EGF in their adult submaxillary glands than do females that were located between female fetuses, indicating that the higher level of androgen during late gestation programs some element involved in EGF regulation in the SMG (9). In EGF receptor knockout mice, the SMG precursor in the newborn shows poor branching of ducts (10).

In the SMG of the normal newborn mouse, EGF receptor mRNA levels rise briefly, then decline after the second postnatal day, whereas EGF binding activity peaks at 10 days, thus clearly showing that the level of receptor mRNA and protein can be regulated separately (11). Granular convoluted tubules develop as the EGF receptor protein appears. After about two weeks, the store of EGF in the male SMG rises sharply, whereas in female mice, the store of EGF rises later and more gradually. The SMG contains 2000 times as much immunoreactive EGF as the kidney does, yet the levels of EGF mRNA in the mouse kidney are similar to those in the SMG. This indicates a great difference in the mRNA stability, translation and/or protein turnover between the two tissues.

Thyroid hormone also regulates EGF mRNA in the adult SMG. If thyroxine is administered to newborn mice during the first week of life, it increases EGF in the kidney (but not in the SMG), whereas if thyroxine is administered during the second week of life, it causes premature development of granular convoluted tubules, and increases EGF and EGF mRNA levels in the SMG (but not in the kidney) (12). This dramatically illustrates how a hormone's effect on EGF expression can shift as organs develop. Giving triiodothyronine to adult hypothyroid mice rapidly raises EGF mRNA levels in the SMG, and nuclear run-off studies indicate this is a transcriptional effect (13). Giving adult females androgen, however, causes a slower response in EGF mRNA, even though it causes a rapid shift of the androgen receptor into the nucleus (14).

MODELS USED IN STUDIES OF THE EGF SYSTEM

Waved-2 Mice

A spontaneous point mutation changing a single amino acid in the murine EGF receptor produces the "waved-2" mouse, an animal model that has helped us understand the signaling and metabolic pathways utilized by EGF. This mutation impairs the binding and tyrosine kinase activities of the receptor, and affects the trafficking of the internalized receptor (15,16). In addition to having a wavy coat, these mice exhibit several other relatively minor defects that contribute to their decreased average life span, including defects in heart valves and in lung, gastrointestinal and breast development (17). Propagation of waved-2 mice is hampered in part because homozygous mothers have impaired lactation, indicating the importance of the EGF system in mammary gland development and function.

Gene Knockout Studies of the EGF System

Mammary gland precursor cells taken from EGF receptor knockout newborns fail to develop when transplanted into nude mice. In contrast, cells taken from wild-type newborns display normal mammary duct development, indicating the importance of the EGF system in breast cell growth (18).

It has proven difficult to obtain informative results from knockout studies of mice in which a single member of the EGF receptor ligand family has been eliminated. Presumably other members of the family are taking over the functions of the lost ligand. Even when the genes for three different ligands (EGF, TGF-α and amphiregulin) were all knocked out, fertile offspring were produced, although such triple knockout mothers displayed impaired lactation and poor development of breast alveoli (19).

THE ROLES OF THE 3' UTR AND OF RNA BINDING PROTEINS IN REGULATING EGF, TGF-α AND EGF RECEPTOR mRNAs

EGF mRNA Metabolism: Polyadenylation and the 3' UTR

Although androgens regulate the level of EGF mRNA in specific tissues, no androgen-responsiveness was detectable in transient transfection studies with up to 7 kb of the 5' untranslated region of the mouse EGF gene (20). This supports other data indicating that androgen has important post-transcriptional actions on EGF mRNA levels (21-24).

A variety of AU-rich elements known to influence mRNA stability and subcellular localization are present in the 750 b that comprise the 3' UTR of EGF mRNA. As outlined below, androgen regulates EGF mRNA polyadenylation site usage, poly-A tail length and the level of the mRNA in the mouse SMG, but not the kidney. However, other genes in the kidney are androgen-responsive (21,23).

Furthermore, androgen regulates several RNA binding proteins that can alter the subcellular localization, translation and/or stability of mRNAs that contain AU-rich elements. The time-courses of the changes in the levels of these RNA binding proteins, as well as their tissue-specific responses to androgen, suggest they are involved in the androgen-dependent changes in EGF.

TGF-α mRNA Metabolism: Alternative Polyadenylation and the 3' UTR

TGF-α mRNA is expressed as 4.5 and 1.6 kb species. The 1.6 kb species results from utilization of an upstream polyadenylation signal that eliminates 2750 bases, including 8 AUUUA elements and several additional polyadenylation signals (25). Post-transcriptional regulation of TGF-α mRNA expression has been demonstrated in various transformed cell lines (26), in response to TCDD in SCC-12F cells (27) and an autocrine feedback pathway is present in KB cells (28).

EGF Receptor mRNA Metabolism: the Role of the 3' UTR

Truncated EGF receptors display longer half-lives than full-length transcripts. The truncated rat liver EGF receptor mRNA is four times more stable than the full length mRNA, which suggests that elements in the deleted 3' region are important for destabilizing full- length transcripts (29). A truncated form of the EGF receptor expressed in A431 cells is twice as stable as the full-length transcript. Both A431 and KB cells contain cytoplasmic proteins that bind to an AU-rich sequence present in the 3' UTR of full length EGF receptor mRNA (30). Treating A431 cells with thyroxine has no effect on transcription of the EGF receptor gene, but drastically decreases the stability of EGF receptor mRNA, with the long form being preferentially degraded (31).

CIS-ACTING AU-RICH ELEMENTS AND SECONDARY STRUCTURES IN THE 3' UTR OF EGF mRNA

Potential Cytoplasmic Polyadenylation Signals in the 3' UTR of EGF

Maternal mRNA in resting oocytes is stored in ribonucleoprotein particles that are protected from degradation by proteins bound to the 3' UTRs that also keep them translationally dormant. Following oocyte activation, the poly-A tails on some of these stored messages change in length as they become translationally active, due to the presence of certain U-rich sequences in their 3' UTRs. EGF mRNAs from four different species contain two such UUUUUAU sequences just upstream of a unique AU-rich 23 b sequence at the end of the 3' UTR.

Potential Stem-Loop Structures in the 3' UTR of EGF

Stem-loops in the 3' UTR can also influence the stability of some mRNAs. The 750 b 3' UTR of murine EGF mRNA would form a complex secondary structure containing several stable stem-loops, if the base-pairing in the stems were not disrupted by RNA binding proteins. Strong stem-loop structures are predicted immediately upstream of two potential polyadenylation signal sites (32).

Poly-A Tail Length and Periodicity in EGF Transcripts

Polyribosomal nuclease activity degrades mRNA in the 3'→5' direction, but if the poly-A tail on a polyribosomal message is long enough to bind at least one molecule of poly-A binding protein (PABP), that message is less sensitive to nuclease degradation. Binding of one PABP molecule protects about 25-30 A's, and the binding of additional molecules of PAPB can protect additional lengths, and thus can produce a 25-30 base periodicity in poly-A tails. Interactions between PABP and other RNA binding proteins confer even more protection from nuclease degradation. Both thyroid hormone and androgen can alter the length and periodicity of poly-A on EGF mRNA in a tissue-specific fashion (33,21,22).

THE KINETICS OF THE RESPONSE OF EGF mRNA AND OTHER GENES IN THE MURINE SMG TO ANDROGEN

Three days after the injection of testosterone, EGF mRNA levels begin to rise in the female mouse SMG (21), yet testosterone causes translocation of androgen receptor to the SMG nucleus within 30 minutes (14) and increases the transcription of other genes rapidly, so it is clear that the lag in the EGF response is not due to a general delay in androgen-responsive genes in that tissue. Furthermore, EGF mRNA in the SMG responds rapidly following triiodothyronine (13), so the lag following androgen administration is not an innate characteristic of EGF gene transcription (33).

Androgen Levels Regulate EGF mRNA Polyadenylation in the SMG but not the Kidney

The state of polyadenylation of an mRNA can affect its stability and/or translation. Androgen levels have been shown to alter poly-A tail length on transcripts of the cystatin-like protein and vasopressin genes (34,35). Using 3' rapid amplification of complementary DNA ends (3' RACE) to assess the effect of testosterone on the polyadenylation of EGF transcripts (21,23), we found that the poly-A tails on EGF transcripts utilizing the terminal polyadenylation site occur in distinct size classes of approximately 20, 50, 70, 100, and 200 A's in the untreated female SMG. Male EGF transcripts, in contrast, have shorter poly-A tails of less clearly defined lengths, ranging from approximately 20-100 A's. Testosterone treatment causes the female poly-A pattern to become more heterogeneous, resembling the male pattern. This change in EGF transcript polyadenylation occurs along with the increase in EGF mRNA levels in the SMG.

The kidney, in contrast, does not display a sex difference in the pattern of polyadenylation. Both male and female transcripts show distinct poly-A tails of approximately 20, 50, 70, 100, and 200 A's. Furthermore, the polyadenylation of EGF transcripts in the kidney is not altered by changing the level of testosterone in either sex. Despite the lack of a sex difference or an androgen response of poly-A tails in the kidney, there is a small sex difference both in the level of EGF mRNA and proteins, and these levels do respond in the kidney, but in the opposite direction to the response in the SMG. Thus androgen levels could be affecting proteins that interact with the poly-A tail, or could be altering poly-A polymerase or a poly-A nuclease activities in the SMG in a tissue-specific fashion.

Androgen Levels Regulate the Level of Proteins that Bind to a Unique AU-Rich Sequence at the End of the 3' UTR of EGF mRNA

A unique AU-rich 23-b sequence is centered on the terminal polyadenylation signal in all published mammalian EGF sequences, suggesting that conserved *trans*-acting factors involved in EGF mRNA metabolism bind to this sequence. We therefore prepared [^{32}P]-RNA containing the 3' terminal 23-b sequence of EGF mRNA plus a short poly-A tail, and incubated it with SMG cytosol. Cytosol retarded the electrophoretic mobility of this RNA, and UV-cross-linking revealed a prominent band of ~47 kDa.

Trypsin abolished both the gel-retarding and cross-linking activities. Cytosol from female SMGs contained approximately 8 times more of these activities than did male cytosol. Injecting testosterone into female mice altered the RNA binding activities in a biphasic fashion, initially increasing them by about 40% at 2 days, possibly involving stabilization of the RNA binding activity involving post-translational modifications and/or changes in a proteolytic activity that could selectively degrade the protein (22,23). After five days of testosterone, however, the RNA binding activities decrease by about 65%, approaching male levels. Adding excess unlabeled 23-b EGF 3' UTR RNA to the [^{32}P]-RNA competed for the cytosolic binding protein, but poly-A was an ineffective competitor, demonstrating the binding was sequence-specific. Testosterone injections changed the level of 47 kDa protein in the SMG before it changed EGF mRNA levels or EGF mRNA polyadenylation, so this cytosolic protein could be a *trans*-acting factor involved in EGF polyadenylation in the SMG.

Kidney cytosol also contained the 47 kDa RNA binding protein, but did not display sexual dimorphism or testosterone-responsiveness. Two weeks following orchiectomy of male mice, mature EGF peptide immunoreactivity levels fell by 96%, and mRNA levels fell by 76% in the SMG. Orchiectomy enhanced the RNA binding activities of the 47-kDa protein in SMG cytosol but not in the kidney: cross-linking to 3' UTR RNA increased 67% following orchiectomy and the amount of RNA showing a shift in gel mobility increased by 47%. The population of EGF transcripts in the SMG with short, heterogeneous poly-A tails (<50 A's) decreased by 31%, whereas transcripts with long poly-A tails of approximately 50, 70, 100, and 200 A's all increased and became more distinct, resembling the pattern found in the normal female. These changes in polyadenylation in SMG did not occur in kidney after orchiectomy. Thus, 2 weeks after orchiectomy, the SMG responses in EGF expression, EGF mRNA polyadenylation, and the activities of the 47 kDa binding protein are the reciprocal of the responses produced by injecting female mice with testosterone for 5 days, providing further evidence that circulating androgens regulate EGF expression post-transcriptionally (21).

Following orchiectomy, as the ~47 kDa band increased, two bands of ~59 and ~72 kDa also appeared, while a third band of ~34 kDa decreased in SMG cytosol, indicating that other proteins that interact with the unique AU-rich 23 b sequence also display androgen-responsiveness (21). Preliminary RNA-protein supershift studies with antibodies to AUF1, hnRNP A1 or hnRNP C1 indicate that they all bind to the 23 b AU-rich element at the end of the 3' UTR of EGF, with anti-AUF1 causing the greatest supershift.

SUMMARY OF SOME ACTIVITIES OF MAJOR AU-RICH RNA BINDING PROTEINS

AUF1 (also called hnRNP D) can exist in four isoforms, produced by alternative splicing of transcripts from a single gene. The isoforms have distinct nucleic acid binding properties: some can accompany mRNA transcripts from the nucleus to the cytosol, but some AUF1 isoforms interact with nuclear proteins as well as ss- and ds-DNA to influence gene transcription and recombination. Isoforms of AUF1 with higher AU-rich binding affinity may facilitate the turnover of AU-rich mRNA (36). HnRNP A1 binds to AU-rich RNA transcripts in the nucleus and modulates exon splicing, but can remain associated with mature mRNA and facilitate the transfer of mRNA to the cytosol (37). Interestingly, hnRNP A1 transcripts themselves contain AU-rich elements that are known to bind AUF1, suggesting that the stability and/or subcellular localization of hnRNP A1 transcripts may be regulated by AUF1 (38). HnRNP C1 is predominantly nuclear and is involved in spliceosome assembly and the splicing of mRNA (37). HuR interacts predominantly with polyadenylated mRNA but also can stabilize partially deadenylated mRNAs that contain AU-rich elements. HuR can protect a U-rich sequence from degradation and can shuttle AU-rich mRNAs between subcellular compartments (39).

Regulation of AUF1 Isoform Expression in Cell Culture Models

AUF1 isoform levels respond differentially to hemin-shock in K562 cells, and to isoproterenol in smooth muscle cells (40,41). p37, the AUF1 isoform with the highest avidity for AU-rich sequences, appears to be especially labile, and may be subject to preferential ubiquitination, since its levels increased dramatically in HeLa cells in response to heat shock or proteasome inhibitors (42).

Regulation of AUF1 Isoforms by Androgens in Murine SMG

We found that androgens differentially regulate the expression of the four isoforms of AUF1 in a characteristic pattern. Cytosol from female SMGs contains two major isoforms (p45 and p40), whereas cytosol from male SMGs contains a prominent p37 and a weaker p42 band. Injecting female mice with testosterone decreased p45 levels by 81% after 7 days, whereas p37 levels increased by 449% at 7 days. Orchiectomy, conversely, decreased p37 levels in the male SMG by 91% while increasing p45 5-fold. Kidney cytosol contains a prominent p37 band in both males and females, which remained unchanged when circulating androgen levels were altered. Dihydrotestosterone (DHT) changed the pattern of AUF1 isoforms in female SMG cytosol more rapidly than testosterone did. In nuclear extracts of female SMG, p45 was predominant, and DHT decreased its level slightly at 24 h. Polysomal extracts from female SMG contained relatively small amounts of p45 but they increased after 24 h of DHT (43).

Regulation of HuR by Androgen in Murine Tissues

Western blot studies indicate that androgen levels can also modulate HuR levels. The levels of HuR in female SMG and kidney homogenates are higher than the levels in male SMG and kidney homogenates. This sexually dimorphic expression of HuR appears to reflect androgen-dependency, since administration of DHT or testosterone to females significantly reduces the levels of HuR in both tissues, and conversely, orchiectomy of males increases HuR level in both the SMG and kidney. Thus HuR levels in the SMG change in the opposite direction to the change in the level of the p37 isoform of AUF1. Changes in the levels of these two proteins in the opposite directions should have an additive effect on the metabolism of AU-rich mRNAs, since a fall in the level of p37 and a rise in the level of HuR both should stabilize AU-rich mRNA (36). The level of HuR in polysomes from female SMG increase transiently following androgen treatment and the levels remain elevated, whereas HuR levels in cytosolic and nuclear fractions fall in both systems. The tissue-response of HuR is different from that of AUF1, since only HuR responds in the kidney, whereas both proteins respond in the SMG (44).

Evidence that AU-Rich RNA Binding Proteins May be Involved in Regulating Other Hormone-Responsive Systems

Parathyroid hormone (PTH) mRNA levels are regulated post-transcriptionally by calcium and phosphate levels in the parathyroid gland. Proteins that bind to the 3' UTR of PTH mRNA are present in many tissues, but apparently, only in the parathyroid is their binding regulated by calcium and phosphate. In an *in vitro* RNA degradation assay, adding excess 3'-UTR destabilized the full-length PTH transcript, indicating cytosolic proteins that otherwise bind and protect PTH mRNA from RNase activity can be removed by binding to the 3' UTR. When brain cytosol was chromatographed on a column containing the 3'-UTR of PTH mRNA, AUF1 was found in the bound fraction, and recombinant AUF1 was also shown to bind to this AU-rich 3'-UTR. Recombinant AUF1 does stabilize PTH mRNA in the *in vitro* degradation assay, but regulation of AUF1 by the level of calcium or phosphate remains to be demonstrated (45).

Adrenergic agonists are known to destabilize mRNAs that encode adrenergic receptors. UV cross-linking studies have detected a ~36,000 protein that binds to A/C+U-rich regions within the 3'-UTR of beta-adrenergic receptor mRNA. Both HuR and hnRNP A1 were detected in proteins extracted from polysomes from hamster smooth muscle cells (46). Although no adrenergic regulation of HuR or hnRNP A1 has yet been reported, there is evidence that the abundance of AUF1 can be regulated by norepinephrine (40).

Post-Translational Modifications of RNA Binding Proteins

The RNA binding activity of several protein-containing fractions may be altered by several kinds of post-transcriptional modification that could be hormone-responsive, including methylation, ubiquitination and phosphorylation. Dephosphorylation has been reported to decrease the binding of proteins to the 3' UTR of angiotensinogen, GM-CSF and cyclooxygenase and EGF mRNAs (22,23), although the specific *in vivo* mechanisms remain to be elucidated. Recently it has been reported that blocking proteasomes or heat shock promotes the stabilization and preferential ubiquitination of the p37 isoform of AUF1 in HeLa cells, suggesting that this post-translational modification could be involved in the preferential degradation of AU-rich mRNAs that are rapidly degraded in normal cells (42).

HnRNPs constitute about 60% of the methylated proteins in the cell, and methylation has been proposed to facilitate the nuclear export. It has recently been reported that the two AUF1 isoforms that contain exon 7, p45/p42, remain in the nucleus, associated with a scaffold-associated factor (SAF-B), following actinomycin D inhibition of active gene transcription. In contrast, p37/p40, which prior to treatment were located both in the nucleus and cytosol, became preferentially located in the cytosol, suggesting that they shuttle between cytosol and nucleus. These authors speculated that methylation of RGG consensus sites that flank exon 7 might regulate binding to SAF-B (47). Methylation may not only affect protein-protein interactions, but also could alter RNA binding properties of the AUF1 isoforms p37/p42 associated with rapid turnover of AU-rich mRNAs (36).

Summary of key concepts

- *The EGF system is crucial for growth, development and repair of many tissues.*
- *Specific hormones regulate the expression of elements of EGF system at different stages during the development.*
- *EGF mRNA levels in the SMG respond more rapidly to thyroid hormone than to androgen.*
- *Knocking out the EGF receptor gene is lethal in mice, but knocking out the EGF gene is not, apparently due to redundancy in ligands.*
- *A hormone can regulate the expression of EGF mRNA in different organs at one developmental stage, but not at another.*
- *Androgen regulates polyadenylation site use, poly-A tail length and the level of EGF mRNA in the SMG but not the kidney of the mouse.*
- *The mRNAs encoding EGF receptor and TFG-α also undergo post-transcriptional regulation.*
- *The terminal polyadenylation signal of mammalian EGF mRNAs is embedded in a highly conserved 23 b AU-rich element.*
- *A 47 kDa protein that binds to the 23 b AU-rich element is androgen-responsive in the SMG but not the kidney.*
- *Other AU-rich RNA binding proteins also display androgen responsiveness in the male SMG.*
- *AUF1 isoforms display a characteristic pattern of response to androgen in the SMG but not the kidney.*
- *HuR levels respond to androgen in both the kidney and the SMG.*

Study Guide Questions

1) If no androgen-responsiveness could be detected in reporter expression studies with 7 kb upstream of the EGF coding sequences and if there is no consensus androgen response element, explain how androgen can regulate EGF gene expression.
2) If two different RNA binding proteins both bind to the same AU-rich sequence in a given mRNA, what are some of the ways that a cell can modulate the function of two binding proteins differentially?
3) Two cells both express the same levels of all the same RNA binding proteins, which all have exactly the same activity in both cells. If the same levels of an mRNA are detected in both cells, using a probe complementary to the coding sequence of the mRNA, what mechanisms could explain why an mRNA can have different half-lives in the two cells?
4) Alternative splicing of transcripts of an AU-rich mRNA binding protein, as well as covalent modification of that binding protein, can alter its interactions with RNAs or proteins, and thus alter its subcellular distribution or metabolism. Explain how such changes in splicing or covalent modification could make an AU-rich mRNA hormone-responsive at one stage in a cell's development and yet make the same mRNA hormone-resistant at another stage in that cell's development (assume that the hormone receptor pathway remains unaltered).
5) The length of the poly-A tail on a transcript can influence mRNA stability, translation and subcellular location. Unfortunately, small differences in poly-A tail length on a long mRNA may be difficult to assess on Northern blots. What more sensitive methods could be used to assess changes in poly-A tail length?

Acknowledgements

Supported, in part, by funds from the Medical Research Service of the Department of Veteran's Affairs.

REFERENCES

1. Singh B., J.M. Rutledge, D.T. Armstrong: 1995, Epidermal growth factor and its receptor gene expression and peptide localization in porcine ovarian follicles, *Mol Reprod Dev* 40:391-9.
2. Vaughan T.J., P.S. James, J.C. Pascall, K.D. Brown: 1992, Expression of the genes for TGF alpha, EGF and the EGF receptor during early pig development, *Development* 116:663-9.
3. Tong B.J., S.K. Das, D. Threadgill, T. Magnuson, S.K. Dey: 1996, Differential expression of the full-length and truncated forms of the epidermal growth factor receptor in the preimplantation mouse uterus and blastocyst, *Endocrinology* 137:1492-6.

4. Kliem A., F. Tetens, H. Niemann, B. Fischer: 1998, Only a truncated epidermal growth factor receptor protein is present in porcine endometrium, *Biol Reprod* 58:1367-71.
5. Threadgill D.W., A.A. Dlugosz, L.A. Hansen, T. Tennenbaum, U. Lichti, D. Yee, C. LaMantia, T. Mourton, K. Herrup K, R.C. Harris:1995, Targeted disruption of mouse EGF receptor: effect of genetic background on mutant phenotype, *Science* 269:230-4, 1995.
6. Chailler P., N. Briere: 1998, Mitogenic effects of EGF/TGF alpha and immunolocalization of cognate receptors in human fetal kidneys, *Biofactors* 7:323-35.
7. Cybulsky A.V., P.R. Goodyer, A.J. McTavish: 1994, Epidermal growth factor receptor activation in developing rat kidney, *Am J Physiol* 267:F428-36.
8. North D., J. Lakshmanan, A. Reviczky, M. Kaser, D.A. Fisher: 1992, Ontogeny of epidermal growth factor, transforming growth factor-alpha, epidermal growth factor receptor, and thyroid hormone receptor RNA levels in rat kidney and changes in those levels induced by early thyroxine treatment, *Pediatr Res* 31:330-4.
9. Brown M.J., G.S. Schultz, F.K. Hilton: 1984, Intrauterine proximity to male fetuses predetermines level of epidermal growth factor in submandibular glands of adult female mice. *Endocrinology* 115:2318-23.
10. Jaskoll T., M. Melnick: 1999, Submandibular gland morphogenesis: stage-specific expression of TGF-alpha/EGF, IGF, TGF-beta, TNF, and IL-6 signal transduction in normal embryonic mice and the phenotypic effects of TGF-beta2, TGF-beta3, and EGF-r null mutations, *Anat Rec* 256:252-68.
11. Gattone V.H. 2d, D.A. Sherman, D.A. Hinton, F.W. Niu, R.T. Topham, R.M. Klein: 1992, Epidermal growth factor in the neonatal mouse salivary gland and kidney, *Biol Neonate* 61:54-67.
12. Salido E.C., J. Lakshmanan, S. Koy, L. Barajas, D.A. Fisher: 1990, Effect of thyroxine administration on the expression of epidermal growth factor in the kidney and submandibular gland of neonatal mice. An immunocytochemical and in situ hybridization study *Endocrinology* 127:2263-9.
13. Fujieda M., Y. Murata, H. Hayashi, F. Kambe, N. Matsui, H. Seo: 1993, Effect of thyroid hormone on epidermal growth factor gene expression in mouse submandibular gland, *Endocrinology* 132:121-5.
14. Kyakumoto S., R. Kurokawa, M. Ota: 1985, Effect of castration and administration of testosterone on cytosol and nuclear androgen receptor in mouse submandibular gland, *Biochem Int* 11:701-7.
15. Fowler K.J., F. Walker, W. Alexander, M.L. Hibbs, E.C. Nice, R.M. Bohmer, G.B. Mann, C. Thumwood, R. Maglitto, J.A. Danks: 1995, A mutation in the epidermal growth factor receptor in waved-2 mice has a profound effect on receptor biochemistry that results in impaired lactation, *Proc Natl Acad Sci U S A* 92: 1465-9.
16. Keegan B.P., L.G. Sheflin, S.W. Spaulding: 2000, The internalization and endosomal trafficking of the EGF receptor in response to EGF is delayed in the waved-2 mouse liver, *Biochem Biophys Res Commun* 267:881-6.
17. Chen B., R.T. Bronson, L.D. Klaman, T.G. Hampton, J.F. Wang, P.J. Green, T. Magnuson, P.S. Douglas, J.P.Morgan, P.G. Neel: 2000, Mice mutant for Egfr and Shp2 have defective cardiac semilunar valvulogenesis, *Nat Genet* 24:296-9.
18. Sebastian J., R.G. Richards, M.P. Walker, J.F. Wiesen, Z. Werb, R. Derynck, Y.K. Hom, G.R. Cunha, R.P. DiAugustine:1998, Activation and function of the epidermal growth factor receptor and erbB-2 during mammary gland morphogenesis, *Cell Growth Differ* 9:777-85.
19. Luetteke N.C., T.h. Qiu, S.E Fenton, K.L. Troyer, R.F. Riedel, A. Chang, D.C. Lee:1999, Targeted inactivation of the EGF and amphiregulin genes reveals distinct roles for EGF receptor ligands in mouse mammary gland development, *Development* 126:2739-50.
20. Fenton S.E., N.S. Groce, D.C. Lee: 1996, Characterization of the mouse epidermal growth factor promoter and 5'-flanking region. Role for an atypical TATA sequence, *J Biol Chem* 271:30870-8, 1996.
21. Sheflin L.G., E.M. Brooks, B.P. Keegan, S.W. Spaulding: 1996, Increased epidermal growth factor expression produced by testosterone in the submaxillary

gland of female mice is accompanied by changes in poly-A tail length and periodicity, *Endocrinology* 137:2085-92.
22. Sheflin L.G., E.M. Brooks, S.W. Spaulding: 1996, Testosterone regulates tissue-specific changes in the binding of a 47-kilodalton protein to a highly conserved sequence in the 3' untranslated region of epidermal growth factor messenger ribonucleic acid, *Endocrinology* 137:2910-7.
23. Sheflin L.G., E.M. Brooks, B.P. Keegan, S.W. Spaulding: 1996, The cytosolic 47-kilodalton protein that binds to the 3' untranslated region of epidermal growth factor transcripts responds to orchiectomy in a tissue-specific fashion, *Endocrinology* 137:5616-23.
24. Pascall J.C.: 1997, Post-transcriptional regulation of gene expression by androgens-recent observations from the epidermal growth factor gene, *J Mol Endocrinol* 18:177-80.
25. Qian J.F., E. Lazar-Wesley, C. Breugnot, E. May: 1993, Human transforming growth factor alpha: sequence analysis of the 4.5-kb and 1.6-kb mRNA species, *Gene* 132:291-6.
26. Berkowitz E.A., Hissong MA, Lee DC: 1996, Transcriptional and post-transcriptional induction of the TGF alpha gene in transformed rat liver epithelial cells, *Oncogene* 12:1991-2002.
27. Gaido K.W., S.C.Maness, L.S. Leonard, W.F. Greenlee 1992, 2,3,7,8-Tetrachlorodibenzo-p-dioxin-dependent regulation of transforming growth factors-alpha and -beta 2 expression in a human keratinocyte cell line involves both transcriptional and post-transcriptional control, *J Biol Chem* 267:24591-5.
28. Nicolini G., M. Miloso, M.C. Moroni, L. Beguinot, L. Scotto: 1996, Post-transcriptional control regulates transforming growth factor alpha in the human carcinoma KB cell line, *J Biol Chem* 271:30290-6.
29. Satoh Y., M. Oomae, Y. Hoshikawa, M. Izawa, T. Sairenji, S. Ichii: 1997, The stability of truncated epidermal growth factor receptor mRNA is higher than that of the full-length receptor mRNA in rat hepatoma cells. *Endocr J* 44:403-8.
30. McCulloch R.K., C.E. Walker, A. Chakera, J. Jazayeri, P.J. Leedman: 1998, Regulation of EGF-receptor expression by EGF and TGF alpha in epidermoid cancer cells is cell type-specific, *Int J Biochem Cell Biol* 30:1265-78.
31. Kesavan P., S. Mukhopadhayay, S. Murphy, M. Rengaraju, M.A. Lazar, M. Das: 1991,Thyroid hormone decreases the expression of epidermal growth factor receptor, *J Biol Chem* 266:10282-6.
32. Brooks E.M., L.G. Sheflin, S.W. Spaulding: 1995, Secondary structure in the 3' UTR of EGF and the choice of reverse transcriptases affect the detection of message diversity by RT-PCR, *Biotechniques* 19:806-15.
33. Sheflin L.G., E.M. Brooks, S.W. Spaulding: 1995, Assessment of transcript polyadenylation by 3' RACE: the response of epidermal growth factor messenger ribonucleic acid to thyroid hormone in the thyroid and submaxillary glands, *Endocrinology* 136:5666-76.
34. Carter D.A., D. Murphy: 1993, Regulation of vasopressin (VP) gene expression in the bed nucleus of the stria terminalis: gonadal steroid-dependent changes in VP mRNA accumulation are associated with alterations in mRNA poly (A) tail length but are independent of the rate of VP gene transcription, *J Neuroendocrinol* 5:509-15.
35. Vercaeren I., J. Winderickx, A. Devos, B. Peeters, W. Heyns: 1992, An effect of androgens on the length of the poly(A)-tail and alternative splicing cause size heterogeneity of the messenger ribonucleic acids encoding cystatin-related protein, *Endocrinology* 131:2496-502.
36. Wagner B.J., C.T. DeMaria, Y. Sun Y, G.M. Wilson, G. Brewer: 1998, Structure and genomic organization of the human AUF1 gene: alternative pre-mRNA splicing generates four protein isoforms, *Genomics* 48:195-202.
37. Dreyfuss G., M.J. Matunis, S. Pinol-Roma, G.C. Burd: 1993, hnRNP proteins and the biogenesis of mRNA, *Annu Rev Biochem* 62:289-321.
38. Bhattacharya S., T. Giordano, G. Brewer, J.S. Malter: 1999, Identification of AUF-1 ligands reveals vast diversity of early response gene mRNAs. Identification of AUF-1 ligands reveals vast diversity of early response gene mRNAs, *Nucleic Acids Res* 27:1464-72.

39. Zhao Z., F.C. Chang, H.M. Furneaux: 2000, The identification of an endonuclease that cleaves within an HuR binding site in mRNA. *Nucleic Acids Res* 28:2695-701.
40. Pende A., K.D. Tremmel, C.T. DeMaria, B.C. Blaxall, W.A. Minobe, J.A. Sherman, J.D. Bisognano, Bristow, G. Brewer, J.D. Port JD: 1996, Regulation of the mRNA-binding protein AUF1 by activation of the beta-adrenergic receptor signal transduction pathway, *J Biol Chem* 271:8493-501.
41. Loflin P., C.Y. Chen, A.B. Shyu: 1999, Unraveling a cytoplasmic role for hnRNP D in the in vivo mRNA destabilization directed by the AU-rich element, *Genes Dev* 13:1884-97.
42. Laroia G., R.G. Cuesta, G. Brewer, R.J. Schneider: 1999, Control of mRNA decay by heat shock-ubiquitin-proteasome pathway, *Science*, 284:499-502.
43. Sheflin LG, S.W. Spaulding: 2000, Testosterone and dihydrotestosterone regulate AUF1 isoforms in a tissue-specific fashion in the mouse, *Am J Physiol Endocrinol Metab* 278:E50-7
44. Scheflin L.G., W. Zhang, S.W. Spaulding. 2001, Androgen regulates the level and subcellular distribution of the AU-rich RNA binding protein HuR both *in vivo* and *in vitro*. *Endocrinology* 142:2361-2368.
45. Sela-Brown A,. J. Silver, G. Brewer, T. Naveh-Many: 2000, Identification of AUF1 as a parathyroid hormone mRNA 3'-untranslated region-binding protein that determines parathyroid hormone mRNA stability, *J Biol Chem* 275:7424-9.
46. Blaxall B.C., A.C. Pellet, S.C., Wu SC, A. Pende, J.D. Port: 2000, Purification and characterization of beta-adrenergic receptor mRNA-binding proteins, *J Biol Chem* 275:4290-7.
47. Arao Y., R. Kuriyama, F. Kayama, S. Kato: 2000, A nuclear matrix-associated factor, SAF-B, interacts with specific isoforms of AUF1/hnRNP D, *Arch Biochem Biophys* 380:228-36. mouse kidney is posttranscriptional, *Biochemistry* 25:1170-5.

16

THE ROLE OF RNA BINDING PROTEINS IN TUMORIGENESIS

Sreerama Shetty
University of Texas Health Center at Tyler, TX

Posttranscriptional regulatory mechanisms control expression of a variety of genes implicated in the pathogenesis of inflammation, fibrotic repair and neoplasia. Posttranscriptional regulation involves specific interactions between specific mRNA binding sequences that interact with mRNA binding proteins. These cis-trans interactions can either stabilize or destabilize the mRNAs. Of the genes regulated at the posttranscriptional level, components of the uPA-uPAR fibrinolytic system are prominently represented and these pathways could importantly influence the pathogenesis of pulmonary inflammation and neoplasia. uPA, uPAR, and PAI-1 are expressed by lung and pleural mesothelial cells and are implicated in the pathogenesis of remodeling of the injured lung and pleural space and in cancers that occur in these locations. It is now clear that the expression of each of these genes is regulated at the posttranscriptional level and that the regulatory mechanisms involve the interaction of these mRNAs with specific mRNA binding proteins that control mRNA stability.

BACKGROUND

Posttranscriptional regulatory mechanisms are integral to diverse pathophysiologic processes. The regulation of mRNA stabilities is particularly important during "stresses" such as starvation, infection, inflammation, exposure to toxins and tissue invasion by cancer cells during neoplastic transformation. Two important factors determine the stability of mRNA: specific sequences (*cis*-elements) in the mRNA determine half-life, and specific binding proteins (*trans*-factors) regulate stability.

The regulation of mRNA stability is thought to play an important role in eukaryotic gene expression by modulating the cytoplasmic abundance of several mRNAs. Different mRNAs have different half-lives in a given cell, and the stability of the same mRNA may be modulated by extracellular stimuli or environmental factors. This scenario raises questions as to why different mRNAs have different half-lives. Obviously, the reason could involve innate structural changes or specific sequences of the mRNA that may be responsible for mRNA lability or stability. Along these lines, a few instability determinants have been identified in either the coding region or the 3' untranslated region (UTR) of diverse mRNAs. Among 3'UTR mRNA stability determinants, AU-rich sequences in various oncogenes, lymphokine and cytokine mRNAs (1-5) and an iron-responsive element in transferrin receptor mRNA (6-7) have been reported.

The Role of Posttranscriptional Regulatory Mechanisms in Cancer Cells

In nonneoplastic tissues, cytokine mRNAs are extremely short-lived, with half-lives ranging from less than 20 minutes to one hour (8-11). Cytokine-induced mRNA instability regulates gene expression and is thought to be mediated at least in part by AU-rich sequences in the 3'-untranslated region of the message. Excess cytokine production may influence growth of cells derived from diverse neoplasms. For example, myeloid leukemic cells, small cell lung carcinoma cells, osteogenic sarcoma, breast cancer and colon cancer cell lines produce and respond to GM-CSF, G-CSF and IL-6 *in vitro*. Cytokines play a role in the maintenance of tumor cells, and an understanding of growth factor production and regulation in tumors is relevant to the biology of cancer. Some tumor cells produce increased amounts of IL-1 and IL-6 mRNAs and IL-1 and IL-6 in turn can stabilize other cytokine mRNAs. AU-rich sequences in the 3'-untranslated region of most cytokine and oncogene mRNAs contribute to the extreme lability of these mRNAs. More recently, prolonged stability of cytokine messages was found to be mediated by alternative factors independent of the AU-rich region (12).

Activation of cytokines, has been implicated in the pathogenesis of transformation and survival of tumor cells (13). Expression of proinflammatory cytokines has been associated with increased potential for growth and metastasis of squamous cell carcinoma and most other types of cancer. Constitutive activation of cytokine genes may also play a role in tumor progression. However, the mechanism(s) involved in activation of cytokine gene expression during tumor progression are, at present, poorly defined. Unrestrained cancer proliferation does not, by itself, appear to lead to metastasis. Rather, it is the tendency of malignant cells to traverse tissue boundries that distinguishes proliferative disorders and carcinomas *in situ* from true malignancy. There are no qualitative differences between cancer cells and normal cells with respect to the basic processes of cell detachment, cell migration, cell invasion and the formation of colonies at distant sites. A critical difference is that cancer cells express the phenotypic traits characteristic for these processes at times and places incompatible with normal cellular behavior. Altered gene expression, determined at the posttranscriptional level can contribute to the neoplastic phenotype.

The Urokinase-Urokinase Receptor System and the Pathogenesis of Neoplasia

Extracellular proteolytic enzymes including serine proteinases such as urokinase plasminogen activator and metalloproteinases have been implicated in cancer metastasis. The basic idea is that the release of proteolytic enzymes in tumors facilitates cancer–cell invasion into the surrounding normal tissue by breakdown of basement membranes and extracellular matrix (14). The suggestion that plasminogen activation may play an important role in these processes goes back several decades (15). There are two types of plasminogen activators, the urokinase-type (uPA) and the tissue-type (tPA). Both are capable of catalyzing the conversion of the inactive zymogen plasminogen to the active proteinase plasmin. Plasmin, in turn, can degrade most extracellular proteins. tPA is mainly involved in intravascular thrombolysis, while uPA generates plasmin in events involving degradation of extracellular matrix. uPA has therefore been studied most extensively in relation to cancer metastasis. There are two main inhibitors of plasminogen activators, PAI-1 and PAI-2, while plasmin is inhibited by a_2-anti-plasmin (aAP). Urokinase receptor (uPAR) is a cell membrane-anchored binding protein for uPA, concentrating plasminogen activation at cell surfaces (14-15). During the past decade, evidence for involvement of the uPA system in cancer metastasis has increased steadily and it is now clear that the uPA-mediated pathway of plasminogen activation is central to this process. Many model systems have indicated that uPA is involved in cancer metastasis. In addition, analyses of many human cancers have shown that high levels of uPA, PAI-1 and uPAR in tumors are correlated with poor clinical prognosis.

Expression of the components of the uPA system is regulated by a large variety of hormones, growth factors and cytokines. Regulatory mechanisms that determine the expression of the components of the uPA system in tumors remain to be fully elucidated. Obviously, the higher expression of the components of the fibrinolytic system in a tumor compared to the corresponding normal tissue could be due ultimately to oncogenic transformation. During cancer cell migration and invasion, composition of the microenvironment of the cell also changes, and interactions of the modified microenvironment with the cells may lead to altered expression of fibrinolytic system. For example, cancer metastasis begins with detachment of cancer cells from their initial location and may require down-regulation of E-cadherin expression and increased uPA expression (16).

Molecular Determinants of mRNA Stability

Cis determinants of mRNA stability: Most investigators have exploited the creation of chimeric mRNA to identify sequences affecting mRNA stability. Segments from genes with unstable mRNAs linked to stable mRNAs in such a way as to maintain the reading frame are transfected into target cells and the stability of the chimeric mRNAs is assayed after blockade of ongoing transcription. Some mRNAs contain multiple stability determinants, any of which might independently formulate part of a decay pathway. The determinants are classified based on their location on mRNA. The poly(A) tail has multiple functions including the regulation of cytoplasmic mRNA stability. The poly(A) tail protects mRNA from rapid degradation (deadenylation) involving a poly(A) binding protein (PABP), as confirmed by decay studies of poly(A) mRNA in a PABP depleted environment (17). Stability of the majority of the mRNAs studied to date have been influenced by sequences in the 3'UTR and specific mRNA binding proteins. Among 3'UTR determinants, a 30 nucleotide histone mRNA 3'UTR stem-loop sequence affects histone mRNA stability. The majority of the cytokines and oncogenes have unstable mRNAs that enable the transient appearance of these proteins associated with specific cellular events. These mRNAs contain AU- or U-rich sequences in the 3'UTR and these sequences can function as mRNA destabilizing signals. The oncogenicity of v-fos gene and c-fos gene lies in the stability determinants encoded in their respective mRNA 3'UTR (18). Destabilization of selected genes can be a very complicated process involving multiple functions depending on the mRNA, the cell type and growth conditions. For example, the mRNAs coding for transferrin receptor and ferritin, both of which regulate iron homeostasis, are regulated posttranscriptionally by a mechanism that is dependent on the intracellular iron concentration. Regulation of mRNA function is achieved through the iron-responsive element, a stem-loop structure and an iron response element binding protein. Similarly, insulin-like growth factor II is regulated at the posttranscriptional level by a stem-loop stability determinant present at the 3'UTR.

Coding region determinants: Four coding region determinants regulating mRNAs have been reported. Some of the coding region stability determinants are also protein-binding sites. *c-fos* mRNA contains at least three mRNA destabilization signals, an AU-rich element (ARE) in the 3'UTR and two other sequences in the coding region (19-21). Destabilization of tubulin mRNA by tubulin monomers and the stabilization of growth-associated protein by nerve growth factor depend on the peptides encoded by their respective coding-region determinants. The c-myc mRNA coding region determinants involve a C-terminal coding region determinant and influence mRNA stability when translation is inhibited.

mRNAses: Most mRNAs are degraded by RNases with little specificity for a particular RNA. However, the stability of mRNA varies from one mRNA species to another and is probably determined by mRNA sequences/structure and mRNA-protein complexes. Some mRNAs are inherently more susceptible to destabilization than others. This variation probably depends on the degree of stability of regulatory RNA-protein complexes. For example, if a poly (A) binding protein were to bind to all mRNA species with equal affinity and protect from deadenylation, the stability of the mRNAs would be identical, which is not the case. Some mRNA binding proteins appear to stabilize the mRNAs to which they bind, while others acts as destabilizers. The stability of a particular mRNA depends on the combined actions of different regulatory factors.

Trans-factors: These proteins can be divided into two groups based on their function including those that bind to mRNA and stabilize the message and others that destabilize the mRNA.

RNA-binding proteins that protect mRNA from degradation: Deadenylation is the first step towards decay of an mRNA. For example, PABP binds to poly(A) mRNA and prevents the mRNA from rapid degradation. Alternatively, one family of proteins have the ability to interact with high affinity to RNAs containing AU-rich or U-rich sequences in the 3'UTR (AUBPs). AUBP binds and protects the mRNA from degradation. These proteins are either located in the cytoplasm or nucleus while some family members travel between two compartments. Most of the AU-rich mRNA are fairly unstable and their half-lives vary. This suggests besides AUBPs that there are other regulatory components. As an example, an iron regulatory mRNA binding protein protects a specific mRNA from degradation. It stabilizes transferrin receptor mRNA by binding to 3'UTR in an iron depleted environment. Additional regulatory proteins in this category are ribonucleotide reductase mRNA-binding proteins, a family of three proteins that interact with ribonucleotide reductase mRNAs and alter their stabilities. There are also c-fos coding region determinant-binding proteins:

two proteins that bind to a purine-rich segment of the c-fos mRNA coding region (22). One, of 64 kDa, and the other, of 53 kDa, binds to a 320 coding region determinant and protects the mRNA from degradation. On the other hand, a 70 kDa protein binds to a 180 nucleotide coding region determinant of c-myc mRNA and protects it from degradation.

RNA-binding proteins that accelerate mRNA degradation: Ribonucleotide reductase mRNA subunit 1 and 2 (RR1 and RR2) stabilty is regulated by three proteins capable of interacting with 3'UTR sequences. Proteins of 57 and 45 kDa bind to RR1 and RR2 mRNAs, respectively. The stability of RR1 and RR2 are inversely correlated with the interaction of RR1 and RR2 with respective binding proteins as confirmed by exposure of cells to phorbol esters (23-25). However, the 75 kDa protein acts as a stabilizer of RR2 mRNA as confirmed by TGF-β-treated cells. Suppression of malignancy by the 3'UTR of R1 and R2 messenger RNAs (26) emphasizes the importance of posttranscriptional regulatory mechanisms. The exact pathway is currently unknown, but the authors indicate that expression of 3'UTRs could bind proteins (R1Bp and R2BP) or other regulators of cell growth to reduce the interaction of these factors with their primary targets.

POSTTRANSCRIPTIONAL REGULATION OF uPAR, uPA AND PAI-1

uPAR: The uPA-uPAR system has, in particular, been linked to altered fibrin clearance and cellular signaling in the context of inflammation and repair as well as neoplasia. Expression of fibrinolytic components such as uPA, uPAR and its major inhibitors (PAI-1 and PAI-2) are now known to be regulated at the posttranscriptional level. Treatment of pleural mesothelial cells or mesothelioma cells (27), lung fibroblasts or lung carcinoma cells (28) with tumor promoters, cytokines and translational inhibitors result in increased mRNA stability. These data indicate that the uPAR gene is regulated at the posttranscriptional level.

One possible mechanism is an interaction between binding proteins and a specific region in the uPAR mRNA to regulate expression of the message and thereby alter uPAR expression at the cell surface. A 50 kDa uPAR mRNA binding protein (uPAR mRNABp) selectively interacts with a 51 nucleotide sequence in the coding region (27). Insertion of 51 nt uPAR mRNABp binding sequence in β-globin cDNA destabilized chimeric β-globin/uPAR/β-globin mRNA and the decay profile of chimeric message was comparable to that of wild-type uPAR mRNA (Fig. 1). This 51 nt protein binding sequence of uPAR mRNA contains regulatory information for message stability.

Tumor promoters and proinflammatory cytokines like TGF-β or TNF-α induce uPAR expression at least in part by stabilizing uPAR mRNA. Tumor promoter compounds, cytokines and translational inhibitors separately or in combination enhance uPAR mRNA stability from 3 h to 6-18 h depending on the agents and cell types. These agents stabilize uPAR mRNA by down-regulating the uPAR mRNA-uPAR mRNABp interaction as shown in Fig. 2. A similar phenomenon occurs in T lymphocyte lymphokine mRNA (29) or ribonucleotide reductase R1 mRNA (24) in mRNA-mRNA binding protein interactions mediated by specific mRNA binding proteins.

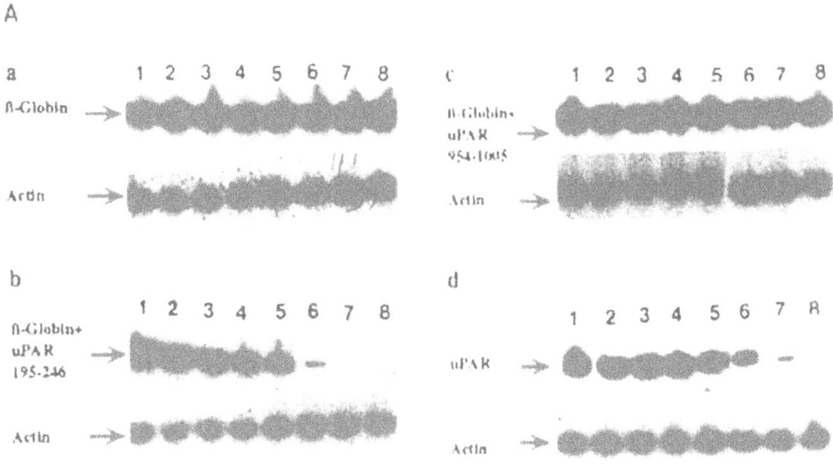

Figure 1. ***Posttranscriptional regulation of uPAR mRNA: demonstration of a uPAR mRNA ccoding sequence that interacts with a uPAR mRNABp to destabilize chimeric β-globin/uPAR/β-globin mRNA.*** *Pleural mesothelioma (MS-1) cells were transfected with β-globin (a), the chimeric β-globin/uPAR/β-globin gene containing the 51 nt (nt 195 to 246) nucleotide uPAR mRNABp binding sequence (b), non-protein binding control sequence of the uPAR coding region (nt 954 to 1005) (c), and the uPAR gene (d) in pcDNA3. Total RNA was isolated at different time intervals after treatment with actinomycin D in serum-free medium and analyzed by RNase protection assay using a ^{32}P-labeled antisense transcript. Lanes 1-8, contain RNA samples after treatment with actinomycin D (10 mg/ml) for 0-24 h.*

The mechanism by which tumor promoters or cytokines induce stabilization of uPAR mRNA remains to be elucidated. The observed stabilization of uPAR mRNA by tumor promotors or cytokines could be due to down-regulation of uPAR mRNABp by either decreased transcription, stabilization, posttranslational modification or a combination of all these putative mechanisms. Translational inhibitors, on the other hand, can stabilize uPAR mRNA by either inhibition of translational or postranslational modification of uPAR mRNABp.

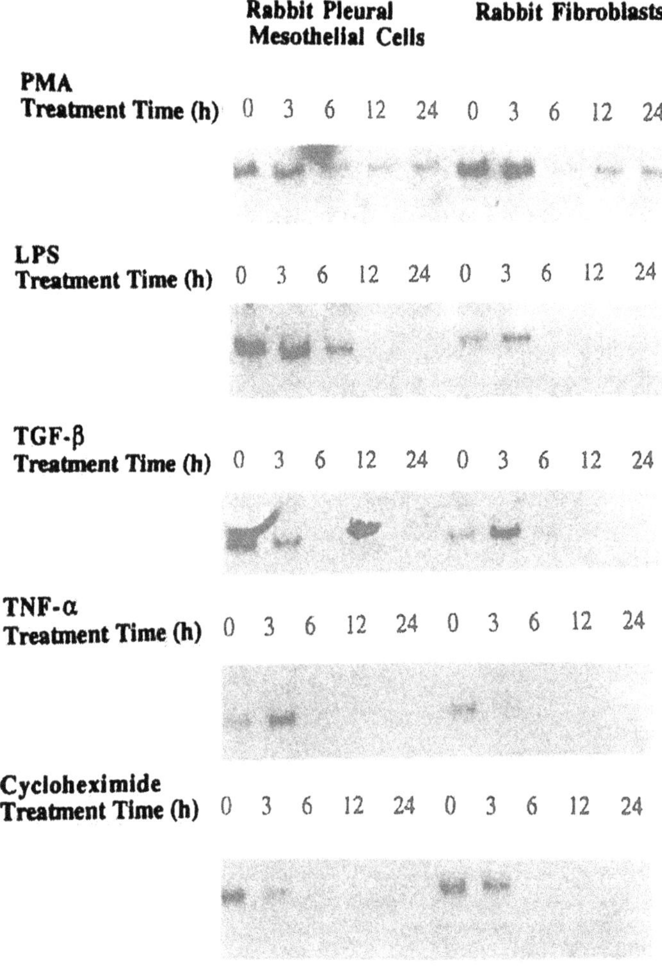

Figure 2. Regulation of the uPAR mRNA-uPAR mRNABp interaction in pleural mesothelial cells and lung fibroblasts by selected stimuli. *Rabbit pleural mesothelial cells and fibroblasts treated with following agents for 0-24 h. Cytosolic extracts were analyzed for uPA mRNA-uPA mRNABp interaction by gel mobility shift assay.*

Overexpression of the chimeric β-globin/uPAR/β-globin sequence containing the uPAR mRNABp binding sequence in stably transfected MS-1 cells increased cell surface uPAR expression, uPA mediated DNA synthesis and cellular migration and invasion (28).

These data confirm that interaction between the uPAR mRNABp and uPAR mRNA regulates uPAR expression through effects on uPAR message stability. These observations also serve to explain at least in part, how expression of uPAR in tumor cells is inversely correlated with poor clinical outcome in many different forms of cancer. These functions contribute to the pathogenesis of neoplastic growth and invasiveness. For example, steady state mRNA analysis indicated that uPAR mRNA is more stable in selected lung carcinoma and malignant mesothelioma. The stability of uPAR mRNA is, at least in part, responsible for overexpression of uPAR at the cell surface and involves posttranscriptional regulation that involves a mechanism in which the uPAR mRNABp participates. Similarly, enhanced message stability of cytokines is also observed in various tumor cells. In a related study of expression of α-tropomyosin, a mechanism involving a 3'UTR determinant suppressed the proliferation, invasion and destruction of muscle tissues (30). In a similar vein, removal of a mRNA destabilizing element from c-fos increased cellular oncogenecity (18). Wang et al. (31) reported that uPAR mRNA stability is also regulated by AU-rich elements in the 3'UTR, representing yet another level of regulation that operates at the posttranscriptional level. Production of cytokines and expression of oncogenes are likewise controlled by 3'UTR AU-rich elements. However, it is not clear as to how posttranscriptional regulation of oncogenes, cytokines and fibrinolytic components are involved in neoplastic transformation, tumor cell proliferation, and extracellular matrix dissolution and invasion.

Fibroblasts isolated from histologically normal lung express relatively less uPAR compared to fibroblasts isolated from fibrotic lungs (32). The higher capacity of fibrotic versus normal lung fibroblasts to express uPAR suggests that fibrotic fibroblasts could augment parenchymal lung inflammation. Differential uPAR expression between fibroblasts isolated from normal and fibrotic lungs is mainly due to a difference in the stability of uPAR mRNA. For example, uPAR mRNABp is present in the cytosolic extracts of normal lung fibroblasts whereas it is absent in fibrotic fibroblasts. Increased uPAR expression by these cells could contribute to the development of fibrosis. For example, the interaction of uPAR and uPA could promote uPA-mediated fibroblast proliferation as well as fibroblast-mediated proteolysis of extracellular matrix. Increased expression of uPAR in fibroblasts that are somehow altered by the fibrotic process suggests a putative role for the receptor in pulmonary fibrosis, a process that recapitulates increased uPAR expression associated with tumorigenesis.

UPA: Another component of the fibrinolytic system; urokinase (uPA), is expressed by phenotypically diverse lung carcinoma as well as non-malignant lung epithelial cells and is also regulated at the posttranscriptional level. uPA mRNA has a short half-life in normal renal epithelial cells (33) and its stability is modulated by protein synthesis inhibition, PKC down-regulation (33-35), calcium ions (36) and a variety of proinflammatory cytokines (*Shetty, unpublished data*), indicating that mRNA stability is as important as transcriptional regulation in the overall control of uPA expression. uPA mRNA stability is also reported to be determined by multiple instability determinants present in the 3'UTR (37). Insertion of the uPA mRNA 3'UTR into the rabbit gene degrades chimeric β-globin/uPA mRNA as rapidly as endogenous uPA mRNA. These results indicate that the uPA mRNA 3'UTR contains most of the information required for its rapid turnover. Further analysis using deletion methodology indicates that the pig uPA mRNA 3'UTR contains three independent determinants of instability. One of these instability determinants is an AU-rich-sequence. For example, uPA mRNA has a short half-life of 70 min in non-metastatic LLC-PK1 cells whereas in metastatic breast cancer MDA-MB-231 cells, message is stabilized for at least 17 h and this is attributed to a regulatory mechanism involving a region in the 3'UTR. Nanbu et al. (38) identified a p40 AU-rich binding protein that interacts with 3'UTR of uPA mRNA and its level is elevated in MDA-MB-231 cells. These observations indicate a correlation between the AU-rich sequence-dependent uPA mRNA stabilization *in vivo* and the binding of p40 to the AU-rich sequence *in vitro*. Henderson et al. (39) reported that the uPA gene is expressed about 3.5-70 fold higher in metastatic cell lines than in non-metastatic rat mammary adenocarcinoma (DMBA-8) cell lines due predominantly to postranscriptional mechanisms.

In addition to an interaction between an AU-rich sequence and AUBp, the message is also regulated by interaction between a 66 nt sequence in the uPA mRNA 3'UTR and a 30 kDa nucleocytoplasmic shuttling protein (40). This uPA mRNA binding protein specifically interacts with uPA mRNA, as confirmed by experiments with non-malignant lung epithelial cells versus malignant lung carcinoma-derived cells. Insertion of this uPA mRNABp binding sequence into the 3'UTR of β-globin mRNA destabilizes chimeric β-globin/uPA mRNA and these results suggest that the 66 nt region of the uPA 3'UTR contains information that influences the rate of uPA mRNA decay. The uPA mRNABp is found in the cytoplasm but not the nucleus of non-malignant epithelial cells. However, the same protein is observed in the nuclear extracts of selected uPA-overexpressing lung carcinoma-derived cells. These results suggest that uPA mRNA expression is regulated at the posttranscriptional level and the nuclear-cytoplasmic distribution of the uPA mRNABp determines the stability. This pathway represents a potential mechanism by which epithelial cells could upregulate uPA gene regulation in lung neoplasia. However, the cellular signaling that regulates complex formation or the mechanism that facilitates nucleo-cytoplasmic shuttling of the uPA mRNABp remains to be elucidated.

PAI-I: The biosynthesis of major inhibitors of plasminogen activation (PAI-1 and PAI-2) appears to be tightly regulated. Expression of the PAI-1 gene is regulated at both transcriptional and posttranscriptional levels in HepG2 cells (41). PAI-1 mRNA exists in two forms; a labile 3.2 kb and a relatively more stable 2.2 kb species. The difference in the stability between the two mRNA forms is attributed to the differences in the 3'UTR, particularly with respect to the variable presence of AU-rich sequences. Insulin and insulin-like growth factor-1 (IGF-1) increase steady state levels of PAI-1 mRNA in HepG2 cells (42). Insulin stabilizes the 3.2 kb PAI-1 mRNA form whereas IGF-1 prolongs the half-lives of both PAI-1 mRNA species. These data indicate that, in addition to AU sequences, other regions may be involved in PAI-1 mRNA turnover. Cyclic AMP (cAMP) destabilizes PAI-1 mRNA and this process involves 134 nt sequences in the 3'UTR of PAI-1 mRNA (43). Tillman-Bogush et al. (44) reported that in rat cells cytosolic mRNA binding multi-protein complexes with molecular weights ranging from 38 to 76 kDa interact with cAMP responsive sequences. These results underscore the complexity of the regulation of PAI-1 gene expression. We recently found that the levels of PAI-1 are significantly increased in selected lung carcinoma cells compared to normal lung epithelial cells. This observation is particularly important, as PAI-1 expression has been related to prognosis in lung carcinoma cells (45-46). Translational inhibitors increased PAI-1 mRNA and the increment is greater in normal lung epithelial cells compared to selected lung carcinoma-derived cells. In these cells, increased stability of PAI-1 mRNA after transcriptional blockade indicated that PAI-1 gene expression could be regulated at the posttranscriptional level. PAI-1 mRNA has a relatively short half-life of 1-2 h in lung small airway epithelial cells (47), a finding consistent with that previously observed in HepG2 cells (41). However, selected cytokines stabilize PAI-1 mRNA for several hours possibly by interfering with the interaction of PAI-1 mRNA and specific mRNA binding protein.

It now appears that malignant lung epithelial cell types exhibit relatively higher basal PAI-1 mRNA levels via a posttranscriptional mechanism that, like uPA and uPAR, involves a mRNA binding protein. We found that a 60 kDa protein that specifically binds to 3'UTR may be involved in the control of PAI-1 mRNA stability (47). The PAI-1 mRNABp-PAI-1 mRNA interaction is readily detectable in small airway epithelial cells but is diminished or undetectable in cytokine-treated cells or malignant epithelial cells over-expressing PAI-1. These observations suggest that interaction of PAI-1 mRNA with its binding protein is likely destabilizing. The RNA-protein complex is inversely correlated with the stability of PAI-1 mRNA. Recently, Maurer et al. (48) found that PAI-2 mRNA is also regulated at the posttranscriptional level and this regulatory pathway involves an AU-rich motif present on PAI-2 mRNA 3'UTR.

Summary of key concepts

- *mRNA stability depends on specific mRNA sequences (molecular determinants) present in the coding or non-coding region.*
- *Interaction between mRNA stability determinants and specific binding proteins stabilize or destabilize the mRNAs.*
- *Alteration in mRNA-protein interaction by cytokines or other agents is involved in the pathogenesis of inflammation, fibrotic repair and neoplasia.*

Study Guide Questions

1) What is the function of the poly (A) binding protein (PABp)?
2) Why do mRNAs with AU- or U- rich sequences have different stabilities?
3) Why do some mRNAs have multiple stability determinants?
4) How do cytokines regulate mRNA stability during cellular differentiation or malignant transformation?
5) What determines the stabilities of uPA, uPAR and PAI-1 mRNAs?

Acknowledgements

This review and part of the work described above was supported by HL 62453 and HL 45018. The author is grateful to Dr. S. Idell for helpful discussions concerning the manuscript.

REFERENCES

1. Akashi, M., Shaw, G., Gross, M., Saito, M. and Koeffler, H.P. (1991) Role of AUUU sequences in stabilization of granulocyte-macrophage colony stimulating factor RNA in stimulated cells. *Blood* 78, 2005-2012.
2. Caput, D., Beutler, B., Hortog, K., Thayer, R., Brown-shimer, S. and Cerami, A. (1986) Identification of a common nucleotide sequence in the 3'-untranslated region of mRNA molecules specifying inflammatory mediators. *Proc. Natl. Acad. Sci. U.S.A.* 83, 1670-1674.
3. Shaw, G. and Kamen, R. (1986) A conserved AU sequence from the 3' untranslated region of GM-CSF mRNA mediates selective mRNA degradation. *Cell* 46, 659-667.
4. Jones, T.R. and Cole, M.D. (1987) Rapid cytoplasmic turnover of c-myc mRNA: requirement of 3'untranslated sequences. *Mol. Cell Biol.* 7, 4513-4521.
5. Wilson, T. and Triesman, R. (1988) Removal of poly(A) tail and consequent degradation of c-fos mRNA facilitated by 3'AU-rich sequence. *Nature* 336, 396-399.
6. Casey, J.L., Koeller, D.M., Ramin, V.C., Klausner, R.D., and Hartford, J.B. (1989) Iron regulation of transferrin receptor mRNA levels requires iron-responsive elements and a rapid turnover determinant in the 3'untranslated region of the mRNA. *EMBO J.* 12, 3693-3699.

7. Mullner, E.W. and Kuhn, L.C. (1988) A stem-loop in the 3'UTR mediates iron-dependent regulaton of transferrin receptor mRNA stability in the cytoplasm. *Cell* 53, 815-825.
8. Yamato, K., El-Hajjaoui, Z., Kuo, J. and Koeffler, H.P. (1989) Granulocyte-macrophage colony stimulating factor: signals for its mRNA accumulation. *Blood* 74: 1314-1320.
9. Akashi, M., Loussararian, A. Adelman, D., Saito, M. and Koeffler, H.P. (1990) Lymphotoxin: stimulation and regulation of colony stimulating factors in fibroblasts. *J. Clin. Invest.* 85: 121-129.
10. Koeffler, H.P., Gasson, J. and Tobler, A. (1988) Transcriptional and postranscriptional modulation of myeloid CSF expression by TNF and other agents. *Mol. Cell Biol.* 8: 3432-3438.
11. Yamato, K., El-Hajjaoui, Z. and Koeffler, H.P. (1989) Regulation of levels of IL-1 mRNA in human fibroblasts. *J. Physiol.* 139: 610-616.
12. Ross, H. J., Sato, N., Ueyama, Y. and Koeffler, H.P. (1991) Cytokine messenger RNA stability is enhanced in tumor cells. *Blood* 77: 1787-1795.
13. Dong, G., Chen, Z., Kato, T. and Waes, C.V. (1999) The host environment promotes the constitutive activation of nuclear factor-kappaB and proinflammatory cytokine expression during metastatic tumor progression of murine squamous cell carcinoma. *Cancer Res.* 59: 3495-3504.
14. Mignati, P. and Rifkin, D.B. (1993) Biology and biochemistry of proteinases in tumor invasion. *Physiol. Rev.* 73, 161-195.
15. Dano, K., Andreasen, P.A., Grondahl-Hansen, K., Kristensen, P., Nielsen, L.S. and Skriver, L. (1985) Plasminogen activators, tissue degradation and cancer. *Adv. Cancer Res.* 44, 139-266.
16. Behrens, J. (1993) The role of cell adhesion molecules in cancer invasion and metastasis. *Breast Cancer Res. Treat.* 24, 175-184.
17. Bernstein, P.L., Herrick, D.J., Prokipcak and Ross, J. (1992) Control of c-myc mRNA half-life in vitro by a protein capable of binding to a coding region stability determinant. *Genes Dev.* 6:642-654.
18. Raymond, V., Atwater, J. A. and Verma, I. M. (1989) Removal of an mRNA destabilizing element correlates with the increased oncogenicity of proto-oncogene fos. *Oncogene Res.* 4: 861-865.
19. Schiavi, S. C., Wellington, C. L., Shyu, A.-B., Chen, C.Y.A., Greenberg, M.E. and Belasco, J.G. (1994) Multiple elements in the c-fos protein-coding region accelerate mRNA deadenylation and decay by a mechanism coupled to translation. *J. Biol. Chem.* 269:3441-3448.
20. Shyu, A-B., Belasco, J.G. and Greenberg, M.E. (1991) Two distinct destabilizing elements in the c-fos message trigger deadenylation as a first step in rapid mRNA decay. *Genes Dev.* 5:221-231.
21. Wellington, C.L., Greenberg, M.E., and Belasco, J.G. (1993) The destabilizing elements in the codin region of c-fos mRNA are recognized as RNA. *Mol. Cell Biol.* 13: 5034-5042.
22. Chen, C.Y., You, Y. and Shyu, A. B. (1992) Two cellular proteins bind specifically to a purine-rich sequence necessary for the destabilization function of a c-fos protein coding region determinant of mRNA stability. *Mol Cell Biol.* 12:5748-5757.
23. Amara, F.M., Chen, F.Y. and Wright, J. A. (1994) Phorbol ester modulation of a novel cytoplasmic protein binding activity at the 3'-untranslated region of mammalian ribonucleotide reductases R2 mRNA and role in message stability. *J. Biol. Chem.* 269:6709-6715.
24. Chen, C.Y., Amara, F.M. and Wright, J. A. (1993) Mammalian ribonucleotide reductase R1 mRNA stability under normal phorbol ester stimulating conditions: involvement of a *cis-trans* interaction at the 3'untranslated region. *EMBO, J.* 12: 3977-3986.
25. Chen, C.Y., Amara, F.M. and Wright, J. A. (1994) Defining a novel ribonucleotide reductase R1 mRNA *cis* element that binds to an unique cytoplasmic *trans*-acting protein. *Nucleic Acid Res.* 22: 4796-4797.
26. Fan H., Villegas, C., Huang, A. and Wright, J. A. (1996) Suppression of malignancy by the 3' untranslated regions of ribonucleotide reductase R1 and R2 messenger RNAs *Cancer Res.* 56:4366-4369.
27. Shetty, S., Kumar, A. and Idell, S. (1997) Posttranscriptional regulation of urokinase receptor mRNA: identification of a novel urokinase receptor mRNA binding protein in human mesothelioma cells. *Mol. Cell. Biol.* 17:1075-1083.

28. Shetty, S. and Idell, S. (1999) Posttranscriptional regulation of urokinase receptor gene expression in lung carcinoma and malignant mesothelioma cells *in vitro*. *Mol. Cell. Biol.* 199:189-200.
29. Brewer, G. (1991) An A+U rich element RNA-binding factor regulates c-*myc* mRNA stability *in vitro*. *Mol. Cell. Biol.* 11:2460-2466.
30. Rastinejad, F., Conboy, M.J., Rando, T.A. and Blau, H.M. (1993) Tumor supression by the RNA from the 3'untranslated region of a-tropomyosin. *Cell* 75:1107-1117.
31. Wang, G.J., Collinge, M., Blasi, F., Pardi, R. and Bender, J.R. (1998) Posttranscriptional regulation of urokinase plasminogen activator receptor messenger RNA levels by leukocyte integrin engagement. *Proc. Natl. Acad. Sci.* 95:6296-6301.
32. Shetty, S., Kumar, A., Johnson, A.R., Pueblitz, S., Holiday, D., Raghu, G. and Idell, S. (1996) Differential expression of the urokinase receptor in fibroblasts from normal and fibrotic human lungs. *Am. J. Resp. Cell Mol. Biol.* 15:78-87.
33. Altus, M. S., Pearson, D., Horiuchi, A. and Nagamine, Y. (1987) Inhibition of protein synthesis in LLC-PK1 cells increases calcitonin induced plasminogen activator gene transcription and mRNA stability. *Biochem. J.* 242: 387-392.
34. Altus, M. S. and Nagamine, Y. (1991) Protein synthesis inhibition stabilizes urokinase-type plasminogen activator mRNA. *J. Biol. Chem.* 266: 21190-21196.
35. Ziegler, A., Knesel, J., Fabbro, D. and Nagamine, Y. (1991) Protein kinase C down-regulation enhances cAMP-mediated induction of urokinase-type plasminogen activator mRNA in LLC-PK1 cells. *J. Biol. Chem.* 266: 21067-21074.
36. Ziegler, A., Hagmann, J., Kiefer and Nagamine, Y. (1990) Ca2+ potentiates cAMP-dependent expression of urokinase-type plasminogen activator gene through a calmodulin- and protein kinase C-independent mechanism. *J. Biol. Chem.* 265: 21194-21201.
37. Nanbu, R., Menoud, P.A. and Nagamine, Y. (1994) Multiple instability-regulating sites in the 3'untranslated region of the urokinase-type plasminogen activator mRNA. *Mol. Cell. Biol.* 14: 4920-4928.
38. Nanbu, R., Montero, L., D'orazio, D. and Nagamine, Y. (1997) Enhanced stability of urokinase-type plasminogen activator mRNA in metastatic breast cancer MDA-MB-231 cells and LLC-PK1 cells down-regulated for protein kinase C--correlation with cytoplasmic heterogeneous nuclear ribonucleoprotein C. *Eu. J. Biochem.* 247:169-174.
39. Henderson, B. R., Tansey, W. P., Phillips, S. M., Ramshaw, I. A. and Kefford, R. F. (1992) Transcriptional and posttranscriptional activation of urokinase plasminogen activator gene expression in metastatic tumor cells. *Cancer Res.* 52:2489-2496.
40. Shetty, S. and Idell, S. (2000) Post-transcriptional regulation of urokinase mRNA: identification of a novel urokinase mRNA-binding protein in human lung epithelial cells *in vitro*. *J. Biol. Chem.* 275:13771-13779.
41. Bosma, P. J. and Kooistra, T. (1991) Different induction of two plasminogen activator inhibitor 1 mRNA species by phorbol ester in human hepatoma cells. *J. Biol Chem.* 266:17845-17849.
42. Fattal, P.G., Schneider, D. J., Sobel, B. E. and Billadello, J. J. (1992) Posttranscriptional regulation of expression of plasminogen activator inhibitor type 1 mRNA by insulin and insulin-like growth factor 1. *J. Biol. Chem.* 267:12412-12415.
43. Heaton, J.H., Tillmann-Bogush, M., Leff, N. S. and Gelehrter, T.D. (1998) Cyclic nucleotide regulation of type-1 plasminogen activator-inhibitor mRNA stability in rat hepatoma cells. *J. Biol. Chem.* 273:14261-14268.
44. Tillmann-Bogush, M., Heaton, J. H. and Gelehrter, T. D. (1999) Cyclic nucleotide regulation of PAI-1 mRNA stability. *J. Biol. Chem.* 274: 1172-1179.
45. Pedersen, H., Brunner, N. Franscis, D., Osterlind, K., Ronne, E., Hansen, H. H., Dano, K. and Grondahl-Hansen, J. (1994) Prognostic impact of urokinase, urokinase receptor, and type 1 plasminogen activator inhibitor in squamous and large cell lung cancer tissue. *Cancer Res.* 54: 4671-4675.
46. Pedersen, H., Grondahl-Hansen, J., Franscis, D., Osterlind, K., Hansen, H. H., Dano, K. and Brunner, N. (1994) Urokinase and plasminogen activator inhibitor type 1 in pulmonary adenocarcinoma. Cancer Res. 54: 120-123.
47. Shetty, S. and Idell, S. (2000) Posttranscriptional regulation of plasminogen activator inhibitor-1 in human lung carcinoma cells in vitro. Am. J. Physiol. 278:148-156.
48. Maurer, F., Tierney, M. and Medcalf, R. L. (1999) An AU-rich sequence in the 3'UTR of plasminogen activator inhibitor type 2 (PAI-2) mRNA promotes PAI-2 mRNA decay and provides a binding site for nuclear HUR. Nu. Acid Res. 27:1664-1673.

17

REGULATION OF G-PROTEIN COUPLED RECEPTOR CYTOSOLIC mRNA BINDING PROTEINS

Kathryn Sandberg, Zheng Wu, Hong Ji, Eric Hernandez, and Susan E. Mulroney

Georgetown University School of Medicine, Washington, DC

The regulation of G-protein coupled receptor (GPCR) expression is an area of intense interest because GPCRs play a crucial role in many key physiological and pathological processes. Indeed, greater than 60% of all pharmaceuticals are targeted towards GPCRs. The ability to regulate GPCR protein expression at the post-transcriptional level provides a mechanism past the transcriptional regulatory point when genes are turned on or off. This allows a finer control of expression for the GPCRs that are critically involved in a myriad of critical cell functions such as growth, differentiation, and development. In the majority of instances of post-transcriptional regulation, the binding of specific proteins to defined sequences and/or structures in the target mRNA, governs splicing, nucleocytoplasmic transport, subcellular localization, translation and mRNA degradation. In this chapter, we focus on the regulation of GPCRs via cytosolic RNA binding proteins that, through interaction with specific sequences or structures in the mRNA, contribute to determining the level of protein expression by altering the rate of mRNA translation or its stability. Highlighted are our studies on RNA binding proteins that interact with the 5' leader sequence of the angiotensin AT_1 receptor, a GPCR that is critical to the control of blood pressure and fluid homeostasis.

INTRODUCTION

In conjunction with transcriptional regulation (synthesis), the rate at which a GPCR mRNA is translated or degraded can markedly affect protein abundance and consequently, receptor function. Although complexities of GPCR transcriptional regulation are becoming better understood, post-transcriptional regulation remains much less so. Post-transcriptional mechanisms are of particular interest since GPCR mRNA expression needs to be tightly controlled by the cell, such as for proto-oncogenes, transcription factors, and key signaling molecules. Therefore, understanding what regulates GPCR mRNAs is central to understanding their overall gene expression. This is still a field in its infancy and consequently, only a few GPCRs have been shown to undergo post-transcriptional regulation at the level of mRNA translation (1) and stability (2). However, the superfamily of GPCRs share many properties in common and thus, it is likely that the types of post-transcriptional mechanisms that are currently being characterized for the angiotensin, lutropin releasing hormone, and ß-adrenergic receptors (ßAR), are mechanisms for regulating GPCRs, in general.

RNA sequences recognized by regulatory proteins are frequently located within discrete regions of the mRNA including the 5' leader sequence (5'LS), coding region (CR) and 3' untranslated region (3'UTR). The classic example of an mRNA controlled by the 5'LS is observed with ferritin (see chapter 12). Recent reports from our lab and others suggest similar mechanisms may also regulate the AT_1 angiotensin receptor (AT_1R). RNA binding proteins (BPs) interacting with the open reading frame, include tubulin, *c-myc* and *c-fos*. The luteinizing hormone/human chorionic gonadotropin receptor is the first reported example of a GPCR in which RNA BPs are shown to interact within the coding region. The best studied RNA-protein interactions in GPCRs are found in the 3'UTR. RNA BPs in the 3'UTR of the ßAR (3-8) have been studied extensively but reports of RNA BPs interacting in the AT_1R 3'UTR (see chapter 6) are not far behind. Highly regulated mRNAs can also contain RNA binding determinants in a combination of regions of the mRNA. For example, *c-fos* and *c-myc* contain regulatory sequences in both the CR and the 3'UTR (9,10) and most recently, we (11) and others (12) have found that sequences in the 5'UTR effect regulatory mechanisms in the 3'UTR of the AT_1R. Based on the exponential growth in the number of reports of post-transcriptional regulation of GPCRs in the last decade, it is clear that control of GPCRs by mRNA BPs is an exciting new field that will yield important information.

THE 5' LEADER SEQUENCE

After post-transcriptional modifications, mRNAs are transported to the cytoplasm where they can be present as translationally inactive (masked) or active (polysomal) forms (13). The masked forms are primarily bound in cytoplasmic RNA-binding protein complexes (RBC). Exchange between active and inactive compartments can modulate mRNA translation (14). Presumably, this is determined by the interaction of mRNA with cytoplasmic

RNA-BPs, such as those responsible for structural organization, translation initiation, specific translational repression or activation and mRNA stability (13,15). Specific protein/mRNA interaction *cis*-elements include the m^7G cap (m^7GpppX, where X is any nucleotide) (15), the initiator codon AUG and surrounding context (e.g., XXXAUGX, where X is any nucleotide) (16), the 5'LS (17), upstream short open reading frames (sORFs) (18), the CR (10,19,20) as well as the 3'-UTR and poly(A+) tail (21). In the 5'LS, the majority of known *trans*-acting proteins act as negative regulators (translational repressors) primarily by inhibiting translational initiation or controlling mRNA entry into polysomes (15). There are also reports indicating that the 5'LS can alter the stability of the mRNA (11,12). Still much remains unknown regarding the molecular events involved in regulating the translational machinery.

One of the best-characterized examples of post-transcriptional regulation by RNA BPs interacting with the 5'LS is the synthesis of ferritin and erythroid 5-aminolevulinate synthase (eALAS). During iron-starvation, the iron responsive binding protein (IRP), a cytosolic protein of 110 kDa, binds to the iron responsive element (IRE) in the 5'LS of ferritin mRNA (see chapter 12). Formation of the IRE-IRP complex prevents the association of the 43S translational pre-initiation complex and results in decreased translation of ferritin. Thus, cellular iron homeostasis is regulated post-transcriptionally through an iron dependent IRE-IRP interaction (22,23). Through the same mechanism, the IRP binds to an IRE located in the 5'LS of eALAS, which is an enzyme involved in heme synthesis and consequently, in repressing the translation of eALAS (24).

Studies in our laboratory indicate that similar mechanisms may govern the AT_1R, which is a GPCR that plays a crucial role in control of blood pressure. The GC content in the 5'LS of AT_1Rs is particularly high and thus, the 5'LS is likely to form stable secondary structures such as hairpins, bulges, internal loops and junctions as predicted by the Zuker algorithm (25), which could serve as a templates for RNA BP recognition (26) (Fig. 1). In fact, RNA electromobility shift assays (EMSA) demonstrate that the 5'LS of the AT_1R (AT_1R-5'LS) binds to proteins in rodent cytosolic extracts yielding an RNA-BPC (Fig. 2). These AT_1R-5'LS BPs were detected in tissues in which the AT_1R is highly distributed, such as brain, pituitary, lung, spleen, adrenal, uterus, vascular smooth muscle cells and kidney (27). Under physiological conditions in which AT_1R binding sites are regulated on the cell surface, the change in AT_1-5'LS BP activity inversely correlated with receptor expression.

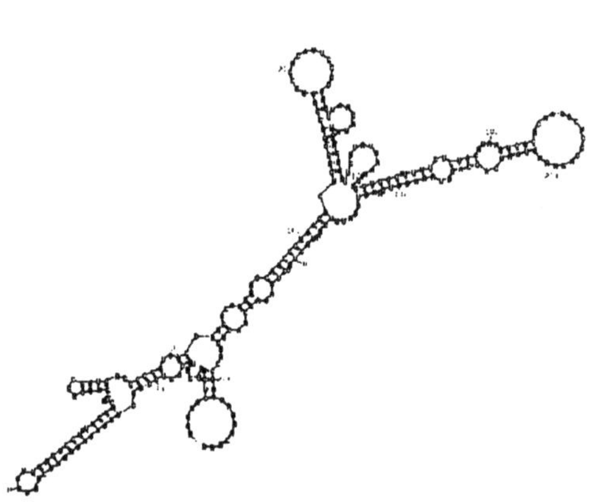

Figure 1. Graphical representation of 5'LS RNA secondary structure predicted by the Stemloop program using the Zuker algorithm. *Secondary structure of the 5'LS of $AT_{1a}R$ (271 nt) shows the formation of hairpin loops, bulges and internal loops that could serve as templates for RNA BP recognition and binding.*

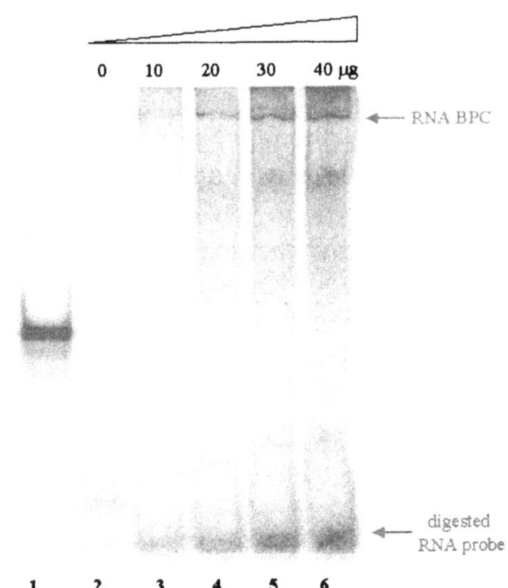

Figure 2. Detecting adrenal cortex AT_1R 5'LS BPC formation. *RNA electromobility shift assay was performed by incubating rat adrenal cortex cytosolic extract with 100,000 cpm of ^{32}P-$AT_{1a}R$ 5'LS at 30°C for 20 min followed by incubating with T1 RNAse and heparin sulfate at 30°C for 15 min. Electrophoresis was carried out at 200 volt for 3 hrs on a 4% polyacrylamide gel. Lane 1, probe alone; Lanes 2 to 6, increasing amounts (0, 10, 20, 30, and 50 µg) of adrenal cortex cytosolic protein; RNA-BPC and digested free probe are indicated by arrows.*

Our previous studies showed that deoxycorticosterone acetate (DOCA) treatment significantly reduced AT_1R-5'LS BP activity in the rat forebrain and hypothalamus, but had no effect on AT_1R-5'LS BP activity in pituitary (27). These results are consistent with reports that DOCA significantly increased AT_1R expression in the brain, but had no effect on receptor expression in the pituitary (27). These data complement our recent studies showing that estrogen up-regulates adrenal AT_1-5'LS BP activity by nearly 2-fold (Fig. 3A) under conditions in which adrenal AT_1R numbers are decreased by 35% in ovariectomized rats (Fig. 3B) (28). The fact that both physiologically mediated up- and down- regulation of AT_1R expression correlates reciprocally with alterations in AT_1R-5'LS BP activity in a tissue-specific manner suggests that 5'LS BPs regulate AT_1R translation. This hypothesis is consistent with *in vivo* studies of polysome distribution analysis in the adrenal cortex. Under conditions in which estrogen significantly increased AT_1R-5'LS BP activity and decreased AT_1R number, the fraction of AT_1R mRNA associated with polysomes decreased and the fraction associated with monosomes increased suggesting that the AT_1R mRNA was translated less efficiently in the adrenal cortex after estrogen treatment (Fig. 4).

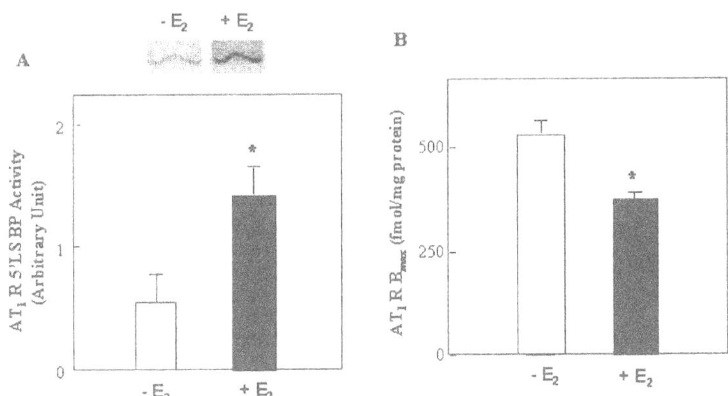

Figure 3. Regulation of adrenal AT_1R 5'LS BP activity and AT_1R expression by estradiol treatment in ovariectomized rats. *(A) RNA EMSA was performed by incubating 100,000 cpm of ^{32}P-AT_1R 5'LS RNA probe with 40 μg of adrenal cortex cytosolic extract prepared from OVX rats ± E_2 (40 μg/kg x 8 days). Autoradiograms (upper panel) were quantitated by densitometry (lower panel). The densitimetric units in the +E_2 group are expressed as a percentage of the average densitometric units in the -E_2 group (2 animals/group, n=3). * $p<0.05$ between ± E_2 treatment groups. (B) Adrenal AT_1R density measured by radioligand saturation binding assays. Membranes were prepared from adrenals of ovariectomized rats injected ± E_2 (40 μg/kg x 8 days) and incubated with ^{125}I-[Sar^1, Ile^8]Ang II in presence of the AT_2 R antagonist PD123319. The Bmax was determined by computerized nonlinear regression analysis. Data were averaged from 3 groups in duplicate and expressed as the mean ± SEM. *$p<0.01$ between ± E_2 treatment groups.*

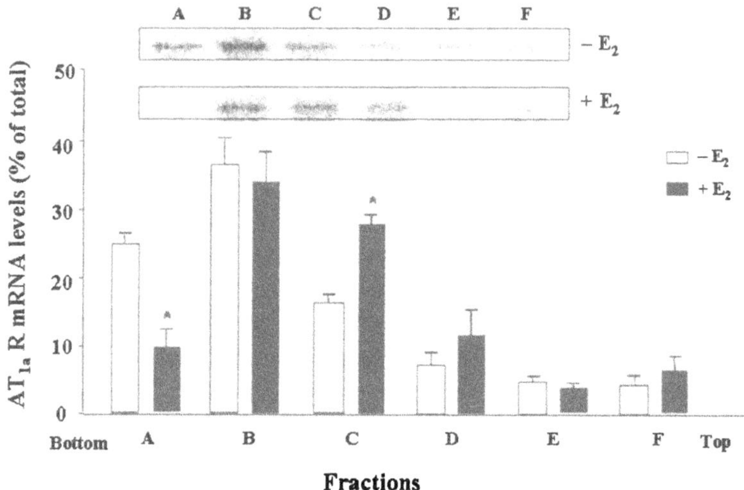

Figure 4. Effect of estrogen on AT_1R mRNA polysome distribution in the adrenal cortex. *Adrenal cortex from ovariectomized rats treated ± estrogen (40 µg/kg/day x 8 days) was homogenized and loaded onto 10-50% sucrose gradients. Six fractions were collected from the bottom to the top of the gradient and RNA was isolated from each fraction. The amount of AT_1R mRNA was determined in each fraction by RNase protection assay. The bottom of the gradient is the heaviest and thus, represents the highest number of polysomes bound to mRNAs while the top of the gradient possesses the smallest number of monosomes. Values are presented as a percentage of the total signal from ± E_2 treated groups (6 animals/group, n = 3). $p<0.05$.*

There are several possible mechanisms that can be envisioned by which 5'LS BPs regulate AT_1R mRNA translation. During translational initiation, the 40S small ribosomal subunit associates with the ternary complex including translational initiation factor eIF-2, GTP and Met-tRNAi to form a 43S pre-initiation complex at the 5' end of the mRNA. The pre-initiation complex migrates along the 5'LS until it reaches the first AUG codon, whereupon a 60S ribosomal subunit joins, the first peptide bond is formed and protein synthesis occurs (Fig 5A). The initiation of translation in eukaryotes is influenced by several aspects of mRNA structure: (a) the 5'm^7G cap; (b) secondary structure formed in the leader sequence; (c) the primary sequence or context and secondary structure surrounding the AUG codon; (d) upstream AUG codons and sORFs; and (e) the length of the leader sequence (29,30). RNA BPC formation at highly stable secondary structures in close proximity to the 5' m^7G cap site could block cap binding, RNA unwinding and installation of the 43S translational pre-initiation complex by steric hindrance (Fig. 5B). Alternatively, RNA BPs interacting close to the 5' cap site could facilitate translational initiation by activating translation initiation factors like eIF4F, which is the initiation factor that regulates the function of the cap. The rate-limiting step in translation is the binding of the 40S ribosome to a mRNA, which is facilitated by the presence of the cap. eIF4F is composed of eIF4A, eIF4E and eIF4G (see Sonnenberg for in depth reviews (31)). eIF4A is an RNA helicase that unwinds RNA secondary structure. eIF4E is the cap binding subunit thought to play a crucial role in regulating translation initiation. eIF4G bridges between eIF4A and eIF4E. RNA BPs could also inactivate translational inhibitors like eIF4E. In this regard, it is interesting that Ang II via its AT_1R in vascular smooth muscle cells stimulates phosphorylation of the translational repressor 4E-BP1, which leads to its dissociation from eIF4E and consequently, relieves translational inhibition (32). If secondary structures are further downstream of the 5' end, then RNA-BPC formation could inhibit translation by blocking the scanning of the 43S pre-initiation complex toward the downstream major open reading frame (AUG) (27,33) (Fig 5C).

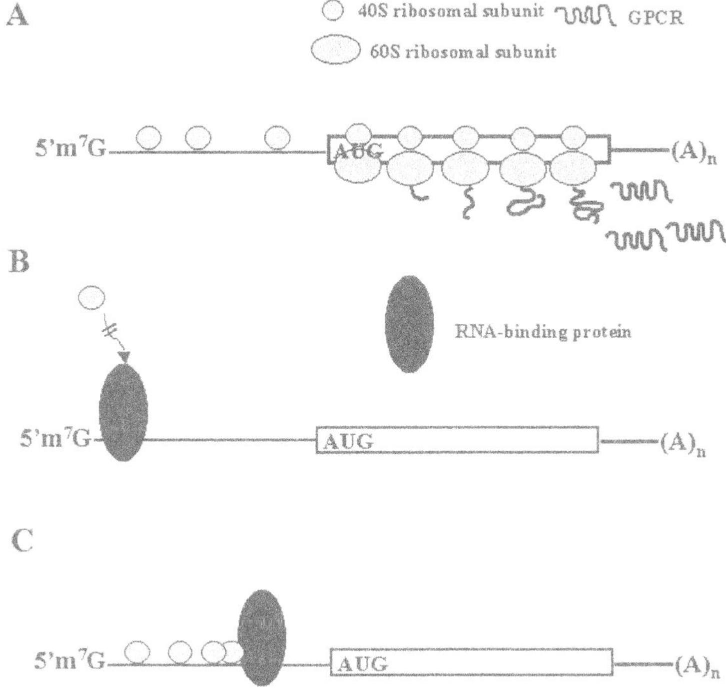

Figure 5. Inhibition of AT_1R translation by proteins that bind to secondary structures in the 5'LS RNA. *Rectangles and lines represent major open reading frame and untranslated regions, respectively. (A) Model of protein translation. (B) When RNA secondary structure exists close to the 5' cap, induced steric hindrance from RBCs could block cap-binding and installation of the translation pre-initiation complex. (C) RNA BPCs at downstream secondary structures in the 5'LS could prevent scanning of the translation pre-initiation complex toward the major open reading frame, thus inhibiting AT_1R translation.*

In addition, RNA-BPC forming near or at upstream sORFs could inhibit translation of the major open reading frame by yet unknown mechanisms, perhaps by encouraging the ribosomes to initiate at the upstream AUGs, thereby inhibiting translation efficiency of the major open reading frame as seen in the well-studied regulation of GCN4, a transcription factor in yeast that controls amino acid biosynthesis. Translation of upstream sORFs in the GCN4 mRNA markedly reduces the translation efficiency of the transcription factor (34). Upstream sORFs (or minicistrons) exist in all the mammalian AT receptors cloned to date (35).

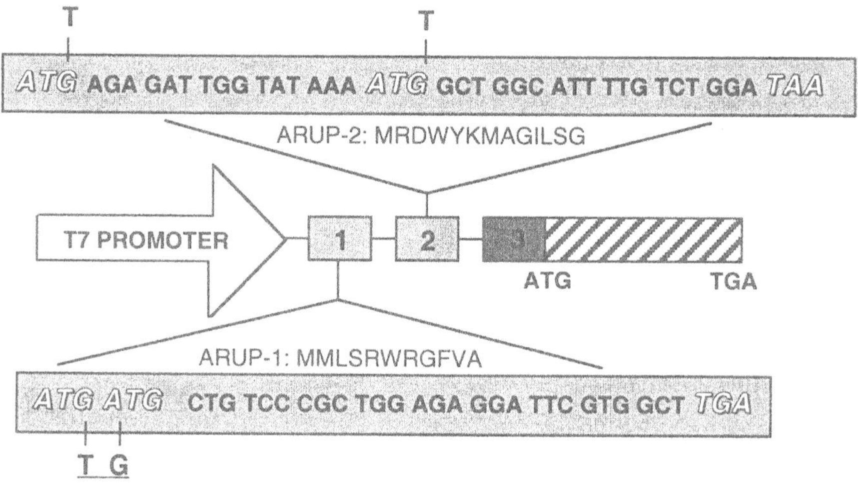

Figure 6: The cDNA structure of the rAT$_{1a}$R QM mutant gene. *The nucleotide sequences of two 5'LS sORFs are shown in boxes. Four upstream ATGs in exons 1 and 2 are inactivated by site-directed mutagenesis. The 3'UTR is not included in the WT or QM mutant cDNAs.*

We have found that disruption of all four upstream AUGs and consequently, the three sORFs in the rat AT$_{1a}$R 5'LS (Fig. 6), increased translation of the receptor *in vitro* by 4-fold (Fig. 7A). Furthermore, while no effects of AUG disruption were observed on AT$_1$R mRNA levels in transiently transfected A10 cells (Fig. 7B), the AT$_1$R number, as determined by Scatchard analysis, doubled compared to the wild type-transfected cells (Fig. 7C). At this point, it is not known if AT$_1$R-5'LS BPs act independently of the role of sORF in suppression of AT$_1$R translation or conjointly. The effect of disrupting sORFs in the AT$_1$R is reminiscent of studies on the type 2 ßAR (ß$_2$AR). Mutational inactivation of the sORF in the ß$_2$AR significantly increased *in vitro* translation of the receptor and also significantly increased receptor expression and translation in transfected COS-7 cells (1). Preliminary studies reveal RNA BPC formation with the

5'LS of the β_2AR, neuropeptide Y receptor, gonadotropin releasing hormone receptor and the corticotropin releasing hormone receptor (*data not shown*).

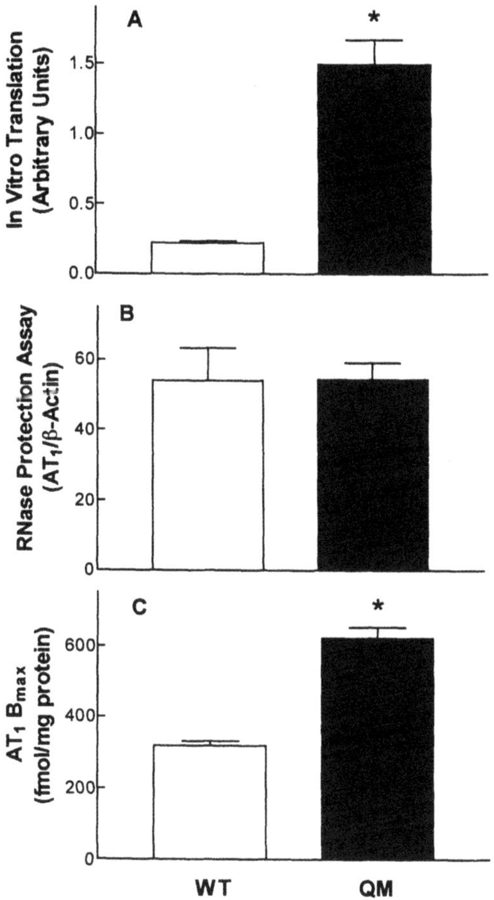

Figure 7: Comparison of QM and WT translation and steady state levels of mRNA. *(A) In vitro translation of WT and QM mRNAs in wheat germ extracts. Total capped RNAs (0.5 µg) were translated for 90 min at 25°C. The translated proteins were analyzed by 15% SDS-PAGE. Autoradiograms of SDS gels were quantified by phosphoimaging and expressed in arbitrary units. The data are the mean ± S.E.M. from 4 experiments. *$p<0.01$. (B) AT_1R RNase protection assays were performed on WT and QM transfected A10 cells. After 48 h transfection, total RNA was extracted from cells and AT_1R mRNA was quantitated by RNase protection assay. Data shown are the ratios of AT_1 vs. β-actin mRNAs. (C) AT_1R number was determined in WT and QM transfected A10 cells. Membrane proteins were incubated with increasing concentrations of ^{125}I-[Sar^1, Ile^8]-Ang II in the presence of a saturating concentration of the AT_2R blocker, PD-123177 in the presence and absence of Ang II (200 nM). The AT_1R B_{max} values were calculated from Scatchard plots. The statistical analysis was done by one-way ANOVA followed by a post hoc Dunnett's t test. *$p<0.01$.*

Studies on the CRH R reveal a similar relationship as in the AT_1R between changes in CRH R expression and binding. Long-term adrenalectomy (ADX) is associated with marked down-regulation of pituitary type-1 corticotropin releasing hormone receptors (CRH-R1) but normal CRH-R1 mRNA levels, suggesting that regulation of receptor levels occurs at post-transcriptional sites. The 5'LS RNA of the CRH-R1 underwent RBC formation with pituitary extracts as observed in RNA electromobility shift assays (36). Interestingly, competition studies and UV-crosslinking analysis suggest that at least some proteins in the $AT_{1a}R$ and CRH-R1 RBCs are common to both receptor mRNAs. Importantly, one week after ADX, CRH-R1 5'LS BP activity in the pituitaries showed a significant 2-fold increase compared to BP activity in pituitaries from SHAM operated animals. This effect was prevented by glucocorticoid replacement (Fig. 8A). In contrast, no differences in the number of $AT_{1a}R$ binding sites or $AT_{1a}R$ 5'LS BP activity were observed between sham and ADX animals, indicating that the effect of ADX was specific for the CRH-R1 mRNA (Fig. 8B). The studies suggest that the binding of protein complexes to the 5'LS of the CRH-R1 is regulated by physiological alterations in the HPA axis and that the consequence is translational regulation of CRH-R1 mRNA. Further insights regarding the role of GPCR RNA BPs will be greatly aided by identification of the specific *cis* elements and secondary structures recognized by the 5'LS BPs as well as purification and cloning of these RBPs.

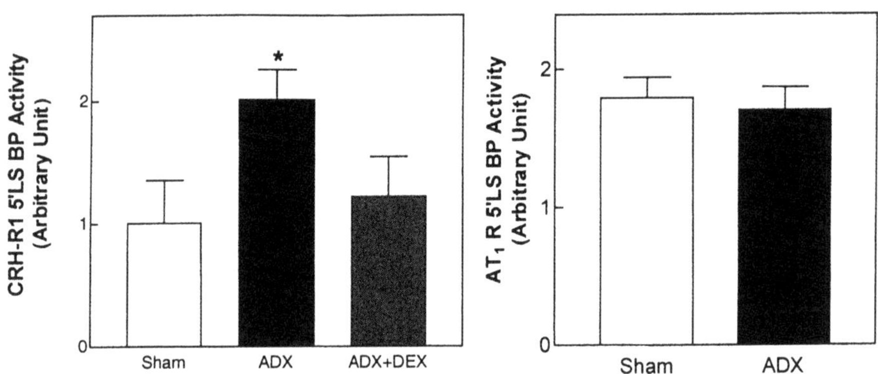

Figure 8. Regulation of CRH-R1 and AT_1R 5'LS BP activity by ADX. *RNA EMSA was performed by incubating pituitary cytosolic extracts with 100,000 cpm of ^{32}P-5'LS RNA probe of the CRH-R1 (A) or AT_1R (B). Autoradiograms were quantitated by densitometry. Values are the mean ± SEM (2 animals/group, n=5). * $p<0.01$ between sham and ADX.*

THE CODING REGION

Sequences in the open reading frame of mRNAs have been shown to regulate mRNA stability. A destabilizing pyrimidine-rich sequence (a.k.a., the coding region determinant, CRD) in the carboxyl-terminal open reading frame of *c-myc* governs its stability (7). When the CRD-BP, a 70 kDa protein, binds to the CRD, it stabilizes the RNA by protecting it from endonucleolytic cleavage (37,38). The rapid autoregulated decay of ß-tubulin mRNA is triggered in the presence of excess tubulin monomers by a signal in the amino-terminal tetrapeptide encoded by the ß-tubulin mRNA (5,39). Interestingly, destabilization of polysomal ß-tubulin mRNA requires translation of the first 41 codons (40,41). There are two destabilizing regions in the open reading frame of *c-fos*, in addition to the AU-rich element in the 3' UTR (9). One of these regions is composed of 320 nucleotides and is located near the middle of the mRNA. This *c-fos* sequence added to α-globin resulted in a significant decrease in stability of the chimeric mRNA (9,42). In addition, a purine-rich segment within this region has been shown to be the site of interaction for two proteins (6). As with tubulin, the induction of *c-fos* mRNA decay requires translation through the open reading frame destabilizing sequences (43).

Similar mechanisms may also be regulating the receptor mRNA for luteinizing hormone (LH) or its human placental counterpart, human chorionic gonadotropin (hCG). The LH/hCG receptor belongs to the family of GPCRs that mediate their effects through Gs and the adenylate cyclase signaling pathway. After an endogenous preovulatory LH surge or the administration of a pharmacological dose of hCG, LH/hCG receptors expressed on ovarian granulosa cells of preovulatory follicles and luteal cells are rapidly down-regulated and which is accompanied by a parallel loss in LH/hCG receptor mRNA (44). Previous studies have shown that this rapid loss in steady state levels of LH/hCG mRNA occurs via a 3-fold decrease in mRNA half-life rather than by a decrease in receptor transcription (45).

Kash and Menon have shown that two RBCs, LRBP-1 and LRBP-2, bind sequences within the open reading frame of the LH/hCG receptor mRNA (46,47). While LRBP-2 is not specific for the LH/hCG receptor mRNA, LRBP-1 partially purified from pseudopregnant rat ovary, displays some selectivity towards a 203-220 region within the coding region (Fig 9). In order to assess the specificity of LRBP-1 towards the LH/hCG receptor mRNA, it will be important in future studies to rigorously test the specificity of this RBC in RNA gel shift and UV-crosslinking competition assays using various regions of the LH/hCG receptor mRNA and other GPCR mRNAs.

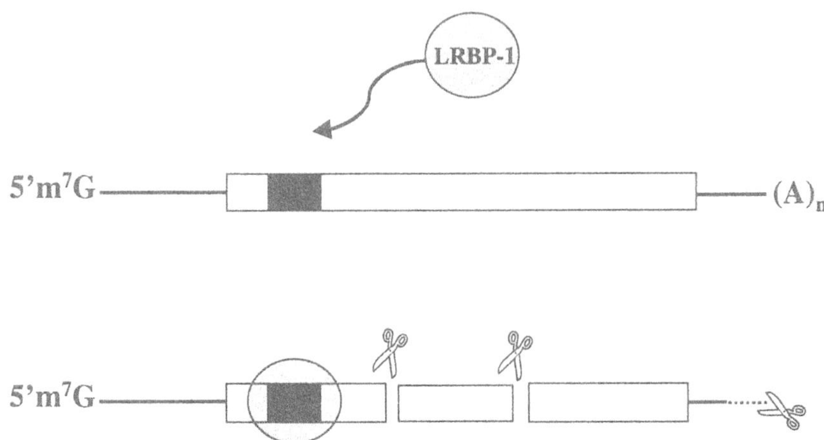

Figure 9. Regulation of LH/hCG receptor mRNA stability by binding of LRBP-1 to the coding region. *Rectangles and lines denote coding region and untranslated regions, respectively. A 50 kDa protein, LRBP-1, binds to the coding region of the LH/hCG receptor mRNA between ribonucleotides 203 and 220 (black box). Upon binding, the LH/hCG receptor mRNA is targeted for degradation by endonucleases thereby increasing the mRNA decay rate.*

There are four splice variants (6.7, 4.4, 2.6 and 1.8 kb) of the LH/hCG receptor mRNA in the ovary (48). The 6.7, 4.4 and 2.6 kb transcripts are variants involving the 3'UTR. These three transcripts all contain the 2.1 kb open reading frame encoding the full-length receptor. Evidence suggests that the smallest transcript (1.8 kb) contains the amino-terminal ligand binding domain and the carboxyl-terminal intracellular tail (49; 50). All four of the transcripts are down-regulated within 12 h and become undetectable by 24 h; by 24-48 h, mRNA expression recovers. The apparent LRBP-1 binding site (203-220 nt) is found in all four ovarian transcripts, which has led to the hypothesis that *control of LH/hCG receptor mRNA stability during hCG-induced down-regulation involves the LRBP-1 RBC*. A 17- nucleotide domain that is approximately 50% cytosine exists within this region. While poly(A), poly(U) and poly(G) were ineffective competitors for binding of LRBP-1 to this region, poly(C) competed with high affinity.

Cytosine-rich RNA sequences are known to play a role in control of mRNA stability. Three cytidine-rich elements in the 3'UTR of α-globin mRNA mediate its stability in erythrocytes (51). Two poly(rC) RNA BPs (αCP-1 and αCP-2) interact with the cytidine-rich region in the α-globin 3'UTR resulting in enhanced mRNA stability (52-54). Removal of these cytidines by site-directed mutagenesis resulted in rapid degradation of the α-globin *in vivo* (53). Tyrosine hydroxylase mRNA is stabilized during hypoxia in PC12 cells by a (U/C)(C/U)<u>CC</u>CU *cis* element in which the most critical nucleotides are underlined (55,56). PolyC BPs, αCP-1 and αCP-2, bind this region in the tyrosine hydroxylase mRNA and are implicated in regulating its stability (57). While a highly homologous sequence to the αCP-1 and αCP-2 binding sites in tyrosine hydroxylase mRNA exists in the critical sequence involved in LRP-1 binding (5'-U_{203}CUC-X_7-UCUCCCU_{220}-3') as determined by RNA foot printing, polyclonal antibodies for αCP-1 and αCP-2 did not result in supershifted RNP-BPCs with the ovarian partially purified fraction. Therefore, it is unlikely that LRBP-1 isolated from rat ovary includes αCP-1 or αCP-2. In fact, the cytidines (C^{217}, C^{218}), which are critical to αCP-1 and αCP-2 binding in tyrosine hydroxylase mRNA are least important for LRBP-1 binding (47).

There is much to be discovered regarding the association of LRBP-1 with the LH/hCG receptor mRNA. The possibility exists that the binding of LRBP-1 to sequences in the receptor open reading frame could influence its mRNA stability, thereby finely controlling LH/hCG receptor expression. It will be critical to purify the protein(s) involved in the LRBP-1 RBC to ultimately understand their role in post-transcriptional mechanisms regulating hCG-induced receptor down-regulation. In this regard, it is important to keep in mind that a RNA-BPC on a non-denaturing gel could include several proteins complexed together. Furthermore, some components could be lost during protein purification and so reconstitution experiments will be necessary to assess LRBP-1 functionality. The caveats also remain that other unidentified proteins or mRNA sequences are involved in LH/hCG mRNA stability and that the LRBP-1 is not a specific regulator of the LH/hCG receptor mRNA especially since, rigorous examination of the LRBP-1 specificity has not yet been completed. Only one region of the LH/hCG receptor (711-1301) was shown not to compete in the gel shift assay. Extensive studies need to be carried out with other regions including deletion of the region considered important and use of the remaining sequence as a competitor. Furthermore, UV-crosslinking studies need to be performed to investigate the potential proteins involved in the LRBP-1 RBC and in conjunction with UV-crosslinking competition assays. Another issue that should be addressed is the tissue-specific expression of the LRBP-1. It is important to note that the gel shift assays merely detect activity and not protein abundance. Therefore, it is possible that LRBP-1 is present in many tissues but that its not able to bind with high affinity either because an inhibitor is present, or some tissues have more protease activity, or more RNase activity. [if the radiolabeled RNA probe or RNA BPs are degraded, then this will radically inhibit RNA-BPC formation]. These concerns are more a consideration in tissues such as the kidney, which is notorious for having high levels of RNase and protease activity. At this point, the evidence is indirect regarding the role of LRBP-1 in mRNA stability. Direct evidence will require cloning and/or protein purification to assess the role of these *trans* factors in regulating LH/hCG mRNA turnover.

THE 3' UNTRANSLATED REGION

Cytosolic proteins that bind to the 3'UTR are predominantly associated with regulation of mRNA stability and turnover. The mRNA decay rate (half-life) is a major determinant of mRNA abundance in organisms from bacteria to mammals. mRNA levels can fluctuate many-fold following a change in mRNA half-life without any changes in gene transcription and these fluctuations affect how a cell grows, differentiates and responds to its environment (58-61). There are several different types of 3'UTR BPs. *Poly(A) BPs:* Deadenylation is the first step in the decay of many mRNAs. The formation of a poly(A)-BP (PABP)-poly(A) RBC at the 3' terminus of mRNAs protects them from rapid destruction by 3' to 5' nuclease activity (62-64) (Fig. 10A). *Proteins that bind to A + U-rich elements (ARE):* The 3'UTR of many unstable mammalian mRNAs contain regions rich in adenosine and uridine. Two domains are identified in the ARE. Domain 1 contains 40-50 nucleotides, is AU rich and includes several AUUUA pentamers. Domain II contains a 20 nucleotide U-rich region. A family of ARE-BPs (ABPs) has been described which are known to destabilize the mRNA (65-67) (Fig. 10B). *Proteins that bind to stem-loop or other secondary structures in the 3'UTR:* For example, a 50 kDa protein was found to bind with high specificity to a stem-loop structure located in the 3' terminus of the histone mRNA and affect mRNA transport, translation and half-life (68) (Fig. 10C). *Binding of proteins with different functions to the same 3'UTR:* Several proteins were identified that bind to the 3'UTR of the mRNAs encoding subunits 1 and 2 of ribonucleotide reductase (RR). Proteins of 57 and 45 kDa bind to RR1 and RR2 mRNAs and are observed to be mRNA destabilizers, while a protein of 75 kDa binds to a different 3'UTR segment in the RR2 mRNA and functions as a stabilizer. It is intriguing that p45 and p75 bind to the same 3'UTR but affect the mRNA half-life in different ways: these proteins work together to regulate the stability of the RR mRNA (69,70) (Fig 10D).

One of the first RNA BPs shown to interact with GPCRs is AUF1, which is also one of the first to be purified, cloned and characterized (71). AUF1 binds to AREs in the AT_1R (see chapter 6 for extensive discussion) and in the ß-AR. In the AT_1R, AUF1 binding is associated with mRNA destability. AUF1 is also known to bind in the 3'UTR of other mRNAs that are tightly regulated such as proto-oncogenes and cytokines. Interestingly, AUF1 may have different functions depending upon the mRNA since in α-globin, AUF1 is part of the mRNA stability complex. Another ARE BP called ßARB, because it binds to the ß-AR, has been purified and is distinct from AUF1 (72). ßARB also binds to other GPCR including the $ß_1$AR (73,74) and the thrombin receptor (75). It is known that ßARB is composed of HuR and hnRNP A1 but it may also possess as yet unidentified mRNA BPs. In this regard, it will be important to characterize the interactions between ßARB and AUF1 in governing GPCR mRNA stabilities. While it is still poorly understood, the accumulating evidence suggests that HuR and hnRNP play a stabilizing role in $ß_2$AR mRNA half-life (72). However, the evidence remains indirect and thus the caveats remain that HuR and/or hnRNP are not directly involved in regulating mRNA half-lives or that these RNA BPs function differently depending upon the mRNA substrate.

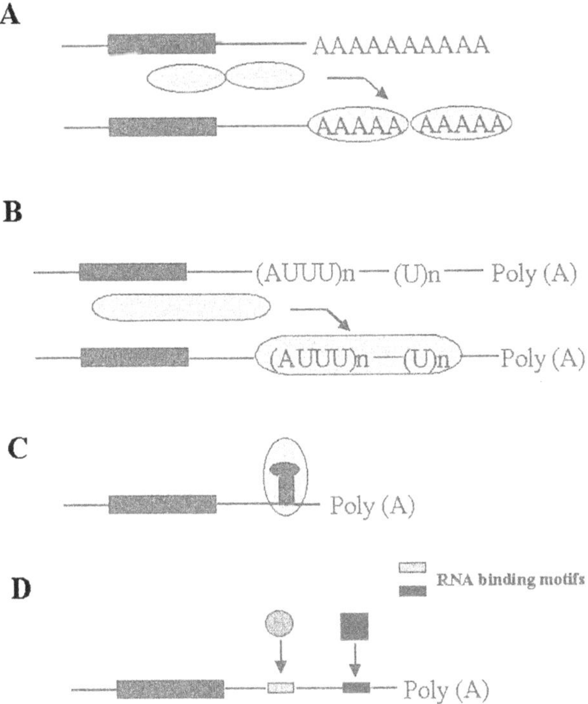

Figure 10. Influence of mRNA stability by 3'UTR BPs. Rectangles and lines denote coding region and untranslated regions, respectively. (A) Poly(A)-BPs protect the poly(A) tract and, thereby, the remainder of the mRNA from rapid degradation. (B) A family of proteins binds to AU- and U- rich elements thought to alter mRNA stability. (C) A protein binds to the stem-loop structure located in the 3' terminus of histone mRNA and influences the transport of the mRNA from the nucleus to the cytoplasm and the translation rate and half-life of the mRNA in the cytoplasm. (D) Two proteins (noted by the arrows) bind to distinct sequences in the 3'UTR of the mRNA encoding ribonucleotide reductase 2. When bound to the mRNA, one of the proteins acts as a stabilizer, while the other acts as a destabilizer. Adapted from Jeff Ross (77).

Recently, Nickenig et al, (76) demonstrated that the AT_1R interacts with at least six polysomal proteins in the 3'UTR within an ARE-rich region (2175-2195 nt). Agonist stimulation enhanced binding of the 50 and 60 kDa proteins and was associated with rapid receptor down-regulation. Thus, it is possible that multiple 3'UTR BPs play a role in regulating the expression of the AT_1R by altering the stability of the mRNA. Once again, the caveats discussed above with the LRBP-1 RNA BPs are particularly relevant for these newly discovered AT_1R 3'UTR RNA BPs and should be considered when evaluating the roles of these proteins in regulating gene expression.

CONCLUSIONS

The discovery of RNA BPs that interact with GPCR mRNAs is an exciting and new field that is bound to be of crucial importance in understanding GPCR gene regulation. It goes without saying that further characterization of these proteins will be greatly facilitated by their purification and cloning, and that questions of specificity and function will remain until direct actions are demonstrated for these GPCR RNA BPs. It is more difficult to isolate and clone new RNA BPs compared to DNA BPs for many reasons. RNA structure is more complex than DNA structure and considerably more variable. While RNA can form similar Watson-Crick base pairing, RNA secondary structure frequently encompasses incomplete intramolecular and intermolecular pairing. This incomplete base pairing leads to a multitude of possible secondary structures from simple base pair mismatches to complicated complex structures like four way junctions and pseudoknots (see Fig. 1). RNA structures can breathe and alternative structures can co-exist with similar free energies. While transcription factors are commonly purified by affinity chromatography using the response element as the fishing hook, affinity chromatography with RNA is more problematic since structure can be just as important as sequence in RNA recognition, and complex structures can be lost on a column. Furthermore, RNase contamination is a chronic problem. Fortunately, there has been an explosion in new techniques devoted to RNA-protein interactions including yeast-three-hybrid systems and purification of RNA protein complexes by immunoprecipitation and identification of the RNA binding elements by PCR-based techniques. [Susan Haynes (78) and Christopher Smith (79) have recently edited excellent books on this subject.] Based on the available evidence thus far in other systems and in the GPCR field, it is entirely likely that cytosolic RNA BPs will generally be multifunctional as is found for the IRP. IRP functions as an aconitase in addition to its role in inhibiting translation of ferritin via an IRE in the 5'LS and in stabilizing transferrin mRNA via several IREs in the 3'UTR. It is exciting to be in the midst of a rapidly growing area of research and in one that is central to our understanding of this important superfamily of GPCRs. In addition to the new insights into GPCR gene regulation, it is also conceivable that studies in this field will lead to the development of novel therapeutic pharmaceuticals that target defects in RNA-BPs.

Summary of key concepts

- *The expression of GPCRs is regulated at the post-transcriptional level by altering the rate of mRNA translation and/or mRNA stability.*
- *Post-transcriptional regulation of GPCRs is achieved by binding specific proteins to defined sequences and/or structures in the 5'LS, CR or 3'UTR of the target mRNA.*
- *Changes in the activity of cytosolic proteins that bind sequences and/or structures in the 5'LS of the adrenal AT_1R and the pituitary CRH-R1 receptor inversely correlates with hormonally-induced changes in AT_1R and CRH-R1 Bmax levels in these tissues. Furthermore, changes in the activity of these RNA BPs parallels changes in the mRNA translational efficiency determined by polysome analysis.*
- *The activity of RNA cytosolic proteins that bind to sequences in the open reading frame of the LH/hCG receptor mRNA is inversely correlated with agonist-induced LH/hCG mRNA degradation.*
- *Several mRNA stability determinants are located in the 3'UTR of GPCRs. Binding of AUF1 and βARB to these determinants in the AT_1R and the β_2AR correlates with altered rates of mRNA degradation.*

Study Guide Questions

1) GPCRs fall into the category of highly regulated mRNAs such as proto-oncogenes, transcription factors, and signaling molecules. Post-transcriptional regulatory mechanisms have been characterized for AT_1Rs, LH/hCG receptors, and βARs. Hypothesize why other GPCRs might also be regulated post-transcriptionally?
2) If identical cytosolic proteins bound to the receptor 5'LS of the AT_1R and CRH-1 R mRNA, how then could you explain the receptor specific regulation of the 5'LS BP activities? How could 5'LS BPs regulate AT_1R expression?
3) The LRBP-1 binding determinant is composed of 50% cytosines and poly(C) is an effective competitor of RBC formation. Could polyC BPs be potential components of the LHBP-1 complex? What factors may influence detecting tissue-specific expression of LRBP-1?
4) In general, 3'UTR BPs regulate mRNA stability rather than translational efficiency. What are some of the pathways by which 3'UTR BPs can regulate GPCR expression?

Acknowledgments

Work done by the authors' laboratories is supported by NIH grants (HL57502 and AG/HL19291) to KS, (DK59652-01) to HJ and (AG18634-01) to SEM. Both KS and SEM are Established Investigators of the American Heart Association.

REFERENCES

1 Parola AL, Kobilka BK. 1994. The peptide product of a 5' leader cistron in the β_2 adrenergic receptor mRNA inhibits receptor synthesis. J. Biol. Chem. 269:4497-4505.
2 Hadcock JR, Wang HY, Malbon CC. 1989. Agonist-induced destabilization of beta-adrenergic receptor mRNA. Attenuation of glucocorticoid-induced up-regulation of beta-adrenergic receptors. J. Biol. Chem. 264:19928-33.
3 Port JD, Huang LY, Malbon CC. 1992. Beta-adrenergic agonists that down-regulate receptor mRNA up-regulate a M(r) 35,000 protein(s) that selectively binds to beta-adrenergic receptor mRNAs. J. Biol. Chem. 267:24103-24108.
4 Theil EC. 1994. Iron regulatory elements (IREs): a family of mRNA non-coding sequences. Biochem. J. 304:1-11.
5 Yen TJ, Gay DA, Pachter JS, Cleveland DW. 1988. Autoregulated changes in stability of polyribosome-bound beta-tubulin mRNAs are specified by the first 13 translated nucleotides. Mol. Cell Biol. 8:1224-1235.
6 Chen CY, You Y, Shyu AB. 1992. cellular proteins bind specifically to a purine-rich sequence necessary for the destabilization function of a c-fos protein-coding region determinant of mRNA instability. Mol. Cell Biol. 12:5748-5757.
7 Bernstein PL, Herrick DJ, Prokipcak RD, Ross J. 1992. Control of c-myc mRNA half-life in vitro by a protein capable of binding to a coding region stability determinant. Genes Dev. 6:642-654.
8 Shaw G, Kamen R. 1986. A conserved AU sequence from the 3' untranslated region of GM-CSF mRNA mediates selective mRNA degradation. Cell 46:659-667.
9 Shyu AB, Belasco JG, Greenberg ME. 1991. Two distinct destabilizing elements in the c-fos message trigger deadenylation as a first step in rapid mRNA decay. Genes Dev. 5:221-231.
10 Wisdom R, Lee W. 1991. The protein-coding region of c-myc mRNA contains a sequence that specifies rapid mRNA turnover and induction by protein synthesis inhibitors. Genes Dev. 5:232-243.
11 Ji H, Wu Z, Lee S, Zheng W, Verbalis JG, Sandberg K. 2000. Translational control in regulation of the renin angiotensin system. Comp. Biochem. Physiol. 126A (suppl 1):132a.
12 Xu K, Murphy TJ. 2000. Reconstitution of angiotensin receptor mRNA down-regulation in vascular smooth muscle. Post-transcriptional control by protein kinase a but not mitogenic signaling directed by the 5'-untranslated region. J. Biol. Chem. 275:7604-7611.
13 Spirin AS, 1996. Masked and translatable messenger ribonucleoprotein in higher eukaryotes. In *Translational Control.* J.W.B. Hershey, M.B. Mathews, N. Sonenberg, eds., Cold Spring Harbor, NY: Cold Spring Harbor Laboratory Press, 319-334.
14 Kozak M. 1989. The scanning model for translation: an update. J. Cell Biol. 108:229-241.
15 Standart N, Jackson RJ. 1994. Regulation of translation by specific protein/mRNA interactions. Biochimie. 76:867-79.
16 Kozak M. 1989. Context effects and inefficient initiation at non-AUG codons in eucaryotic cell-free translation systems. Mol. Cell Biol. 9:5073-5080.
17 Kozak M. 1991. A short leader sequence impairs the fidelity of initiation by eukaryotic ribosomes. Gene Expr. 1:111-115.

18 Geballe AP, 2000. Translational control by upstream open reading frames. In *Translation Control.* N. Sonenberg, J.W.B. Hershey, M.B. Mathews, eds., Cold Spring Harbor, NY: Cold Spring Harbor Laboratory Press, 595-614.
19 Mondino A, Jenkins MK. 1995. Accumulation of sequence-specific RNA-binding proteins in the cytosol of activated T cells undergoing RNA degradation and apoptosis. J. Biol. Chem. 270:26593-26601.
20 Cao J, Geballe AP. 1996. Coding sequence-dependent ribosomal arrest at termination of translation. Mol. Cell Biol. 16:603-608.
21 Adam SA, Nakagawa T, Swanson MS, Woodruff TK, Dreyfuss G. 1986. mRNA polyadenylate-binding protein: gene isolation and sequencing and identification of a ribonucleoprotein consensus sequence. Mol. Cell Biol. 6:2932-2943.
22 Rouault TA, Hentze MW, Haile DJ, Harford JB, Klausner RD. 1989. The iron-responsive element binding protein: a method for the affinity purification of a regulatory RNA-binding protein. Proc. Natl. Acad. Sci. U. S. A. 86:5768-5772.
23 Klausner RD, Rouault TA, Harford JB. 1993. Regulating the fate of mRNA: the control of cellular iron metabolism. Cell 72:19-28.
24 Gray NK, Hentze MW. 1994. Iron regulatory protein prevents binding of the 43S translation pre- initiation complex to ferritin and eALAS mRNAs. EMBO J. 13:3882-3891.
25 Zuker M. 1989. Computer prediction of RNA structure. Methods Enzymol. 180:262-288.
26 Kozak M. 1991. An analysis of vertebrate mRNA sequences: intimations of translational control. J. Cell Biol. 115:887-903.
27 Krishnamurthi K, Zheng W, Verbalis AD, Sandberg K. 1998. Regulation of cytosolic proteins binding cis elements in the 5' leader sequence of the angiotensin AT1 receptor mRNA. Biochem. Biophys. Res. Commun. 245:865-870.
28 Krishnamurthi K, Verbalis JG, Zheng W, Wu Z, Clerch LB, Sandberg K. 1999. Estrogen regulates angiotensin AT1 receptor expression via cytosolic proteins that bind to the 5' leader sequence of the receptor mRNA. Endocrinology 140:5431-5434.
29 Kozak M. 1987. An analysis of 5'-noncoding sequences from 699 vertebrate messenger RNAs. Nucleic Acids Res. 26:8125-8148.
30 Kozak M. 1999. Initiation of translation in prokaryotes and eukaryotes. Gene 234:187-208.
31 Sonenberg N, Hershey JWB, Mathews MB, 2000. Translational control of gene expression. Cold Spring Harbor: Cold Spring Harbor Laboratory Press.
32 Fleurent M, Gingras AC, Sonenberg N, Meloche S. 1997. Angiotensin II stimulates phosphorylation of the translational repressor 4E-binding protein 1 by a mitogen-activated protein kinase- independent mechanism. J. Biol. Chem. 272:4006-4012.
33 McCarthy JE, Kollmus H. 1995. Cytoplasmic mRNA-protein interactions in eukaryotic gene expression. Trends Biochem. Sci. 20:191-197.
34 Hinnebusch AG. 1994. Translational control of GCN4: an *in vivo* barometer of initation-factor activity. Trends Biochem. Sci. 19:409-414.
35 Sandberg K. 1994. Structural analysis and regulation of angiotensin II receptors. Trends Endo. Metab. 5:28-35.
36 Wu Z, Aguilera G, Zheng W, Sandberg K. 2000. Adrenalectomy regulates corticotropin-releasing factor receptor expression by regulating mRNA binding proteins (New Orleans, RNA Society 5th Annual Meeting), pp. 708.
37 Herrick DJ, Ross J. 1994. The half-life of c-myc mRNA in growing and serum-stimulated cells: influence of the coding and 3' untranslated regions and role of ribosome translocation. Mol. Cell Biol. 14:2119-2128.
38 Prokipcak RD, Herrick DJ, Ross J. 1994. Purification and properties of a protein that binds to the C-terminal coding region of human c-myc mRNA. J. Biol. Chem. 269:9261-9269.

39 Gay DA, Sisodia SS, Cleveland DW. 1989. Autoregulatory control of beta-tubulin mRNA stability is linked to translation elongation. Proc. Natl. Acad. Sci. U. S. A. 86:5763-5767.

40 Pachter JS, Yen TJ, Cleveland DW. 1987. Autoregulation of tubulin expression is achieved through specific degradation of polysomal tubulin mRNAs. Cell 51:283-292.

41 Schiavi SC, Wellington CL, Shyu AB, Chen CY, Greenberg ME, Belasco JG. 1994. Multiple elements in the c-fos protein-coding region facilitate mRNA deadenylation and decay by a mechanism coupled to translation. J. Biol. Chem. 269:3441-3448.

42 Shyu AB, Greenberg ME, Belasco JG. 1989. The c-fos transcript is targeted for rapid decay by two distinct mRNA degradation pathways. Genes Dev. 3:60-72.

43 Lu DL, Menon KM. 1996. 3' untranslated region-mediated regulation of luteinizing hormone/human chorionic gonadotropin receptor expression. Biochemistry 35:12347-12353.

44 Hoffman YM, Peegel H, Sprock MJ, Zhang QY, Menon KM. 1991. Evidence that human chorionic gonadotropin/luteinizing hormone receptor down-regulation involves decreased levels of receptor messenger ribonucleic acid. Endocrinology 128:388-393.

45 Lu DL, Peegel H, Mosier SM, Menon KM. 1993. Loss of lutropin/human choriogonadotropin receptor messenger ribonucleic acid during ligand-induced down-regulation occurs post transcriptionally. Endocrinology 132:235-240.

46 Kash JC, Menon KMJ. 1998. Identification of a hormonally regulated luteinizing hormone/human chorionic gonadotropin receptor mRNA binding protein. J. Biol. Chem. 273:10658-10664.

47 Kash JC, Menon KM. 1999. Sequence-specific binding of a hormonally regulated mRNA binding protein to cytidine-rich sequences in the lutropin receptor open reading frame. Biochemistry 38:16889-16897.

48 Wang H, Ascoli M, Segaloff DL. 1991. Multiple luteinizing hormone/chorionic gonadotropin receptor messenger ribonucleic acid transcripts. Endocrinology 129:133-138.

49 Koo YB, Ji I, Slaughter RG, Ji TH. 1991. Structure of the luteinizing hormone receptor gene and multiple exons of the coding sequence. Endocrinology 128:2297-2308.

50 Hu ZZ, Buczko E, Zhuang L, Dufau ML. 1994. Sequence of the 3'-noncoding region of the luteinizing hormone receptor gene and identification of two polyadenylation domains that generate the major mRNA forms. Biochim. Biophys. Acta. 1220:333-337.

51 Weiss IM, Liebhaber SA. 1995. Erythroid cell-specific mRNA stability elements in the alpha 2-globin 3' nontranslated region. Mol. Cell Biol. 15:2457-2465.

52 Holcik M, Liebhaber SA. 1997. Four highly stable eukaryotic mRNAs assemble 3' untranslated region RNA- protein complexes sharing cis and trans components. Proc. Natl. Acad. Sci. U. S .A. 94:2410-2414.

53 Kiledjian M, Wang X, Liebhaber SA. 1995. Identification of two KH domain proteins in the alpha-globin mRNP stability complex. EMBO. J. 14:4357-4364.

54 Kiledjian M, DeMaria CT, Brewer G, Novick K. 1997. Identification of AUF1 (heterogeneous nuclear ribonucleoprotein D) as a component of the alpha-globin mRNA stability complex. Mol. Cell Biol. 17:4870-4876.

55 Czyzyk-Krzeska MF, Dominski, Z., Kole, R. and D.E. Millhorn. 1994. Hypoxia stimulates binding of a cytoplasmic protein to a pyrimidine-rich sequence in the 3'-untranslated region of rat tyrosine hydroxylase mRNA. J. Biol. Chem. 269:9940-9945.

56 Czyzyk-Krzeska MF, Beresh JE. 1996. Characterization of the hypoxia-inducible protein binding site within the pyrimidine-rich tract in the 3'-untranslated region of the tyrosine hydroxylase mRNA. J. Biol. Chem. 271:3293-3299.

57 Paulding WR, Czyzyk-Krzeska MF. 1999. Regulation of tyrosine hydroxylase mRNA stability by protein-binding, pyrimidine-rich sequence in the 3'-untranslated region. J. Biol. Chem. 274:2532-2538.

58 Carter BZ, Malter JS. 1991. Regulation of mRNA stability and its relevance to disease. Lab. Invest. 65:610-621.
59 Peltz SW, Brewer G, Bernstein P, Hart PA, Ross J. 1991. Regulation of mRNA turnover in eukaryotic cells. Crit. Rev. Eukaryot. Gene Expr. 1:99-126.
60 Hargrove JL, Schmidt FH. 1989. The role of mRNA and protein stability in gene expression. Faseb J. 3:2360-2370.
61 Hargrove JL. 1993. Microcomputer-assisted kinetic modeling of mammalian gene expression. Faseb J. 7:1163-1170.
62 Bernstein P, Ross J. 1989. Poly(A), poly(A) binding protein and the regulation of mRNA stability. Trends Biochem. Sci. 14:373-377.
63 Bernstein P, Peltz SW, Ross J. 1989. The poly(A)-poly(A)-binding protein complex is a major determinant of mRNA stability in vitro. Mol. Cell Biol. 9:659-670.
64 Peltz SW, Jacobson A. 1992. mRNA stability: in trans-it. Curr. Opin. Cell Biol. 4:979-983.
65 Burd CG, Dreyfuss G. 1994. Conserved structures and diversity of functions of RNA-binding proteins. Science 265:615-621.
66 Caput D, Beutler B, Hartog K, Thayer R, Brown-Shimer S, Cerami A. 1986. Identification of a common nucleotide sequence in the 3'-untranslated region of mRNA molecules specifying inflammatory mediators. Proc. Natl. Acad. Sci. U. S. A. 83:1670-1674.
67 Chen CY, Shyu AB. 1995. AU-rich elements: characterization and importance in mRNA degradation. Trends Biochem. Sci. 20:465-470.
68 Pandey NB, Williams AS, Sun JH, Brown VD, Bond U, Marzluff WF. 1994. Point mutations in the stem-loop at the 3' end of mouse histone mRNA reduce expression by reducing the efficiency of 3' end formation. Mol. Cell Biol. 14:1709-1720.
69 Chen FY, Amara FM, Wright JA. 1993. Mammalian ribonucleotide reductase R1 mRNA stability under normal and phorbol ester stimulating conditions: involvement of a cis-trans interaction at the 3' untranslated region. EMBO J. 12:3977-3986.
70 Amara FM, Chen FY, Wright JA. 1993. A novel transforming growth factor-beta 1 responsive cytoplasmic trans- acting factor binds selectively to the 3'-untranslated region of mammalian ribonucleotide reductase R2 mRNA: role in message stability. Nucleic Acids Res. 21:4803-4809.
71 Zhang W, Wagner BJ, Ehrenman K, Schaefer AW, DeMaria CT, Crater D, DeHaven K, Long L, Brewer G. 1993. Purification, characterization, and cDNA cloning of an AU-rich element RNA-binding protein, AUF1. Mol. Cell Biol. 13:7652-7665.
72 Blaxall BC, Pellett AC, Wu SC, Pende A, Port JD. 2000. Purification and characterization of beta-adrenergic receptor mRNA- binding proteins. J. Biol. Chem. 275:4290-4297.
73 Mitchusson KD, Blaxall BC, Pende A, Port JD. 1998. Agonist-mediated destabilization of human beta1-adrenergic receptor mRNA: role of the 3' untranslated translated region. Biochem. Biophys. Res. Commun. 252:357-362.
74 Pende A, Tremmel KD, DeMaria CT, Blaxall BC, Minobe WA, Sherman JA, Bisognano JD, Bristow MR, Brewer G, Port J. 1996. Regulation of the mRNA-binding protein AUF1 by activation of the beta- adrenergic receptor signal transduction pathway. J. Biol. Chem. 271:8493-8501.
75 Tholanikunnel BG, Granneman JG, Malbon CC. 1995. The M(r) 35,000 beta-adrenergic receptor mRNA-binding protein binds transcripts of G-protein-linked receptors which undergo agonist-induced destabilization. J. Biol. Chem. 270:12787-12793.
76 Nickenig G, Michaelsen F, Muller C, Vogel T, Strehlow K, Bohm M. 2001. Post-transcriptional regulation of the AT_1 receptor mRNA. Identification of the mRNA binding motif and functional characterization. Faseb J. 15:1490-1492.
77 Ross J. 1996. Control of messenger RNA stability in higher eukaryotes. Trends Genet. 12:171-175.

78 Haynes SR. 1999. RNA-protein interaction protocols. In *Methods in Molecular Biology*. J.M. Walker, ed. Vol 118, Totowa: Humana Press.
79 Smith CWJ. 1998. RNA:Protein Interactions. In *The Practical Approach Series*. B.D. Hames, ed., Oxford: Oxford University Press.

Index

angiotensin 287
androgens 261
AREs (*A-U rich elements*) 104, 167
AT_1R (*angiotensin receptor type 1*) 287
AUF1 62, 107, 125, 167, 265, 299

β-globin 197, 245, 281

collagen-α1 55
CPEB (*cytoplasmic polyadenylation element binding protein*) 38, 43
CRHR (*corticotropin-releasing hormone receptor*) 294
CYPs (*cytochrome P450 enzymes*) 227

DRBM (*ds-RNA binding motif*) 176

EGF (*epidernmal growth factor*) 257
eIF4E (*eukaryotic initiation factor 4E*) 4, 40, 43, 92, 291
ELAV (*embryonic lethal abnormal visual system*) 163
estrogen 289

GAP-43 (*growth-associated protein-43*) 165, 166

hnRNP A1 (*heterogeneous nuclear ribonucleoprotein*) 265
hnRNP K 18, 156
hnRNP L
hsps (*heat shock proteins*) 122
HuD 163
HuR 164, 265

IGF-II (*insulin-like growth factor*) 241
IRE (*iron response element*) 216, 217, 287
IRES (*internal ribosomal entry site*) 2, 56, 253
IRP1 (*iron response protein*) 218, 287
IRP2 218

KH domain 18, 60

LDH (*lactate dehydrogenase*) 196
LDH-A mRNA 3'UTR binding proteins 199
LRBP (*LH/hCG receptor binding protein*) 296

masked mRNA 35
mTOR 73

ODC (*ornithine decarboxylase*) 89

p70^{s6k} 74
PAI-1 (*plasminogen activator inhibitor*) 137
PAI-RBP1 (*plasminogen activator inhibitor binding protein*) 149
PASR (*PKA stabilizing region*) 201
PCSR (*PKC stabilizing region*) 201
PCBP (*poly(rC) binding proteins*) 15, 53
PHAS-I (*phosphorylated heat and acid stable protein that is regulated by insulin*) 72
picornaviridae 2, 94
PKA (*protein kinase A*) 199
PKC (*protein kinase C*) 199
PABP (*poly(A) binding protein*) 262, 298
polypyrimidine 13
polysomes 37, 130, 164, 286

uPAR (*urokinase receptor*) 273
uPAR mRNABp (*urokinase receptor binding protein*) 276

www.ingramcontent.com/pod-product-compliance
Ingram Content Group UK Ltd.
Pitfield, Milton Keynes, MK11 3LW, UK
UKHW020104200726
13856UKWH00002B/374

* 9 7 8 1 4 7 5 7 6 4 4 7 5 *